Out of the Earth Into the Fire

SECOND EDITION

RELATED TITLES
published by
The American Ceramic Society

The Extruder Book
Daryl E. Baird
©2000, ISBN 1-57498-073-4

Glazes and Glass Coatings
Richard A. Eppler and Douglas R. Eppler
©2000, ISBN 1-57498-054-8

The Magic of Ceramics
David W. Richerson
©2000, ISBN 1-57498-050-5

Boing-Boing the Bionic Cat
Larry L. Hench, Illustrated by Ruth Denise Lear
©2000, ISBN 1-57498-109-9

Setting up a Pottery Workshop
Co-published by The American Ceramic Society,
Westerville, Ohio, USA and A&C Black, London, England
Alistair Young
©1999, ISBN 1-57498-106-4

Glazes for the Craft Potter, Revised Edition
Co-Published by The American Ceramic Society,
Westerville, Ohio, USA and A&C Black, London, England
Harry Fraser
©1998, ISBN 1-57498-076-9

Answers to Potters' Questions II
Edited by Ruth C. Butler
©1998, ISBN 1-57498-085-8

Great Ideas for Potters II
Edited by Ruth C. Butler
©1998, ISBN 1-57498-068-8

Answers to Potters' Questions
Edited by Barbara Tipton
©1990, ISBN 0-934706-10-7

Great Ideas for Potters
Edited by Barbara Tipton
©1983, ISBN 0-934706-09-3

Glaze Projects
Richard Behrens
©1971, ISBN 0-934706-06-9

Potter's Wheel Projects
Edited by Thomas Sellers
©1968, ISBN 0-934706-04-2

Decorating Pottery
F. Carlton Ball
©1967, ISBN 0-934706-05-0

Brush Decoration for Ceramics
Marc Bellaire
©1964, ISBN 0-934706-02-6

Ceramic Projects
Edited by Thomas Sellers
©1963, ISBN 0-934706-08-5

Throwing on the Potter's Wheel
Thomas Sellers
©1960, ISBN 0-934706-03-4

Underglaze Decoration
Marc Bellaire
©1957, ISBN 0-934706-01-8

Copper Enameling
Jo Rebert and Jean O'Hara
©1956, ISBN 0-934706

For information on ordering titles published by The American Ceramic Society, or to request a catalog, please contact our Customer Service Department at 614-794-5890 (phone), 614-794-5892 (fax), customersrvc@acers.org (e-mail), or write to Customer Service Department, 735 Ceramic Place, Westerville, OH 43081, USA.

Subscribe to Clayart!
Clayart is the "electronic voice of potters worldwide." Subscriber-initiated discussions range from questions/answers on materials and techniques to business advice and philosophical debate, sponsored by The American Ceramic Society. Visit the website and subscribe at www.ceramics.org/clayart.

Out of the Earth Into the Fire

SECOND EDITION

A Course in Ceramic Materials
For the Studio Potter

By Mimi Obstler

Edited by **Robina Simpson**
Curator of Department of Earth and Environmental Sciences
of Columbia University

Photographs by **Anthony Israel**

The
American
Ceramic
Society

Published by The American Ceramic Society
   735 Ceramic Place
   Westerville, Ohio 43081
   www.ceramics.org

The American Ceramic Society
735 Ceramic Place
Westerville, Ohio 43081

04 03 02 01 00   5 4 3 2 1

ISBN: 1-57498-078-5

Executive Director and Publisher
W. Paul Holbrook

Senior Director, Publications
Mark Mecklenborg

Acquisitions
Mary J. Cassells

Developmental Editor
Sarah Godby

Marketing Assistant
Jennifer Brewer

Production Manager
John Wilson

Design by Paula John, Boismier John Design, Columbus, Ohio.
Photography by Anthony Israel, except where noted.
Cover image: Jar by Bill Lau. Glazed by Sue Browdy after potter's death. Used by permission. See p. 18.

Library of Congress Cataloging-in-Publication Data
Obstler, Mimi.
    Out of the earth, into the fire: a course in ceramic materials for the studio potter/by
  Mimi Obstler; edited by Robina Simpson; photographs by Anthony Israel.--2nd ed.
      p. cm.
    Includes bibliographical references and indexes.
    ISBN 1-57498-078-5
     1. Ceramic materials. 2. Pottery. I. Simpson, Robina. II. Title.

TP810.5 .O28 2000
666'.3--dc21   00-065067

For more information on ordering titles published by The American Ceramic Society or to request a publications catalog, please call (614) 794-5890 or visit our online bookstore at <www.ceramics.org>.

In memory of Bill Lau.

The second edition of *Out of the Earth Into the Fire* proudly displays on its cover one of the last pots by master potter and teacher, Bill Lau. Bill made this pot during a demonstration class at the N.Y. 92nd Street Y shortly before his untimely death from cancer. After his death, Sue Browdy glazed the interior of the pot and it was then fired in the 92nd Street Y stoneware kiln. On behalf of all of us who were privileged to know and love Bill, I have dedicated this book to his memory.

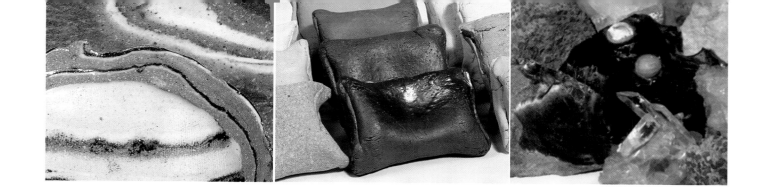

Brief Contents

Contents

CHAPTER 2

CLAYS and CLAYBODIES

CHAPTER 3

AUXILIARY MELTERS

CHAPTER 4

AUXILIARY SILICA and ALUMINA MINERALS

APPENDIXES

Preface

"...the lives of men are short beside the history of rocks and stone..."[1]

SECOND EDITION

A second edition immediately brings up the question of what is new. In this second edition, although basic content, format, and structure remain unchanged, there are three important changes.

First, wherever possible, technical data has been updated, and origin, availability, and current manufacturing source duly noted. In doing this update, I was struck with the incredible changes that have taken place over the last eight years on the manufacturing scene. Of the 30 or more ceramic materials described in this book, one-third are no longer manufactured under the same company name. Tracking down their corporate identities often required the abilities of a sleuth. In contrast, the names and chemical structures of the ceramic materials have changed very little in the last eight years. Only a few materials have become ghosts of the past. This fact, plus the slight variations in chemical and mineral structures, bears witness to their welcome stability.

Second, Appendix A1, Glaze Calculation Techniques, has been revised for the purpose of clarification.

Finally, and most important, the second edition adds 50 new photographs. Pictures describe test results far more eloquently than words. Short of actually holding the test result in your hands, a picture is the best way to visualize the ceramic surface. In selecting the objects for the photographs, aesthetic values of form, color, and technical virtuosity had to give way to the paramount consideration, namely, the visual clarity of the ceramic surface under discussion. The objective was not to create a photographic portfolio of ceramic wonders, but rather to provide a concrete visualization of the text. In this way I hope to deepen your understanding of the resulting ceramic surface and spur you on to bigger and bolder experiments with materials of your craft.

FIRST EDITION

As a potter and teacher of glaze chemistry, ceramics has been a part of my life for over thirty years. Early on it became clear to me that I was more interested in the raw materials of ceramics than in the finished product, and that the challenge of experimental surfaces held more interest than the successful repetition of glazes. My fascination with ceramics began when glazes unexpectedly turned traitor and displayed strangely blemished surfaces. A desire to know why this occurred led me to explore the raw materials of claybodies and glazes.

The search for this knowledge led inevitably to a geology course at Columbia University. As a result of this course, the close relation between the formation of minerals of the earth's crust, and the claybodies and glaze surfaces of ceramic pots became a self-evident truth. Studying the connection between the the raw materials of the crust of the earth and the surface of a ceramic form gave ceramics a new meaning for me. From this point on, it was this connection that was emphasized in my glaze class.

This book seeks to restore the geological context from which our ceramic materials have been extracted. In this way, I hope to connect the potter once again with the eternal, awe-inspiring processes which produce our ceramic materials. Many of our ceramic materials are hundreds of millions of years old. As the potter handles these materials in their ignoble, powdered state, surely some sense of their ancient, majestic origins should stir the soul. Hence, the focus of this book is not only upon the final ceramic result but also upon the minerals and rocks which create our ceramic materials. The ultimate purpose is to retrieve our earth materials from the grey shadows of indifference, and give them the attention and respect which they so richly deserve.

---

[1]A. S. Byatt, "Dragons' Breath," *The Djinn in the Nightingale's Eye* (New York: Vintage International, Vintage Books, 1997), 79–80.

Acknowledgments

I am greatly indebted to Robina Simpson, Curator of Earth and Environmental Sciences of Columbia University. In addition to editing the geological material and providing the rock samples, she gave invaluable assistance with respect to style, clarity, and overall organization of both editions. She also provided the sorely needed clarification of Appendix A1. My thanks to Stefan Larson for masterminding the cleanup and also to Raymond Woodhead, my Australian correspondent, whose careful reading helped to pinpoint some of the confusions in Appendix A1.

To Sue Browdy, with whom I shared a studio for many years, I owe more than I can say. Her quiet intelligence, superb craftsmanship, and masterful handling of the kiln firings have brought to life many of the gemstone surfaces described in this book. She also played a key role in creating the photographic record for the text.

My thanks to Jacqueline Wilder and Peggy Bloomer for taking time out from their busy schedules to provide me with important cone 5/6 tests. Their spirited contributions to the photographic record in this book show clearly the rich potential of cone 5/6 oxidation surfaces.

I am very grateful to Jane Hartsook, Barbara Knapp, and Harold Obstler for their careful reading and excellent suggestions with respect to the first edition and for their unflagging encouragement and support.

I thank Peggy Bloomer for her mercurial computer skills and infinite patience in the numerous revisions of the first edition. Mary Cassells of The American Ceramic Society initiated the second edition project, and I am most grateful for her support and assistance. All my thanks go to Eileen Handelman, whose eagle eye found some remaining typos and errata.

Grateful acknowledgment is made to the technical service representatives of the manufacturers and suppliers listed in this book. They have been most generous with their time and expertise. In particular, my thanks to Robert Kistler, Charles Frame, R. J. Brotherton and Dr. H. Steinberg of U.S. Borax Inc. for their informative letters and memoranda on Colemanite and Gerstley Borate in the first edition. Ragnar Hess deserves a special mention for obtaining primary information for me on these materials. Tom Wilhelm of U.S. Borax Inc., Derek McCracken of F & S International, and Di Stonhill of Bikita Minerals were outstanding in their generous response to my numerous queries. Tommy Kilgore of ECC International, Sara Robinson of NYCO Sales, David Morse of Industrial Minerals Section U.S.G.S., J. David Sagurton of Dry Branch Kaolin Sales, Dorna Isaacs of Hammill & Gillespie, K. C. Rieger of R. T. Vanderbilt Company, and John Cowen and Lee Watroba of Sheffield Pottery Inc. were most helpful in my search for current technical data, as were most of the other manufacturers whom I contacted.

To Janet Bryant a special thanks for starting me out on my lifelong exploration of ceramic materials.

Above all I thank my students. Their thirst for knowledge and their tireless testing have truly created this book.

Notes

This book does not consider *barium carbonate* or *lead materials* because of their toxic properties. These toxic properties made it inadvisable for students to perform tests with them in our ceramic studio environment. The contribution of these materials to claybodies and glazes is indisputable. For example, barium carbonate is an essential ingredient for earthenware, casting slips, and certain glaze colors at all firing temperatures. The honey-colored, golden glow of pure lead oxide when applied to an earthenware claybody cannot be duplicated. However, nature provides us with many choices for melters in the production of superb ceramic surfaces. The primary focus of this book is on those materials that when used with proper safeguards would not expose the maker or the consumer to undue risk.

Glaze colorant and opacifying materials like iron oxide, rutile, ilmenite, copper oxide, copper carbonate, cobalt oxide, and cobalt carbonate, tin oxide, and zirconium materials, together with their fixed percentage limits, are not discussed separately in this text. The reasons for their exclusion are twofold; firstly, a definitive account of the development of specific colors, together with their suggested percentage limits, is already available for the reader in *The Ceramic Spectrum*, by Robin Hopper (Hopper 1984, 138–44). Secondly, to the extent that percentage lists of colorants imply fixed results, they can be extremely misleading. This book is focused on the crucial trinity of glassmaker, adhesive, and melter oxides that underlies all glaze and claybody materials. This oxide trinity (subject, of course, to the variables of firing temperature, kiln atmosphere, and underlying claybody color) ultimately determines the final surface color produced by the colorant material. Consequently, colorant materials are presented within the context of a specific glaze base, firing temperature, atmosphere, and claybody. The point is not that 1% to 3% of a particular coloring oxide creates a blue, green, or yellow color; the emphasis is as follows:

1. On the *kind* of green or blue that results from a particular glaze base when it is applied to a light- or dark-colored claybody and fired to high or low stoneware oxidation and/or reduction temperatures

2. How various glaze bases encourage or inhibit the development of certain colors

The lab tests, in particular, are designed to show different effects produced by the colorant in a specific glaze as the existing ratio of glassmaker-adhesive-melter oxides is altered.

A FURTHER NOTE

Unless otherwise indicated:

1. The primary sources for the characteristics of the ceramic materials described are the test results obtained from glaze chemistry classes at the 92nd Street Y, Greenwich House Pottery, Parsons School of Design, Crafts Students League, North River Pottery, and various private ceramic studios, all of which are located in New York City.

2. The claybody that underlies the glaze tests is the stoneware claybody of the 92nd Street Y, prior to the demise of Jordan clay. The porcelain claybody that underlies the porcelain tests is the Temple porcelain claybody (see p. 113).

A FINAL NOTE

Throughout the period of preparation for this book, I have attempted to obtain the most current chemical and mineralogical data from ceramic manufacturers and suppliers. The keynote of the manufacture of ceramic materials, particularly in the United States, is *change*. As the following chapters will show, mineralogical compositions of ceramic materials are not always constant and may undergo some change from time to time. Similarly, manufacturing corporate identities frequently change, as one company merges with another or leaves the field to another manufacturer. Hence, you are urged to not rely exclusively on the technical data presented here; instead, you should periodically consult the supplier or manufacturer for up-to-date chemical and mineralogical information concerning each ceramic material used in a particular glaze and claybody.

Introduction

This book presents a twofold approach to the study of claybodies and glazes that is both empirical and historical in nature. It is empirical because it seeks to create and understand ceramic surfaces in terms of a hands-on experience with the primary minerals of our earth. It follows a historical approach in its focus on a single mineral as the core of the glaze or the claybody. The earliest stoneware glazes and claybodies evolved in precisely this manner over 2,000 years ago. The incomparable surfaces of the Chinese Song Dynasty still serve as the basic model for many of the glazes and claybodies used today by studio potters. *Figure Intro.2, p12.*

This dual approach does not deny the validity of more technologically advanced methods, such as molecular calculation procedures, and oxide percentage analysis. These procedures now greatly simplified by the introduction of computer glaze software programs, provide essential information about the skeletal structure of glazes and claybodies. An oxide percentage analysis for each ceramic material appears in each section; in addition, oxide percentage analyses are included for various test glazes. Appendix A describes procedures for obtaining molecular and oxide percentage analyses without the benefit of computer software programs such as Hyperglaze Software. These molecular and oxide percentage procedures are relegated to the Appendix because this approach can have meaning only *after* the potter is familiar with the working properties of the ceramic materials of his or her glaze. Too often, glaze calculation techniques are introduced before the student has a good understanding of the ceramic material under analysis. The result can be bewilderment and frustration. The seductive speed of the computer in the analysis and creation of glazes and claybodies can lead to a heavy dependence on its technological prowess. To the extent that complex technology supplants hands-on testing procedures, our estrangement from ceramic materials can only deepen. In this book, glaze calculation techniques are viewed as useful diagnostic tools for the solution of glaze and claybody problems or substitutions, rather than as an essential means for the creation or understanding of ceramic surfaces. This does not mean that molecular and oxide percentage analyses cannot always be successfully used for the creation of excellent ceramic surfaces with a minimum of time and effort. We live in an age of computer technology, and many potters make use of this technology to obtain superb glaze surfaces. (See Richard Burkett 1992, "Ceramics and computers;" Gary Hatcher 1992, "Software for DOS claypersons;" Richard Zakin 1992, "Computer aided thinking about clay," *Studio Potter* (June): 81–87. See also Rick Malmgren, "Glaze calculation software," *Ceramics Monthly* 1992 (January): 29–33; "A look at glaze calculation software," *Ceramics Monthly* 1998 (June July August): 38–45). However, all potters need first to understand the current mineralogical structure of their ceramic materials, if only to ensure the accuracy of the chemical data that feeds the computer programs. The important fact is that chemical analysis is not the only prediction of how a material behaves in a particular ceramic fusion. The chemical structure of two materials may be identical, yet their physical appearance, particle size, heat reactions, and reactions to other minerals in the glaze magma may be totally different. Only hands-on testing will produce this kind of information. Finally, the exclusive use of computer programs to analyze and originate ceramic surfaces will only make our understanding of ceramic materials more abstract. A firsthand, concrete understanding of ceramic materials without the interposition of complex technological machinery is an essential part of the ceramic process. We need to restore that lost connection between our earth materials and ceramic surfaces. If we lose this connection, no matter how successful the end product, we have lost the true romance of the potter's craft.

THE GEOLOGICAL CONNECTION[1]

The primary rocks and minerals of the earth's crust crystallize from magma, or molten rock, which lies deep within the earth. This magma includes the elements of silicon, oxygen, aluminum, sodium, potassium, calcium, iron, and magnesium, plus water and carbon dioxide.

Our clay and glaze materials are made up of these same materials. Every kiln firing re-creates the igneous and metamorphic processes of the formation of the earth's crust. The re-creation of igneous processes occurs when the heat of the kiln melts the glaze materials on the walls of the clay form and transforms these materials into their liquid stage, or magma. The metamorphic process is reenacted when the heat of the kiln fire bakes the soft claybody minerals into an impermeable material.

Thus, on a miniature scale, a universal drama of the earth's formation is reenacted with every kiln firing. In this connection, it is significant to note that the heat of the earth's volcanic gases that spew forth rock-forming magma, known as lava, may reach temperatures between 2000°F and 2300°F—the final temperature of the stoneware kiln fire.

The extraordinary variety of the rocks of the earth's crust results from complex geological processes that are, even today, not fully understood. The following processes are briefly described here because of their relevance to the formation of ceramic surfaces.

DIFFERENT RATES OF SOLIDIFICATION, OR THE LAW OF FRACTIONAL CRYSTALLIZATION

Some of the minerals of the earth's crust form from the crystallization of erupting magmas. Different minerals solidify at different temperatures.

"As the crystallization process continues, the composition of the melt (liquid portion of a magma, excluding any solid material) continually changes. For example, at the state when about 50% of the magma has solidified, the melt will be greatly depleted in iron, magnesium, and calcium, because these elements are found in the earliest formed minerals. But at the same time, it will be enriched in the elements contained in the later-forming minerals, namely sodium and potassium. Further, the silicon content of the melt becomes enriched toward the later states of crystallization" (Tarbuck and Lutgens 1987, p. 55).

Iron, magnesium, and calcium crystallize out of the magma first, at the highest heat. Hence, these elements appear in the earliest formed minerals, such as the minerals of the olivine group, which are rich in iron and magnesium. Sodium is the next to crystallize. As the cooling continues, the liquid portion will be low in iron, magnesium, calcium, and sodium and rich in potassium and silicon. The last minerals to crystallize are potassium feldspar and quartz (silicon). These are the minerals that make up a large portion of the granitic rocks.

Newly formed minerals react with the remaining magma to form new and different minerals. Earlier formed calcium-rich feldspars will react with sodium in the magma and form sodium-rich feldspar. In addition, solidified particles are heavy and settle to the bottom of the melt. The molten portion, which is lighter and more fluid, flows into cracks and crevices of solidified minerals to create yet another kind of mineral.

The question is to what extent does the solidification of the glaze magma follow this pattern of fractional crystallization of the earth's magma? If the same principles are indeed applicable, then the study of fractional crystallization of minerals has much to tell the potter about the formation of glaze surfaces during a kiln firing. For example, in the early solidification of iron-rich minerals, such as olivine, their greater weight and consequent separation from the rest of the melt could account for the sometimes green spotted surface of a black/brown saturated iron glaze. Olivine minerals are generally green in color. Similarly, separation of solid and liquid components during the melting of the glaze could explain the following ceramic phenomena:

a) the occasional difference between the inside, outside, upper, and lower surfaces of a bowl; and

b) the different appearance of the same glaze on different forms and surfaces, especially heavily carved and textured surfaces that reproduce the cracks and crevices mentioned above.

RATE OF COOLING OF MAGMA

Slow cooling of the earth's magma beneath the earth's surface results in minerals that display a large-grained crystalline surface. This slow cooling produces crystals that are large in size and therefore visible to the naked eye. Magma deposited nearer to the earth's surface cools faster. The resulting rock surface will be fine to medium grained and will contain many more crystals than in the large-grained crystalline surface. The crystals may or may not be visible to the naked eye. Molten rock ejected into the air during a volcanic eruption cools almost immediately. The result is a glassy rock surface with almost no crystalline structure.

It follows from this that two rocks or minerals with totally different textures can, in fact, have the same mineral composition—the difference is due entirely to the cooling rate. For example, the igneous rocks, granite and rhyolite, have an identical mineral composition but are very different in appearance. Rhyolite exhibits a dense, fine-grained texture with a small interlocked crystalline network. Its color can be light gray, buff, or pink. Its surface is smooth and sometimes contains glassy fragments. Granite, on the other hand, is visibly coarse grained with glassy, lustrous quartz crystals, pink or white rectangular feldspar crystals, and some dark grains of mica. Granite crystallizes from a magma that cools slowly beneath the earth's crust. Rhyolite forms when this same magma is hurled onto or near the earth's surface during a volcanic eruption and thus cools quickly. The extraordinary difference in color and texture of the two rocks results from the different rate of cooling of the magma. *Figure Intro.1, p11.*

This same kind of variation in the rate of cooling exists in the formation of ceramic surfaces and could explain why certain glazes produce very different surfaces from one kiln firing to the next. It is a well-known fact that certain glaze surfaces will appear shiny if the kiln is cooled quickly. Conversely, a surface that is usually glossy can become dull and stony if the kiln cools slowly. Even in the same kiln firing, different results are possible; the position of a pot in the kiln could subject it to a different temperature or a different cooling rate from another pot on a higher or lower shelf. *Figure 3. 14, p236.*

As we shall see from the chapters that follow, the varied spectra of ceramic glaze surfaces, ranging from a transparent gloss surface to a semi-opaque, satin-matt surface to an opaque, stony-matt surface owe their existence to the geological principles previously set forth. The important question is, If "one magma can generate several different igneous rocks" (Tarbuck and Lutgens 1987, p. 57), can one glaze magma generate different glaze surfaces even on the same pot? The answer, depending on the form and texture of the pot, the position of the pot in the kiln, the speed and temperature of the firing, and the kiln's cooling rate, surely is yes. Thus, the connections between the igneous geologic formations of our earth and the ceramic process move closer and closer—the study of one enriches the other and reveals an underlying harmony. It is this connection between geologic and ceramic processes that I will emphasize in the exploration of the ceramic materials that follows.

OBJECTIVES

The Song Dynasty potters (960–1278 A.D.) produced pots and glazes that we, for all our technical achievements, have yet to surpass. The ancient Chinese potters did not have many resources available to them. They crafted their pots and glazes with but three or four materials found in their native soil. These incomparable vessels were a simple combination of limestone, clay, wood ash, and an altered granite.[2] All of these materials were part of the familiar landscape in which the Song Dynasty potter lived. *Figure Intro.2, p12.*

Today, we live in an age of superabundance of materials. Innumerable clays and glaze materials offer us a bewildering array of choices. Far from understanding these materials as familiar rocks, feldspars, and clays, each with unique personalities of their own, we know them only as white, gray, or brown powders neatly packaged in uniform bags. Consequently we beg, borrow, and steal glaze and claybody formulas that "work." Many of us purchase our claybodies from suppliers who guard the secret of their claybody formulas. Even if we should know their formulas, we are often ignorant of the properties of our clay and glaze materials. We spin our creations in a veil of mystery, and, when that mystery stops working for us, we move on to an equally mysterious formula or material until that also ceases to work. We never fully understand the causes for the failures. Our ignorance is born of the very richness of our choices and the technical triumphs of our age. Raw materials come to us from every corner of the earth in a purified and refined state. We no longer live at the source of our ceramic materials as did the potters of ancient times. It now becomes a Herculean feat to know or understand in any meaningful way the flood of ceramic materials at our disposal. Fortunately, it is not necessary to have an intimate understanding of these hundreds of ceramic materials. Ancient potters created their masterpieces from three or four ceramic materials, and, if we similarly narrowed our choices, we could also achieve extraordinary results.

This book is a record of the many years I have spent exploring the relatively few ceramic materials that make up the bulk of stoneware claybodies and glazes. It presents a simple method of organizing the hundreds of ceramic materials known as THE CHEMICALS, which can be applied to any firing temperature and atmosphere. The results of this method are set forth with a threefold aim:

First: I hope to familiarize you with the basic materials of claybodies and glazes. I will summarize the results of past explorations and provide a series of simple tests for you to follow in your own studio or classroom.

Second: I will present a classification system based on empirical tests and percentage oxide analyses of our basic ceramic materials, which charts their principal functions, so that you can make intelligent choices and substitutions with some understanding of the ultimate consequences.

Third: Most important of all, I hope that the results of points one and two will enable you to formulate new glazes and claybodies that are personal and unique. It is my fervent belief that if you conscientiously follow this course to the end, you will indeed be rewarded, not just with claybody and glaze formulas, but with something far more important: namely, the confidence to make the necessary adjustments and substitutions to ongoing formulas; the freedom to experiment with new materials so that you can create new glazes and claybodies that are a truly unique expression of your ceramic efforts; and, lastly, a close and knowledgeable connection with our ancient natural resources.

TRIAL-AND-ERROR METHOD

For more than 25 years, with the help of adventuresome and hardworking students, I have explored minerals that are basic to the stoneware claybody and glaze. We used a trial-and-error approach—that is, we tested these minerals alone and in different combinations over and over again. We fired them in an oxidation atmosphere to 2190°F, which is the beginning of stoneware, and then again to 2340°F in a reduction atmosphere. In this way, we came to recognize the special contribution of a mineral to a particular glaze or claybody. We inadvertently stumbled upon glaze surfaces that took our breath away. These surfaces seemed both unique and staggeringly simple; they clearly revealed their inspired source, namely the minerals crusting the earth's surface. Hence, we were able to throw away that discouraging concept of glazes and claybodies as mathematical combinations of uniform powders. We saw instead a vision of earth materials transformed by fire into robes of melted stone. *Figure Intro.3, p13.*

All the world's a kiln. According to the most commonly accepted theory, the earth was formed by a cloud of particles and stellar temperature gases that consolidated, solidified, and cooled for millions of years. The center of the earth is still a molten mass. The outer shell of the earth's continental crust consists mainly of granites, which are the original source of all of our ceramic materials. As a result of physical forces, granites disintegrate into feldspars, quartz, and mica minerals. The forces of chemical weathering decompose the feldspar minerals and cause them to release a portion of their silica, soda, potash, and calcia. When water molecules replace the released minerals, the clay mineral forms. These granitic deposits, which supply us with clay and glaze materials, form and reform into different kinds of minerals in an ongoing process that never ceases.

Scientists are still discovering new minerals and estimate that there are now about 2,500 known minerals. Despite the large number of existing minerals only a very few (10 minerals) make up the bulk of the earth. Feldspars alone account for 50% by volume of the earth's crust and are the primary rock-forming mineral. Feldspars are also a major constituent of both the earth's upper mantle and some meteorites as well.

Feldspar is the core of the stoneware glaze. Limestone,

which is composed principally of the mineral calcite (calcium carbonate), helps the feldspar to melt and flow at the temperature of the stoneware kiln. Clay improves the physical suspension of the liquid glaze mixture, and, if the clay contains iron, functions as a colorant. If we ask, *What is a glaze?* our answer would be: a glaze is a combination of a few earth minerals that, when heated up to a particular temperature, will form a glassy coat on the clay structure. At stoneware firing temperatures (from about 2190°F–2400°F) the primary earth material of the stoneware glaze is feldspar. *Figure 1.3, p80.*

Thus, it sounds simple to understand and create a glaze, at least at stoneware temperatures. The catch is that there are many different kinds of feldspars, limestones, and clays, all with different properties and different firing consequences. Furthermore, the availability of any one kind of material can change as time elapses, so that if one becomes familiar with one kind of feldspar, in a matter of months it may no longer be available. For this reason, the simple trial-and-error approach, which was so natural and successful for ancient potters, is not enough for us. We need to organize these ceramic materials into a system that will permit us to make intelligent choices and substitutions. This means that we must now look at the earth materials in terms of their chemical structure.

CLASSIFICATION SYSTEM: THE CHEMICAL STRUCTURE OF GLAZES AND CLAYBODIES[3]

Atoms are the building blocks of the earth. Every atom consists of a heavy nucleus[4] surrounded by clouds of one or more negatively charged particles that we call electrons. The heavy nucleus of hydrogen is unlike other nuclei, as it contains one proton, it has no neutron, and its mass is the proton. Its nucleus is surrounded by a cloud of one electron. All other nuclei of atoms contain both protons and neutrons. It is interesting to know that in 1925 the positively charged proton was thought to be a positive electron. All attempts to create atoms from protons and electrons alone proved to be impossible. Then in 1932, I. Curie-Joliot and F. Joliot (Paris)

and J. Chadwick (London) discovered the neutrally charged neutron (no charge). Thus, we are now able to build up all the atoms in the periodic table of the elements from the electron, proton-neutron atomic construct.

Scientists are still learning about the structure of the materials that make up our earth.

"After 2500 years or more, atomicity of nature has been established. As the year 2000 approaches, of the three constituents of matter, which we have discussed, only the electron remains fundamental i.e. indivisible. Both the proton and neutron are found to be composites of even more basic constituents called quarks, proposed by M. Gell-Mann (USA) & G. Zweig (Cern) in 1964, and gluons, observed in 1979 at Deutsches Elektronen Synchrotron [DESY] (Hamburg). Thus we arrive at the present with wonderment still reserved for the future" (Larson, Stefan, May 1999).

Eight of the 90 natural elements (elements 93 through 105 are artificially prepared by bombardment reactions) make up 99% of the earth's crust by weight. These eight elements are oxygen, silicon, aluminum, iron, sodium, potassium, calcium, and magnesium. Most of the elements do not exist in nature in a pure state. Oxygen—the first element in the above list—frequently combines with the other elements on this list to form oxides. The oxides in turn combine with each other or with other compounds and elements in a specific three-dimensional pattern to form the bulk of the 2,500 existing minerals.

If we probe deeper into our question, *What is a glaze?* our answer will now be: a glaze is a ratio of the glassmaking oxides, structural adhesive oxide, and melting oxides.

I. GLASSMAKING OXIDE: SILICA (SiO_2)

The classic glaze is a layer of glass bonded by fire to a clay form. The essence of glass is the property of transparency. Transparency means that light may travel freely through the glassy layers to reveal whatever lies beyond. Glass is transparent because the atoms of glass, unlike the atoms of most other materials, do not arrange themselves into an orderly pattern. The atoms of glass crowd into a random pattern that allows light rays to flow toward the underlying surface without the obstructions presented by an orderly crystalline structure.

The atoms of most earth materials regroup into their characteristic three-dimensional crystalline structure after being heated and cooled in a potter's kiln. Thus, they do not form a glass at the various kiln temperatures. Silica is one of the few minerals that does have this glass-forming property. Because of the high viscosity of cooling silica magma, its atoms are unable to resume their original three-dimensional pattern. Instead, they flow into a random arrangement, which we see as transparency or glass. On the other hand, when silica was heated in the fires of the earth's formation and cooled during a time span of thousands of years, its atoms had the time to form a characteristic pattern. The result is the mineral form of silica known as quartz.

Hence, most forms of glass are manufactured and do not exist in nature. Two natural silica glasses that are found in nature are obsidian and pumice (a frothy glass).

In terms of sheer weight, silica makes up the bulk of the earth's crust. The earth's crust consists of approximately 59% silica. Feldspar contains 60%–70% silica; the majority of stoneware glazes and claybodies contain 60% silica.

Silica requires a fire of more than 3000°F to reach a melted state. It naturally forms eutectic bonds with certain other materials, which cause the resulting alloy to meet at lower firing temperatures.

Silica is primarily responsible for the transparency and gloss of the glaze surface and for the fit of the glaze coat with the claybody.

II. ADHESIVE OXIDE: ALUMINA (Al_2O_3)

Alumina provides the glue that holds the molten glaze materials onto the walls of the clay form during the molten phase of the glaze firing. In large amounts, alumina is primarily responsible for a stony, opaque, nonfluid glaze surface. In combination with silica, alumina provides the basic structure and strength of the underlying stoneware claybody.

The earth's crust contains about 15% alumina. Alumina constitutes 15%–25% of many feldspars and about the same amount of many stoneware glazes.

Alumina, together with the glassmaker silica, some melt-

ing oxides, and substantial amounts of water, make up the clay minerals.

The melting point of alumina is almost 4000°F. After cooling, its atoms speedily regroup into its characteristic dense crystalline structure. Alumina retards the formation of large crystals by the other materials present in the glaze magma.

III. MELTING OXIDES: SODA (Na_2O), POTASH (K_2O), CALCIA (CaO), MAGNESIA (MgO), LITHIA (Li_2O), BORIC OXIDE (B_2O_3), ZINC OXIDE (ZnO)

The high melting temperatures of both silica and alumina require the addition of a third oxide, which will lower the melting temperatures of silica and alumina to that attainable in the potter's kiln. A material that has this power to lower the melting temperature of another material is called a "melter." This ability of certain materials to lower the melting temperature of another material is referred to as a "eutectic power." The particular proportion of any two materials that results in their fusion at a lower temperature than the fusion point of each material separately is known as their "eutectic ratio." It is interesting to note that the dynamics of the eutectic power is not completely understood, yet without it, there would be no ceramics industry. Soda, potash, and calcia are the primary melting oxides, both in the feldspar minerals and in the stoneware glazes and claybodies.

Now, if we ask for a third time, *What is a glaze?* this time our answer will be: a glaze is a ratio of silica, alumina, and melter oxides. Silica forms a glassy melted surface on the walls of the clay form when combined with the right proportion of melter oxides. The presence of a low amount of alumina adheres the melted glass to the clay walls and prevents it from flowing off the sides of the clay form.

And if we ask, *What is a claybody?* our answer will be: it, too, is a ratio of these same three oxides, but now the ratio has changed. Alumina oxide has increased, and the melter and glassmaking oxides have decreased. The reason for this is that our goal has now become plasticity, workability, and fired strength. Hence, the clay minerals now take precedence over the feldspar minerals and constitute the core of the structure. This change in priorities is reflected in the changed oxide ratio. Although the proportion of the oxide trinity changes, the structure of a claybody is still a ratio of these three kinds of oxides. Consequently, a claybody can change into a glaze and a glaze into a claybody simply by altering this ratio.

All of our ceramic materials, the 200 or more uniform powders known as the *chemicals*, can now be understood and classified in terms of these three kinds of oxides. *Each of the chemicals functions either as a glassmaker, an adhesive glue, a melter, or a combination of any of these three. The function of each of the chemicals may change, depending on the firing temperature and whether it appears in a glaze or claybody.*

CONCLUSION

In conclusion, note that the two definitions of a glaze as a ratio of silica, alumina, and melter oxides or, alternately, as a combination of a few earth materials, such as feldspar and limestone, are actually one and the same. It is simply a matter of whether the question is approached from the standpoint of internal chemical structure or from the empirical standpoint of earth minerals. This twofold approach, which relies on mineralogical, molecular, and oxide percentage analyses, as well as on hands-on trial-and-error tests, is the best way to understand the dynamics of ceramic materials.

FIRING TEMPERATURE AND ATMOSPHERE

Thus far we have approached ceramic materials in terms of trial-and-error tests, together with some consideration of internal chemical oxide structure. There is yet another factor to consider that vitally affects the final result: namely, the kiln atmosphere and heat of the firing, which together make up the environment of the kiln.

The 2,500 or more minerals of the earth form in response to different environments that exist on or within the earth's crust. Magmatic, metamorphic, and hydrothermal environments may occur *within* the earth's crust. Other environments exist at the surface of the earth, or in its shallow depths. Our ceramic surfaces form in response to equally complex environments contained within the four walls of the kiln. At various points of time during a firing, the chamber of a fuel-burning kiln recreates the hydrothermal and metamorphic environments. Inside the kiln an inferno rages. Flames, together with hot reactive gases of oxygen, nitrogen, carbon monoxide, sulfur dioxide, and water vapor, swirl around the still, receptive clay forms. Air is the primary source of oxygen and nitrogen. Incomplete combustion of hydrocarbons and other fuels releases hydrogen gas, carbon monoxide gas and water vapor. Glaze and claybody materials will also release varying amounts of sulfur, carbon, and various gases as they are transformed by fire into permanent, immutable objects. These extraordinary transformations of glaze and claybody materials are accomplished by means of complex chemical and physical reactions, which take place in response to the heated, gaseous atmosphere of the kiln. The material covered in the following chapter requires the reader to have some understanding, however incomplete, of these reactions, and of the meaning of the terms oxidation and reduction atmospheres. What follows, therefore, is a simplified description of the above. For a more detailed account, the reader is referred to Hamer and Hamer 1977, 172–179; Lawrence 1972, 118–120, 129–140; and Brodie 1982, 111–144. These texts provide the basis for the following account of reduction and oxidation atmospheres.

REDUCTION AND OXIDATION ATMOSPHERES

A reduction atmosphere is one in which carbon monoxide and hydrogen gases are released into the kiln. A kiln fired by combustibles such as gas, oil, coal, or wood will release these gases into the kiln if the fuel is not completely burned off. The amount of the released gases in the kiln at any one time depends on the ratio of the fuel to the air and can be controlled by the potter to achieve the desired reduction effects upon the claybody and the glaze. Changes in the claybody and glaze surface occur as a result of the search by carbon and hydrogen gases for more oxygen in order to become carbon dioxide (CO_2) and water vapor (H_2O). Old partnerships dissolve and new ones form as the rising heat (1652°F) loosens the iron-oxygen partnership of iron oxide ($2Fe_2O_3$) contained in the claybody. Iron oxide surrenders two oxygen atoms to the carbon monoxide gas and changes from $2Fe_2O_3$ to $2FeO$. This reaction causes a change in claybody color from pink, red-bronze, or orange to gray, the color of reduced iron. The process is known as "body reduction"; it takes place at anywhere from 1652°F–1815°F. During the later stages of the stoneware firing, when the glaze materials have begun to enter the melt (2190°F), carbon and hydrogen gases attack the reactive glaze materials for their oxygen content. The increased heat has now begun to loosen the bonds of the various metal oxides contained in the glaze materials, such as tin, copper, zinc, nickel, cobalt, iron, and manganese. In a prescribed order of reduction, they obligingly give up oxygen to the invading gases. The changed state of the metal oxides creates a host of color and surface effects in the fired glaze surface. These specific color and surface effects will vary in accordance with the rest of the glaze ingredients present in the melt.[5] For example, 1/2%–2% red iron oxide in a high feldspar, low alumina, calcium melter glaze (the basic celadon glaze) will color the glaze surface blue-green-gray as a result of its transformation from iron oxide (Fe_2O_3) to reduced iron oxide (FeO). A pale yellow color will result if the iron retains its oxygen content. *Figure Intro.7, p17.* A higher concentration of iron oxide in the same glaze base (6%–10%), when transformed to reduced iron oxide, will produce a brilliant black-rust color, sometimes flecked with orange or green spots, depending on the dispersion of the iron oxide in the melt. A brown color of varying strength will result if the iron retains its full oxygen content (Fe_2O_3) because of an oxidation atmosphere throughout the kiln firing. If there is only partial reduction in the kiln atmosphere, so that the reduction process is not completed, an intermediate form of iron oxide will form (Fe_3O_4) and a duller, dark brown glaze surface will be the result. A low proportion of copper oxide (1/2%–2% Cu_2O) in the same feldspathic glaze base can color a glaze red-purple in a reduction firing and pale yellow-green or colorless in an oxidation firing. *Figure Intro.7, p17.*

At the end of the stoneware reduction firing, air is introduced into the kiln atmosphere, causing it to become fully oxidized. Most of the transformed metal oxides are now fused with the rest of the glaze and claybody materials. Only those metal oxides in the surface layers are still susceptible to change. The iron in the surface of the exposed claybody, which is not protected by the fused glaze coat, has not entered into a deep bond with the vitrified claybody materials. Hence, this iron will respond to the heated oxygenated atmosphere and retrieve its oxygen from the atmosphere. The surface of the exposed claybody will change from the gray color of reduced iron to a warm toast red-brown, the color of reoxidized iron. *Figures Intro.5, p15; Intro.8, p18.*

This entire process from start to finish is known as reduction firing. It is a natural consequence of a fuel-burning kiln because hydrocarbons are contained in the fuel itself. It is the oldest method of firing, and until the 1930s it was the predominant method. At that time, the electric kiln became available for potters. (Jane Hartsook, Director of Greenwich House Pottery from 1945 to 1982, recalls the excitement produced by the building of the first electric kiln at Alfred University, New York, in the early 1940s). The atmosphere in electric firing remains basically oxygenated, and carbon-free. However, carbonaceous and sulfurous material in a claybody or glaze can result in the release of reducing carbon monoxide and sulfur dioxide gases into an oxidation atmosphere and create partial reduction effects. "Kiln gases even in electric kilns can never be considered exclusively oxidizing" (Hamer and Hamer 1977, 175).

Subject to the above exception, iron oxide in the oxidation glaze and claybody produces different visual effects from its reduction counterpart. Oxidized iron oxide may color a glaze or claybody surface red, orange, yellow, brown, or black depending upon the amount of its concentration, the glaze and claybody ingredients, and the heat of the firing. *Figure Intro.6, p16; Figure Intro.7, p17; Figure 3.8, p230.*

There are even more profound differences than color. Reduced iron functions as a melter as well as a colorant. Therefore, in an oxidation firing of an iron-bearing claybody or glaze, there is less fusion and less interaction between claybody and glaze than exists in a reduction atmosphere. This is true despite the fact that reduction and oxidation kilns are ostensibly fired to the same temperature for the same period of time. *Figure Intro.5, p15.*

TIME AND TEMPERATURE

Even more important than atmosphere is the time and temperature of the firing. Heatwork in ceramics, as in cooking, is determined by time as well as temperature. Potters measure these two factors—time and temperature—by using cones.[6]

Cones are a precisely balanced ratio of earth materials that will soften and slump at the time and temperature that the clay and glazes reach maturity in the kiln. They are mounted in a wad of clay at a recommended 8 degree angle from the vertical. When cones are viewed through the kiln's peephole, their posture (angle of deformation) will indicate the right moment to turn off the kiln. The temperatures given for the various cones throughout this book are based on the Edward Orton Jr. Ceramic Foundation Equivalents for Orton Standard Pyrometric Cones. These are the cones used by domestic potters to measure the time and temperature of their ware. The Orton chart shows a variation in the given temperature for a particular cone depending on the following factors:

(1) SPEED OF FIRING

The longer and slower the firing, the lower will be the temperature of maturation. Thus, if during the last several hundred degrees of temperature rise, the kiln temperature increases at the rate of 108°F per hour, large cone 9 will slump at 2300°F. If the rise of temperature accelerates and reaches the rate of 270°F per hour, large cone 9 will not slump until 2336°F.

> ". . .as the heating rate increases, the temperature required to reach the end point also increases. This type relation holds for most ceramic bodies and glazes. Also it is a good example of why temperature alone is not adequate in reporting the maturing or firing conditions of ceramic bodies or glazes"
>
> (Fronk and Vukovich, Jr. 1973, 156).

(2) SIZE OF CONES

Both the speed of the firing and the temperature of maturation are considerably higher for the small cones than for the large cones. *The large cones are used for the temperatures given throughout this book.*

(3) SETTING HEIGHT AND SETTING ANGLE OF THE CONE WILL AFFECT THE DEGREE OF CONE DEFORMATION

Inconsistent mountings will produce misleading deformation positions and make difficult an accurate evaluation of the heatwork within the kiln. The lower the height at which the cone is mounted in its clay wad, the lower the degree of its deformation. Similarly, a cone set into its clay wad at an angle less than the recommended 8 degrees produces less deformation; if set at an angle greater than 8 degrees, it will show more deformation. Care should be observed in the mounting of the cones, in order to produce controlled results (Ibid, 157).

Cones measure the maturation points of individual ceramic materials, as well as mixtures of various ceramic materials. The maturation point of a specific ceramic material is known as its pyrometric cone equivalent (P.C.E.), and it customarily appears as the last item in the chemical structure table. (Note that the P.C.E. of ceramic materials is dependent on particle size. For example, the P.C.E. of fine-grained Nepheline Syenite (A400) is cone 3; the P.C.E. of medium-grained Nepheline Syenite (A270) is cone 5; the P.C.E. of coarser-grained Nepheline Syenite (A200) is cone 6.[7]

It is obvious that vastly different results will occur depending on the atmosphere and on the length of time and temperature of the firing. High-fired stoneware pots are, of course, very different in visual appearance, feel, and ring than lower fired ware. Although pots that are fired to about 2150°F–2232°F (cone 5/6) are considered "stoneware," the true nature of stoneware appears only at about 2300°F–2380°F (cone 9/10). The higher silica and alumina content of such ware makes them hard, acid-resistant, and, thus, relatively impervious to the assault of time. They are truly wares of stone.

In addition, the higher and longer the fire, the more the glaze material bonds with the claybody. This bonding creates a middle layer between glaze coat and claybody in which both are merged as one, known as the interface. The glaze does not merely coat the surface of the claybody as it does in earthenware (1922°F–2012°F), but actually fuses with a section of the claybody itself. The interface is characteristic of all high-fired stoneware. This means that in all high-fired pottery, the color and the texture of the underlying claybody becomes an integral part of the final glaze coat. At stoneware temperatures, it is necessary to understand the claybody materials as well as the glazes; together they constitute the total glaze surface. *Figure Intro.5, p15; Intro.6, p16.*

The internal structure of higher fired stoneware glazes and claybodies is different from lower fired ware. The ratio of the three kinds of oxides changes in accordance with the temperature of the firing. Silica and alumina oxides are refractory and have a high melting temperature. For this reason, in order for them to melt at earthenware temperatures, smaller amounts of silica and alumina oxides and a higher amount of melting oxides are necessary. The melting oxides must now do the work that a higher fire would have done. In other words, the lower the fire, the longer the list of additional melters; the higher the fire, the fewer the glaze materials because the increased heat provides most of the melting action. *It is, therefore, preferable to begin a study of glazes with the high-fire stoneware glazes because they contain fewer materials and are therefore easier to understand. Once you have some understanding of how higher temperature glazes work, you will be prepared for the more complex combinations required for lower temperature glazes.* A comparison of the internal structures of earthenware glazes (cone 04–cone 02) and lower fired stoneware glazes (cone 5/6) with high-fire stoneware glazes (cone 9/10) does, in fact, show this difference. When the temperature of the fire is markedly increased or decreased, there should be a corresponding change in the glaze or claybody materials. This will readjust the ratio of the three internal oxides. Otherwise, the fire itself will transform the materials of a claybody into a glaze or, conversely, the glaze and claybody materials into a dry, opaque, and immature surface. *Figure Intro.4, p14.* The fire is always a primary consideration. We must understand the reactions of the glaze and claybody materials to our various firing temperatures and kiln atmospheres. Only this understanding will enable us to make full use of the great potential contained in our ceramic materials.

Notes

[1] The following geological material derives from: Tarbuck and Lutgens 1987, 50–61; Physical Geology, taught wonderfully by Peter Bower and Robina Simpson.

[2] Nigel Wood states that the earliest Chinese glazes were made up of the body clay, wood ash, and burnt or crushed limestone. The core of the later Chinese glazes was altered granitic rock materials, which contained potash mica and quartz. (Wood 1978, 7–13). The Technical Manager of Goonvean and Rostowrack China Clay Co. Ltd., the manufacturer of Cornish Stone, stated in his presidential address of 1977 to the Cornish Institute of Engineers that the altered granitic rock known as "petuntse" was first used in Chinese porcelain about 700 A.D.

[3] Source: Lawrence 1972, 17: Kingery and Vandiver 1986, 214–217; Rhodes 1958, 1–4, 53–56, 61–70; Sanders 1974, 14–18. Larson, Stefan, *Theoretical Physicist*, 1999; conversations with author.

[4] E. Rutherford in 1911 (London) discovered through statistics and experiment, heavy nuclei. We use the term *heavy* nucleus because the electron, which had its charge to mass (e/m) ratio determined by J. J. Thomson in 1897 (London), is 1:1836 the mass of the proton. The mass of the neutron and the mass of the proton is similar (Larson, Stefan, May 1999).

[5] See Kingery and Vandiver 1986, 214–217 for description of scientific phenomena that underlie these different color effects.

[6] Edward Orton Jr. Ceramic Foundation, Fronk, and Vukovich, Jr. 1973, 156–157.

[7] Indusmin Syenite in ceramic whitewares.

Figure Intro.1 (Collection of Department of Earth and Environmental Sciences, Columbia University, New York).

Left: Granite. Slow-cooled, coarse-grained, igneous rock containing 25% quartz, 50% feldspar (mostly potash in this sample), some muscovite, biotite and/or amphibole.

Right: Rhyolite. Fast-cooled, fine-grained igneous rock with the same chemical composition as granite.

Figure Intro.2 Bowl with raised lotus panels on exterior. (Asian Art Museum of San Francisco, The Avery Brundage Collection B60 P120+.)

Late Southern Song (1127–1279), Zejiang (longquan).

Figure Intro.3a Stoneware test bottles.

Cone 9–10 reduction.

Left: Potash feldspar 90%, Whiting 10%, Red iron oxide 1/2%. Rim dip in Albany Slip 100, Fire Island purple sand 10, Red iron oxide 6.

Right: Saturated Iron glaze over potash feldspar 90%, Whiting 10%, Red iron oxide 1/2%.

Back center: Nepheline Syenite, 50%, Albany Slip 50%.

Figure Intro.3b Stoneware Egg jar by Sue Browdy.

Cone 9–10 reduction.

Potash feldspar 90%, Whiting 10%, Red iron oxide 1/2%. Over dip of Albany Slip 100, Red iron oxide 6, Fire Island purple sand 10.

Figure Intro.4 Albany Slip as:

Descending order:
1. Raw clay
2. Claybody, cone 05–06 oxidation
3. Stoneware glaze, cone 5/6 oxidation
4. Stoneware glaze, cone 9–10 oxidation

14

Figure Intro.5

Descending order:
1. Shard of stoneware pot, cone 9–10 reduction; note glassy layer (interface) between claybody and glaze.
2. Shard of stoneware pot, cone 9–10 oxidation, interface layer barely visible.
3. Shard of stoneware pot, cone 5–6 oxidation, interface not visible.

Figure Intro. 6 Jacky's Clear glaze.

Cone 5–6 oxidation.

Left to right:
1. Porcelain claybody.
2. Black claybody.
3. Stoneware claybody.
4. Stoneware claybody, cone 9–10 reduction.

Figure Intro.7 Tests (porcelain claybody).

From left counterclockwise:
1. Potash feldspar 90%, Wollastonite 10%, Red iron oxide 1/2%. Cone 9–10 oxidation.
2. Same glaze cone 9–10 reduction, copper-red blush from vaporization of copper-red neighbor.
3. Same glaze, cone 9–10 reduction.

Background Rocks: (Collection of Department of Earth and Environmental Sciences, Columbia University, New York.)
Left: Granite with potash feldspar.
Right: Wollastonite.

Figure Intro.8 Jar by Bill Lau.

Stoneware, cone 9–10 reduction. Exterior unglazed. Interior glaze: Charlie D Black. Black Mason Stain 3%, Manganese Dioxide 3%, Cobalt oxide 5%. Glazed by Sue Browdy after death of potter.

LAB I Test Procedures

The laboratory sections at the end of each chapter outline a work plan for the student. The proof is always in the doing. One can read about ceramic materials for years without truly understanding how ceramic materials function in glazes and claybodies. Knowledge comes first and foremost from the testing process.

Before you plunge in, however, consider the problem of the test pots themselves. How many of us have spent hours measuring out our test glazes only to apply them on broken shards or nondescript forms that someone else has made? Then there is, of course, the opposite situation in which we spend much time on the clay form only to cover it with an unknown, white-colored glaze liquid. In either case, the final result will be difficult to read. It is for this reason that you should give much thought and care to the test pot form and the method of glaze application, as well as to the constituents of the glaze and claybody. In this respect, the following procedures may prove helpful.

GENERAL TEST PROCEDURES RULES

1. The test pot should reflect, on a small scale, the basic style and personality of the potter. The shape and form of the test pots should constitute the identifying signature of the maker. Wherever possible, the test pot should be a miniature replica of the ultimate form on which the glaze will appear. Ideally, the test forms should display interior, exterior, vertical, and horizontal flow of the glaze magma, as these different surfaces often produce different appearances. Thus, a vertical form such as a tall cylinder or bottle combined with a shallow bowl or plate might well be necessary for accurate evaluation of the final result.

2. Each test should be large enough so that it can be easily divided in half. One half will contain the control (the original glaze) and the other half will contain the test materials. The bottom of the test should be smooth and wide so that the contents of

the test may be clearly written with a mixture of iron oxide and water. Position your writing so that you will be able to identify and distinguish each test mixture from the control. Because some of the test materials may cause the glaze to run, each test should have an attached foot that can catch the glaze overflow.

3. There is often more variation in results between thick and thin glaze coats than there is between different glaze formulas! Each test should show the result of a thin (1 dip), medium (2 dips), and thick (3 dips) coat.

4. Similar principles apply to changes in the thickness, scale, and shape of the test pot form. If thickness, scale, or form of the pot changes, the glaze surface may also appear changed. For this reason, test pots in any one test series should be as uniform as possible. The claybody for the tests should remain constant throughout the testing series and should not be changed. It should consist of the primary claybody used by the potter for his or her wares.

5. Different firing conditions will also cause far-reaching changes in the glaze surface. There can be considerable heat differences between the top, middle, or bottom of your kiln. *Figure 3.8, p230.* If you make three tests of each test mixture and place one of them on each of these three positions, you will be able to eliminate the misleading variable of uneven firing conditions. The more test samples that you place in a single kiln firing, the more accurate will be your reading of the final result.

6. Try to fire the same test in as many different firing temperatures and kilns as are available. In all cases, make at least two identical samples of each test mixture so that you can fire each test in at least two different firing temperatures.

7. Save all leftover test mixtures and place each in a separate jar, clearly labeled with the name of the test and date of mixing. You will use these residues for comparative tests and for repetition of tests when needed. It is to be hoped that during the course of your test efforts, you will find some of your test

LAB I

results interesting enough to repeat on a larger scale. Hence, this practice of saving test mixtures will also enable you to evaluate the storage life of the various test mixtures, and determine the necessity for the addition of deflocculants, or suspension materials, should you mix up any of these test materials in larger quantities for further use.

8. Make 30 test pots the size of a lemon to be used in the first, highly experimental testing stage. Make 30 more test pots the size of an average teacup for the second stage of retesting. An experienced potter is well aware that a test pot is but the first stage in a long progression of testing and retesting the test formula on different shapes and sizes and in different firing temperatures and kiln atmospheres. Only then can one fairly evaluate the test result.

SUGGESTED READING AND MUSEUM TRIPS

Leach, Bernard. 1975. *The Potter's Challenge.* Chapter I, 15–16. New York: E.P. Dutton & Co.

Palissy, Bernard. 1975. "A renaissance glaze manuscript." Trans. J. David Townsend. *Ceramics Monthly* (September): 29–33.

American Museum of Natural History, New York, N.Y. The Hall of Gems.

The Metropolitan Museum of Art, New York, N.Y. Chinese ceramics collection.

Chapter 1 | GLAZE CORES

FELDSPARS AND ROCKS

Figure 1.3 p80.

INTRODUCTION

An analysis of certain beautiful Song Dynasty porcelain glazes revealed that a single feldspathic rock material (Petuntse) provided the core of the glaze. (Presidential Address of C. V. Smale 1977. Hetherington 1978, 4; Wood 1978, 17–21.)[1] This single material contained nearly the right proportion of glassmaker, adhesive, and melter oxides. Only small amounts of wood ash and limestone materials were added to improve the color and melt of the glaze. *I believe that this is still the most meaningful way to approach the stoneware glaze, or any glaze or claybody for that matter.* The objective is to locate one single earth material that alone almost provides the desired surface, and then to add as few additional materials as possible. The high heat of the stoneware fire makes it possible to use this empirical approach in the search for glaze surfaces because most of nature's minerals will melt at these temperatures with the addition of but a very few materials. I call this primary material, which almost achieves the desired glaze surface, a "glaze core." The list of glaze cores is long and disparate and includes feldspars, mica, granitic rocks, some clays, volcanic ash, wood ash, boron minerals, and the artificial manufactured frits. The key characteristic of these materials is their combination of glassmaker, adhesive, and melter functions.

Feldspars and feldspathic rocks contain a complex structure of silica, alumina, and the melter oxides of sodium, potassium, and calcium. Their complex structure makes them ideal glaze cores at stoneware temperatures. Important ceramic texts stress the glaze core role of feldspar. Susan Peterson gives feldspars their due as "Nature's Glazes"

(Peterson 1992, 148, 309; see also Rhodes 1958, 59–73). On the other hand, technical data issued by the ceramics and glass industry describes feldspar as a melter or flux and as an important alumina source.

"... the role of feldspar in glazes is similar to that in (clay)bodies. It again is used for its fluxing action . . . ."
(*Ceramic Industry* 1998,104)

"Unimin's F-4 Feldspar provides a constant, reliable, and economic source of alumina....Our F-4 Feldspar is an excellent flux and contributes to obtaining the desired vitrification."
(*Technical Data* 1988, May)

The kinds of ceramic materials available to potters are mainly determined by industrial ceramics, and it is, therefore, not surprising to find that many teachers of nonindustrial ceramics in colleges and ceramic schools follow industrial nomenclature and view the role of feldspar as a flux. If, however, the study of ceramic material is approached from a purely empirical standpoint, it becomes difficult to understand how feldspar can function solely as a flux in a glaze magma. Mix powdered feldspar with water—apply this mixture to a clayform—fire it to stoneware temperatures, and there will appear a glossy, white surface on the clayform. *Figure 3.2, p224.* Thus, feldspars and feldspathic rocks with their complex chemical structure of silica, alumina, and melter oxides of sodium, potassium, and calcium possess the unique ability to form an "almost" acceptable glaze surface at stoneware firing temperatures. Ceramic materials that have this potential are described throughout this book as *glaze cores*. Because feldspars and feldspathic rocks are the major ingredient of many stoneware glazes, much time will be spent in exploring their potential.

ORIGIN

Feldspars and feldspathic rocks are natural glazes. Their glassmaker, melter, and adhesive oxides form a tightly bonded network that is barely disturbed by the short-lived fires of a stoneware kiln. There is no comparison between a crystalline feldspar formed within the earth and a combination of the same oxides that have been fused together by human effort. Attempts to recreate the fired surface of a potash feldspar by firing a combination of silica (the glassmaker), potassium carbonate (the melting oxide), and kaolin (the adhesive) to stoneware temperatures resulted in a stiff, white, unmelted surface that in no way resembled the crystalline sparkle of the potash feldspar surface. Although the oxide structure of the recreated or synthetic feldspar was almost identical to the crystalline feldspar, neither the fired surface nor the duration of the fire was comparable. Similarly, when equivalent amounts of silica, potassium, and alumina hydrate were substituted for the potash feldspar in a stony, white, orange-brown-flecked stoneware reduction glaze, the final fired result was an unmelted, stiff, whitish coat. Once again, the oxide structure of both glazes was identical—only the source of the oxides and the firing process were different. Here was dramatic proof that a stoneware kiln fire was unable to fuse these materials into the tightly bonded structure of a crystalline feldspar. *Figure 1.1, p78.*

Throughout earth's history, violent upheavals have forced silica-rich magma up toward the earth's outer layers. Under these outer layers, the magma cooled slowly for thousands of years to form the large-grained crystalline rocks known as granite. The process of granitic formation is highly complex and is not fully understood even by geologists (Bates 1969, 27). When exposed on the earth's surface, granites are subjected to two types of weathering. Mechanical weathering (physical disintegration of granites by expansion of water, tree roots, groundwater, animal footsteps, etc.) causes the granites to be broken down into their various minerals—mainly feldspars, quartz, and micas. Chemical weathering (chemical reaction of the granites to the air (atmosphere), living beings (biosphere), earth (geosphere), and water on the earth's surface and atmosphere (hydrosphere) causes some feldspar and mica minerals to further decompose into clay minerals.

Granites are the basis of most of our ceramic materials and make up 75% of the earth's crust. *Figure Intro.1, p11.* They are rocks, which by definition are mixtures of one or more minerals. Granites consist of over 50% potash and soda feldspar and up to 25% quartz. They also contain as much as 20% mica and lesser amounts of magnesium-iron minerals, such as hornblende (Tarbuck and Lutgens 1984, 59). Some granites, if crushed to a fine particle size, will make exciting glaze surfaces at high stoneware temperatures. The material known as Cornwall Stone (composed of feldspar, mica, kaolin, and quartz) makes a fine glaze on a porcelain claybody when it is combined with but 10% calcium carbonate. Cornwall Stone is an altered granite in which chemical weathering has caused the granite to partially decompose into kaolin (clay). Although granite contains different minerals, the primary mineral is feldspar. More variable in structure than the feldspars, Cornwall Stone and Nepheline Syenite are "granitelike" materials that become superb glaze surfaces at stoneware temperatures. *Figure 1.3, p80; 1.4, p81; 1.8, p85.* These materials, like the feldspars, contain potash and soda melters in a relatively insoluble form. Unlike feldspars, Cornwall Stone and Nepheline Syenite are not minerals, but rather are a combination of minerals that can be physically separated into their component parts. Hence, these materials are rocks. Granite is a common example of a rock. Cornwall Stone is described by its manufacturer as a "lightcolored, naturally occurring granite" (*Technical Data*, 1989). Nepheline Syenite is a quartz-poor "granitelike" rock. The different minerals that a rock contains can vary in proportion from time to time. Therefore, a rock may be more variable and less stable in performance than the more structurally intact feldspar minerals.

GENERAL CHARACTERISTICS OF FELDSPARS

Feldspar includes an assortment of minerals of varying composition. Despite this range, the feldspars commonly used by potters tend to follow a fairly recognizable pattern when fired to stoneware temperatures.

1. The most striking characteristic of a feldspar that is fired to stoneware temperatures is the formation of a glassy, white surface. The heat of the stoneware kiln fire, combined

with the fusion power of the feldspar's soda and potash melter oxides (14%–15%) have transformed its considerable silica content (60%–70%) into glass. The white color is a happy consequence of the selection of atoms by size—the atoms of the coloring minerals such as iron and copper are too large to fit into the feldspathic structure. The result is a relatively pure white material to which colorants can always be added.

2. The melting action of the feldspars has a very long range. They begin to melt at about 2138°F (Cone 4) and continue their fusion well beyond the Cone 10 temperature range (2345°F–2381°F). This range makes feldspar a desirable material for a stoneware kiln because the temperature range of the bottom and the top of the kiln is often lower or higher than the middle sections.

3. Melted feldspars possess a high surface tension because of their considerable alumina content (17%–25%); they crawl and flow unevenly on the walls of the clay form. This is especially noticeable with a thick coat of feldspar.

4. The surface of melted feldspars contains an intricate network of fine cracks alternately described as "crazes" if considered a glaze defect and "crackle" if considered aesthetically desirable. It is interesting to note that some feldspars have a highly individualistic craze/crackle network. Their characteristic pattern springs to life when the feldspar combines with 10% calcium melter and is fired to high stoneware temperatures. (See Comparative Tests of Feldspathic Glaze Cores, pp. 42–43.)

The network of G-200 feldspar (a predominantly potassium feldspar mined in Monticello, Georgia) tends to resemble flowerlike petals. Custer Feldspar (another predominantly potassium feldspar mined in Custer, South Dakota), on the other hand, produces a network that is filled with tiny glassy bubbles.

Melting oxides, contained in the oxide structure of the feldspar, are responsible for the craze/crackle network that appears on the surface of the melted feldspar. These melting oxides are for the most part sodium and potassium. Sodium and potassium oxides undergo a high rate of expansion when heat converts them from a solid into a liquid state. As they cool and change back again into a solid state, these oxides will contract at a corresponding rate. This high rate of contraction occurs at a point when the clay form has hardened and is no longer soft and pliable. Therefore, the cooled glaze surface breaks up into a fine network of crazes and/or cracks and in this way reduces the tension and pressures of the underlying claybody.

5. Feldspars do not remain evenly dissolved in the liquid glaze mixture. Within hours, the feldspathic powder settles at the bottom of the glaze bucket, forming a dense, rocklike substance that defies even the most vigorous attempts at disbursement. No matter how feverishly one stirs the glaze, the feldspathic particles remain at the bottom of the liquid mixture in a sodden mass. This makes it difficult to apply an even coat of high feldspathic glaze to the wall of the clay form.

It must now be apparent that although feldspar provides the basic core of a stoneware glaze, it does present certain problems for the potter. We can solve these problems by adding small quantities of three or four minerals to the feldspathic glaze. *The result is the formula for a standard stoneware glaze.*

The most serious problem to correct is the crawl and uneven flow of the melted feldspathic glaze. What this means is:

1. The heat of the stoneware kiln is not high enough to completely melt all of the materials contained in the feldspar so that the magma will flow evenly over the clay form.

2. The high alumina content of the feldspar creates a viscous magma of high surface tension, which tends to bead up on itself.

Additions of limestone or calcium minerals will increase the melt at stoneware temperatures and thus quicken the flow of the feldspathic glaze.

Additions of the glassmaker (silica) will eliminate the craze/crackle network, should this be desired. Silica, unlike the sodium and potassium melters, has a minimal rate of contraction upon cooling, and thus inhibits the high contraction rate of these melters.

Physical suspension of the feldspar in the liquid glaze may be improved by adding 10% or more clay materials such as kaolin or ball clays. The addition of the clay materials will also toughen the raw glaze coat and help it withstand the handling that takes place when the kiln is stacked. Suspension will be further improved by the addition of 2%–3% superplastic clay (bentonite) or even smaller amounts of soda ash or Epsom salts (magnesium sulfate).[2]

Minerals, such as copper, iron, or cobalt, may be added in oxide or carbonate form to achieve color. In the case of the celadon glaze, which is fired in the oxygen-reduced atmosphere known as "reduction," small amounts of iron oxide (1/2%–2%) will produce the gray-blue-green color of the ancient Chinese celadon glazes. Small amounts (1/2%–2%) of copper oxide carbonate added to this same celadon glaze will produce the copper red glaze if the firing conditions are perfect. Increased amounts of iron oxide will produce the black-rust, saturated iron glaze. *Figure Intro.7, p17; 1.7, p84.*

This combination of materials spawns a broad range of standard stoneware glazes. Although a specific stoneware glaze formula may show four or even five ingredients in its recipe, in most cases the core of the glaze is the feldspar. The rest of the materials are present in order to cure the problems contained in the feldspar.

The role of feldspar is clearly illustrated by five glaze dissection tests made with Herbert Sanders' celadon glaze. In each test we omitted one of the five glaze ingredients.

SANDERS CELADON		
Figure 1.1, p78; 1.5, p82.		
		Transparent. Gloss. Gray-green.
Kona F-4	44%	
Flint (silica)	28	
Whiting (calcium carbonate)	18	
EPK kaolin	10	
Barnard Clay	5	Colorant

Although this glaze recipe includes five ingredients, its central ingredient or core lies in Kona F-4 feldspar. This is clearly demonstrated by the test in which the feldspar was omitted. The surface changed from a transparent, gray-green gloss to a dull, opaque, yellow-green matt. With the exception of the test where the additional melter calcium carbonate was omitted, the separate omission of the other glaze ingredients did not change the glaze surface as drastically as did the omission of the feldspar. *Figure 1.1, p.78.*

Similar results appeared in a series of tests performed on a second celadon glaze in which the feldspar was only 1% higher than the silica content.

RHODES PORCELAIN GLAZE		
Figure 4.3, p267.		
		Transparent. Gloss. Gray-blue.
Custer feldspar (potash)	33%	
Flint (silica)	32	
Whiting (calcium carbonate)	20	
EPK kaolin	15	

As in the Sanders Celadon test series, the test that omitted Custer feldspar displayed an opaque, yellow-brown, stony-matt surface in strong contrast to the gray-blue, transparent gloss of the original glaze.

The following tests display further persuasive evidence of the glaze core function of feldspar.

A combination of 90% potash and/or soda feldspar, 10% calcium carbonate, 2% bone ash, and 1/2% red iron oxide creates a sparkling blue or blue-green glaze surface at cone 9/10 reduction temperatures.

At the cone 5/6 oxidation temperatures, 70% Kona F-4 and 30% Wollastonite creates a creamy, satin-matt surface. See also Nepheline Syenite 80%, Wollastonite 20%, *Figure 4.7, p271.*

The oxide structure of a feldspar explains why it constitutes the central ingredient core of a stoneware glaze. As stated previously, the term feldspar includes a vast array of feldspathic minerals that have a variable oxide structure. Despite this variation, the oxide structure of most feldspars tends to follow a general pattern. Most feldspars contain about 60%–70% silica (the glassmaker), 17%–25% alumina (the adhesive), and 10%–15% sodium, potassium, and/or calcium oxide (the melters). The feldspar that appears in Sanders Celadon, known as Kona F-4, follows this general pattern. The general oxide structure of Kona F-4 at the time of the tests was as follows:[3]

Silica (the glassmaker) .67.2%
Alumina (the adhesive) 19.5%
Melter oxides (sodium, calcium, potassium) . . .13.0%

Compare this oxide structure with the oxide structure of Sanders Celadon itself:

Silica 69%

Alumina 14%

Melter oxides (sodium, calcium, potassium) . . .17%

The difference between the oxide structures of the feldspar and Sanders Celadon result from the corrective additions of the minerals described above. Additions of the calcium carbonate melter, Whiting, increases the melt; additions of the silica glassmaker, Flint, reduces the craze/crackle network. These additions create an oxide structure with increased melter and glassmaker oxides. The alumina oxide is correspondingly lower in amount. Thus with the corrective additions of but a few minerals, a feldspar is transformed into the celadon glaze.

SODA AND POTASH MELTERS
Figure 1.2, p79.

Ours is a world of plenty, and there are many different kinds of available feldspars. Feldspars are classified according to the predominate melter contained in their oxide structure. In ceramics there are two basic categories of feldspars: potash feldspars in which the primary melting oxide is potassium and soda feldspar in which the primary melter oxide is sodium. In order to understand their special character traits, it is helpful to consider the properties of soda and potash melters.

Pure sodium oxide (soda) and potassium oxide (potash) are powerful melters. Their melting point is low (1652°F, soda; 1291°F, potash).[4] Soda was one of the earliest melters used by the potters of ancient Egypt and is the primary melter in the turquoise blue Egyptian glazes that originated some thousands of years ago. Sodium, in the form of salt (NaCl), creates the peculiar, orange-peel texture known as "salt glaze," which was perfected by the German potters in the fourteenth century and introduced into England at the end of the seventeenth century. Soda and potash are both important melters in the production of glass. They share the following characteristics:

1. Soda and potash have the highest thermal expansion and contraction rate of all the ceramic melter oxides (Hamer and Hamer 1986, 356) This means that a sizeable amount of either melter[5] in a glaze will make the glaze coat shrink more than the claybody; a network of craze-crackle lines will thus appear on the surface of the glaze.

2. Soda and potash promote color brilliance and luster at most firing temperatures. In addition, soda and potash materials encourage specific color results. A robin's egg blue color results from the combination of a high-soda feldspathic glaze core with copper oxide (1%) and a calcium melter, fired in an oxidizing atmosphere. If the reducing atmosphere is sufficient, a high-potash feldspathic glaze core in a celadon glaze with 1/2% red iron oxide will produce a blue-gray color. The substitution of a high-soda glaze core will change the color to gray-green. Note the opposite result attributed to soda in Hamer and Hamer 1986, 177. This contradiction underscores the difficulty of making conclusions that hold true for all situations. In each case, the conclusions are based on test results obtained from different materials, glaze formulas, and kiln firings. The presence of considerable amounts of alumina in the available soda materials in our tests may well have influenced the green cast of the final result. In the final analysis, each potter must be guided by his or her own test results.

3. Soda and potash applied alone to the surface of the stoneware claybody produce startling and dramatic surface effects. They singe the surface of an iron-bearing claybody deep orange or yellow-gray-brown, bringing flashes of color to a static oxidation body and deepening the red-orange color of a reduction claybody.

Of the two oxides, soda is the stronger and more reactive melter. Unlike potash, soda may volatilize above temperatures of 2192°F. It is more soluble than potash and is even slightly soluble in feldspathic compounds (Hamer and Hamer 1986, 250, 298–299). Soda reddens a claybody more deeply and consistently than potash in both oxidation and reduction atmospheres. It creates a higher gloss surface on the claybody by melting the silica contained in the clay surface. In a reduction atmosphere, a clay surface streaked with soda may appear glassy and a black-orange in color. The black color results from the unique attraction of soda to the carbon particles in the reduction atmosphere, which it traps within the glossy surface layer. This special "carbon trap" quality of soda is important to many glaze surfaces. For example, in a Shino "Carbon Trap" glaze, soda plays a crucial role in producing the typical orange-red-black colors. *Figure 1.10, p87*. If the soda is omitted from the Shino glaze, the surface changes to a dull white.

Potash fired alone creates a duller, more matt surface, interspersed with puddles of glass. Its reddening power seems not to be as strong or as consistent, and it tends to create a grayer color (yellow-brown in oxidation) on the stoneware clay surface in an oxidation atmosphere compared to the red mahogany hue of soda. On the other hand, our tests of potash showed that it flowed through the claybody with more force than did soda, which seemed to alter only the surface layer of the claybody. Note that the initial melting temperature of potash (1291°F) is lower than that given for soda (1652°F). Despite this fact, once soda begins to melt, its faster melting power creates glassier surfaces than that of potash. Soda and potash feldspars reflect these differences. Once again, soda feldspar melts at a higher temperature than the potash feldspar; however, the actual flow of the soda feldspar, once it begins, is more fluid and less viscous than its potash feldspar counterpart. Hence, after a kiln firing, the soda feldspar displays a shinier and more melted surface.

4. Both soda and potash counteract the settling or flocculating tendency of high feldspathic glazes. They accomplish this result by neutralizing the electrical charge that exists between the molecules of a liquid feldspathic glaze; this causes the molecules to slip past each other instead of coagulating into nonfluid groups (Hamer and Hamer 1986, 100). We found that a 5% addition of soda ash was sufficient to keep a high feldspar glaze in suspension for days. When the soda ash was replaced by calcium carbonate, the glaze materials settled out into a gummy, sodden mass at the bottom of the container.

5. Both potash and soda melters create a glaze magma with low surface tension (Hamer and Hamer 1986, 307–308). A glaze magma with low surface tension flows freely over the surface of the clay form; a glaze magma of high surface tension will crawl or bead up on itself, leaving bare patches of exposed claybody. Potash has a slightly lower surface tension than soda and has the lowest surface tension of all materials used in ceramics. Thus, it would seem that a high potash material such as potassium carbonate[6] (deemed to be highly toxic) is a desirable addition for the purpose of curing a crawled glaze surface. (With this theory in mind, I added 2.5% potassium carbonate to a glaze that had often produced a crawled surface due to the high surface tension of the feldspathic glaze core. However, the test result did not reveal the superior workings of potassium carbonate. The glaze without the potassium addition produced a surface as smooth as the test glaze! Further test firings with increased additions of potassium carbonate are needed in order to make a conclusive evaluation of the test results.)

6. The only pure source of soda and potash that is readily available to the potter is found in water-soluble materials such as soda ash and potassium carbonate. When a water-soluble material becomes part of a liquid glaze, it dissolves into the water and travels with the water into the open pores of the unfired claybody. Once a powerful melter flows into the open pores of the claybody, it can drastically lower the firing temperature of the ware and cause it to become bloated, brittle, and overfired. It may also flow in the other direction towards the surface of the glaze where it will form unsightly bubbles and eruptions. In addition, the loss of some of the liquid content of a glaze that contains water-soluble materials will cause a change in the balance of the glaze ingredients. The chemist transforms these water-soluble melters into an insoluble state by bonding them to low amounts of silica and alumina. The result of this fusion is an artificial, laboratory-produced, nonsoluble glaze core known as a "frit." Silica, alumina, and water-soluble materials are fired in a kiln to the temperature that will fuse them into a glass. The glass is ground into a powder and is thus ready for use as an insoluble soda and/or potash glaze material. (See Chapter 3, p193.)

The laboratory method of producing insoluble soda and potash is of course expensive—a far cheaper source lies at hand. Feldspars contain sizeable amounts of these melters in a comparatively insoluble form and are, therefore, natural frits.[7] Nature thus provides us with the heart and soul of the stoneware glaze.

POTASH FELDSPARS

TRADE NAME	Custer, G-200, K200
GEOGRAPHICAL SOURCE	**Custer Feldspar** Custer, South Dakota: Pacer Corporation
	G-200 Feldspar Monticello, Georgia: Feldspar Corporation, Zemex Industrial Minerals
	K200 Feldspar Kings Mountain, North Carolina: Franklin Industrial Minerals
GEOLOGICAL SOURCE	Orthoclase and microcline found in coarse-grained granitic rock (pegmatite).

MINERALOGICAL STRUCTURE[†]

Primarily orthoclase and microcline
(potash feldspars)
Some Albite (soda feldspar) and
Anorthite (calcium feldspar).
Free Quartz (0.4%–26.7%)
Minor amounts of iron oxide (0.04%–0.15%)
Magnesia—trace

CHEMICAL STRUCTURE

	Custer[*]	G-200[**]	K200[***]
Silica (SiO_2)	68.50%	66.30%	67.70%
Alumina(Al_2O_3)	17.00	18.50	18.00
Iron Oxide (Fe_2O_3)	0.15	0.082	0.07
Potash (K_2O)	10.00	10.75	10.0
Soda (Na_2O)	3.00	3.04	3.2
Calcia (CaO)	0.30	0.81	0.14
Magnesia (MgO)	Trace	Trace	Trace
L.O.I.[****]		0.16	0.30
M.P. (Melting Point)[†]	2014°F–2640°F		

[*]   Pacer Corporation, 1998.
[**]   The Feldspar Corporation, Zemex Industrial Minerals, 1998
[***] Franklin Industrial Minerals, 1998.
[****] Loss on ignition
[†]   (*Ceramic Industry* 1998, 102, 104.)

POTASH FELDSPARS

INTRODUCTION

Creations of nature are rarely found in a purified state. Thus, granites contain both soda and potash feldspars; the feldspars mined for industrial use usually contain a mixture of both potash and soda. A feldspar is labeled "potash" feldspar if the greater proportion of its melting oxides consists of potassium. Over the years, there have been many potash feldspars on the market—Clinchfield, Buckingham, Oxford, and Kingman are but a few ghosts from the past. In view of this high mortality rate, it is important to understand the industrial history of your particular potash feldspar so you can be sure of its stability. In this connection, note that the manufacturer of Custer Feldspar is an old timer on the ceramic scene. For 70 years Pacer Corporation has mined high-quality potash feldspar from pegmatite rock located in Black Hills, Custer, South Dakota. Different grades of Custer Feldspar from over 700 mines are blended together to produce a standard product. The daily output is over 100 tons. There are no minimum order requirements. Although this feldspar is used primarily in the manufacture of porcelain insulators, "the ceramic industry is very important to Pacer" (Letter of June 16, 1989, Technical Service Manager, Pacer Corporation; *Technical Data*, Pacer Corporation, 1998). These facts point to the stability of this feldspar and are the kind of background information that helps a potter make an intelligent selection from the vast array of available feldspars.

CHARACTERISTICS OF POTASH FELDSPARS

1. Effect on Fusion of Glaze Surface

According to industrial research, the presence of potash feldspar in a glaze or claybody has a more refractory effect on the ceramic surface compared to equivalent amounts of soda feldspar. Although potash feldspar actually begins its melt at a lower temperature than soda feldspar (Cardew 1973, 306), once the melt begins, the formation of leucite crystals causes a slower and more viscous flow (*Ceramic Industry* 1998, 102, 104). In addition, potash feldspar is said to dissolve silica more effectively than soda feldspar and generally creates a stronger and more stable surface (Ibid.).

In practice, our tests of feldspars fired alone bear out the more refractory nature of potash feldspar compared to those feldspars that are higher in soda. This result is clearly evident at lower stoneware temperatures in an oxidation atmosphere. On the other hand, once a feldspar is combined with a calcium melter, or is incorporated into a total glaze batch, we found inconsistent results. *Ceramic Industry* reports that in a claybody, potash feldspar combined with calcium carbonate shows more fusion than comparable tests with soda feldspar (Ibid, 102). Our comparative tests of soda feldspar Kona F-4 and G-200 feldspar, with 10% additions of Whiting (calcium carbonate), did not show any clearly discernible difference in fusion, possibly because Kona F-4 contains a considerable amount of potash. A comparable test of Soda Spar #56 (now extinct), which contained more soda than Kona F-4, did produce a less-glossy surface compared to its potash counterpart. The results are clouded by the fact that we do not use pure soda or pure potash feldspars in ceramics. As the chemical analyses clearly show, both kinds of feldspars contain soda and potash in their oxide structure. Hence, it is not surprising that we obtain varied results in our glaze core substitution tests depending on the particular oxide ratio of the total glaze in each case. In many tests, the substitution of Kona F-4 for the potash feldspar caused little change in the glaze surface. Of all the glaze cores discussed in this book, potash and soda feldspars produce the least difference when substituted for each other; they can often be interchanged without causing drastic changes in the glaze surface.

2. Effect on Craze-Crackle Network

The high expansion and contraction rate of the potassium melter in the potash feldspar causes the glaze surface to craze. In this connection, it should be remembered that different potash feldspars have correspondingly different craze/crackle patterns. We tested five potash feldspars with 10% calcium melter (Whiting) and found that of the five, G-200 feldspar had the most pleasing, flowerlike, craze/crackle network filled with tiny white bubbles of gas. These results were consistent and repeated in different kiln firings.

3. Effect on Color

Potash feldspar combined with low amounts of iron oxide and calcium minerals and fired to stoneware temperatures in a proper reduction atmosphere creates an exquisite, sky-blue glaze surface. No other kind of feldspar produces a purer shade of blue. A faint, orange-red line occasionally appears at the edge of a high-potash feldspar glaze surface in both oxidation and reduction atmospheres. This is especially visible on a light-colored oxidation body or a porcelain claybody. Potash feldspars usually contain 3%–4% soda, and this could account for the occasional red-orange edge of a high-potash feldspathic glaze.

4. Variation of Different Potash Feldspars

Although new potash feldspars occasionally appear on the scene, they can usually substitute for each other in most glaze formulas without producing major changes in surface, provided the silica and alumina content are not too different. One exception to this rule is the high-iron glaze. The substitution of G-200 potash feldspar (lower silica and higher potash content) for Custer feldspar produced changes in both the color and surface of a high-iron glaze (see Appendix B1). It should be possible to compensate for the different silica content of the substituted feldspar by adding or removing silica (Flint) from the glaze formula. In any case, before making large-scale substitutions of one potash feldspar for another, one should compare the oxide structure of both feldspars and recompute the percentage oxide analysis of the glaze with the substituted feldspar (see Appendix A1, Glaze Calculation Techniques).

SODA FELDSPARS

Kona F-4 feldspar is an important "soda" feldspar for industry, individual ceramic studios, and schools. (Unimin Corporation has dropped Kona from its name and simply refers to it as F-4.) This feldspar has remained on the market for a long period of time and is widely used in the whitewares and ceramics industries (*Technical Data*, Unimin Corp. 1998). Kona F-4's importance to industry guarantees the future stability of its performance. Note that Kona F-4 also contains a fair amount of potassium oxide (5.0%), and that its total sodium content (6.8%) is not as high as the total content of potassium in potash feldspars (10%). Kona F-4 is a hybrid that incorporates some qualities of both potash and soda feldspars. This is especially evident when Kona F-4 is compared to stronger sodium materials, such as the now-extinct soda feldspar #56 or Nepheline Syenite. Hence, it is often possible to substitute the Kona F-4 feldspar for potash feldspars without causing a dramatic surface change. Nevertheless, Kona F-4 feldspar is classified as a "soda" feldspar and commonly appears in stoneware glaze formulas that require the presence of soda feldspar.

CHARACTERISTICS OF SODA FELDSPARS

1. Effect on Fusion of Glaze Surface

Soda feldspar fired alone produces a slightly more melted glaze surface than potash feldspar. This result is clearly visible at the lower stoneware firing temperatures in an oxidation atmosphere. Kona F-4 also produces this result. However, these differences are obscured once the Kona F-4 feldspar is incorporated into a glaze. The results become inconsistent and vary, depending on the oxide ratio of the particular glaze formula and the presence of the calcium melter.

2. Effect on Craze/Crackle Network

There are subtle differences in the craze/crackle network of soda feldspar combined with 10% calcium carbonate as compared to potash feldspar. The craze/crackle network of Kona F-4 often appears smaller and finer without the distinct, flowerlike, craze network that appears in some

---

SODA FELDSPARS

TRADE NAME	Maxum® Sodium Feldspar (Kona F-4); NC-4
GEOGRAPHICAL SOURCE	Spruce Pine, N.C. Kona F-4—Unimin Corp. NC-4—The Feldspar Corp. Zemex Industrial Minerals.
GEOLOGICAL SOURCE	Decomposition of granitic pegmatite (coarse-grained granite rock)

MINERALOGICAL STRUCTURE*

PRIMARY MINERAL	**Soda Feldspar** Albite ($6SiO_2 \cdot Al_2O_3 \cdot Na_2O$)
LESSER MINERALS	**Potash Feldspar** Orthoclase or Microline ($6SiO_2 \cdot Al_2O_3 \cdot K_2O$)
	Calcium Feldspar Anorthite ($2SiO_2 \cdot Al_2O_3 \cdot CaO$)
	Iron oxide 0.04%–0.15% **Free quartz** 0.04%–26.7% (Kona F-4 < 8.0%)** **Magnesium** Trace

CHEMICAL STRUCTURE

	Kona F-4** Maxum®	NC-4***
Silica (SiO_2)	67.90%	68.15%
Alumina (Al_2O_3)	19.00	18.85
Potash (K_2O)	4.80	4.10
Soda (Na_2O)	6.70	6.82
Calcia (CaO)	1.60	1.40
Magnesia (MgO)	Trace	Trace
Iron Oxide (Fe_2O_3)	0.05	0.07
L.O.I.	0.20	0.09
M.P.	1958°F	2014°F–2640°F

*   *Ceramic Industry* 1998, 102.
**   *Technical Data*, Unimin Corporation, 1998
***   *Product Data*, the Feldspar Corporation, Zemek Industrial Minerals, 1998.

potash feldspars such as G-200 and the extinct Buckingham feldspar. *Figure 1.3, p80; 3.2, p224*. Note that a fine network of craze/crackle lines indicates a higher contraction rate than a network of larger, more visible lines. This indicates that Kona F-4 has a greater expansion and contraction rate than its G-200 potash counterpart, despite its higher silica content. Because soda has a higher contraction rate than potash, we would expect to find a stronger craze/crackle pattern in a surface created with a soda feldspar than with a potash feldspar.

3. Effect on Color

A high-soda feldspathic glaze, fired to stoneware temperatures will singe a claybody orange or deep red. This effect is caused by the volatilization of some of the soda at temperatures of 2192°F (Hamer and Hamer 1986, 298). Although as mentioned previously, the vaporization from a potash feldspar will also redden and singe a claybody, a high-soda feldspar will produce this effect more consistently and with greater force and effect. This reddening effect is especially noticeable on a light-burning claybody. Kona F-4 does not consistently produce this result.

Soda feldspars appear to be more sensitive to atmospheric copper vapor than their potash counterpart and will display a pinkish-gray cast whenever copper vapor is present in the same kiln firing. For this reason, copper red glazes often contain a soda feldspar glaze core.

Some iron glazes may produce a different surface color when soda feldspar replaces a potash feldspar as the glaze core. A grayer celadon resulted when a soda feldspar with higher alumina content (soda feldspar #56, no longer available) replaced a potash feldspar glaze core in a celadon (low-iron) glaze. Changes appeared in high-iron glazes when Kona F-4 replaced the potash feldspar glaze cores. The black color of a high-gloss, celadon base glaze changed to dark green-brown with the substituted soda feldspar glaze core in at least two different glazes.

Oxidation cone 5–6 glazes show color changes in response to soda or potash glaze cores. A unique robin's-egg blue was produced with the help of a soda feldspar in a high-calcium, satin-matt, copper glaze. Subtle color changes occurred in a high-iron, gloss glaze when the Kona F-4 glaze core was replaced by a potash feldspar. (For the specific color changes that occurred in each of the above test glazes, see Appendix B2.)

> **Note:** The interchange of the potash and soda feldspars within a glaze magma usually produces some changes in the alumina-silica ratio, and this changed ratio contributes to the color and surface changes as much as the exchanged soda or potash content. For this reason, it is advisable to compute the percentage oxide analysis of the glaze both with and without the glaze core substitution in order to make the necessary adjustments to the total batch recipe (See Appendix A1).

In conclusion, although there are subtle differences between the soda and potash feldspars described above, on the whole, and especially with the Kona F-4 feldspar, these differences are not dramatic. With the exceptions mentioned above, substitutions may be made without creating substantial changes in the glaze surface. The same is not true of the glaze cores discussed in the following sections.

FELDSPATHIC ROCKS

NEPHELINE SYENITE[8]
Figure 1.3, p80; 1.4 p81.

Nepheline Syenite is a special part of the complex igneous rock formations known as the syenites that take their name from their ancient Egyptian quarry site on the Nile river in Syene (Aswan), Egypt (*Encyclopedia Britannica* 1968, 21:555; *Glossary of Geology* 1974, 717). As in the case of granites, the syenites contain soda and potash feldspars together with minor amounts of iron minerals. Unlike granites, the syenites contain little or no quartz, and, hence, are low-silica rocks. Nepheline Syenite is made up of the minerals nepheline (22%); accessory iron minerals (biotite, muscovite, and magnetite, removed during manufacturing process; 4%); soda feldspar (albite; 54%); and potassium feldspar (microcline; 20%). Nepheline is a low-silica, high-soda, high-alumina mineral, which forms from a magma too deficient in silica to create a granite. The word Nepheline derives from the Greek word nephele, which means cloud and is a poetic reference to its color when plunged into acid (Fyre 1981, 677).

Nepheline Syenite is one of our most ancient ceramic materials; our current ceramic source originated in the Blue Mountain region of Ontario, Canada, almost 1.3 billion years ago. The manufacturer is Unimin Corporation, formerly Indusmin Ltd. Tonnage reserves are estimated to be in the millions. Thus, an almost unlimited supply of this valuable ceramic material exists for the potter.

The United States glass industry provides an important market for Nepheline Syenite (Taylor 1989). The high alumina content of Nepheline Syenite both strengthens the glass and decreases its flow during its molten and final stages. In addition, the high soda content of Nepheline Syenite provides the glass with a powerful flux that is cheaper and more concentrated in soda than soda ash.

Once again, the important industrial function of this ceramic material ensures its stability. The chemical structure of Nepheline Syenite, as revealed in the chemical data sheets, has remained remarkably uniform over the years. The greatest degree of variation in Nepheline Syenite occurs with a change in grain size. The finer grinds produce increased melt

NEPHELINE SYENITE

TRADE NAME	(Nepheline Syenite)*
	Spectrum® (Minex®)
GEOGRAPHICAL SOURCE	Nephton, Blue Mountain, Ontario, Canada: UNIMIN Corporation
GEOLOGICAL SOURCE	Medium-grained, low silica, igneous, feldspathic rock formation (Syenites), 1.3 billion years ago (Precambrian)

MINERALOGICAL STRUCTURE

Syenite Rock	78%
Nepheline	22%

Composition:

Albite (soda spar—$Na_2O \cdot Al_2O_3 \cdot 6SiO_2$)	54%
Microline (potash spar—$K_2O \cdot Al_2O_3 \cdot 6SiO_2$)	20%
Magnetite and Iron Minerals**	2%
Muscovite	2%
Nepheline	22%

CHEMICAL STRUCTURE

Silica (SiO_2)	60.20%
Alumina (Al_2O_3)	23.60
Iron oxide (Fe_2O_3)	0.08
Calcia (CaO)	0.35
Magnesia (MgO)	0.02
Potash (K_2O)	4.80
Soda (Na_2O)	10.50
L.O.I.	0.42
P.C.E.***	1868°F

* Sources: Taylor 1989, #5; Indusmin, Inc. 1967, "Mining and Milling Nepheline Syenite"; Indusmin Syenite in ceramic whitewares; Unimin Corporation Technical Data 1998; Bates 1962, 230–231; *Ceramic Industry* 1998, 138; Hammill & Gillespie Technical Data 1985.
** Biotite and Hornblende
*** Pyrometric Cone Equivalent. See p. 9.

and shrinkage rate. The various grinds of Nepheline Syenite are produced at a sophisticated processing complex in Nephton, Ontario. Minor amounts of iron impurities (magnetite and mafic materials such as hornblende) are removed by the milling process. The final product is a pure, uniform, and relatively nontoxic material that stands alone and unequalled as a natural glaze core at stoneware firing temperatures.[9]

Kinds of Nepheline Syenite

Nepheline Syenite is available in various particle sizes ranging from coarse to very fine. They present a wide range of choices and make it imperative that you understand the consequences of the grade number on your supply of Nepheline Syenite. The fluxing power and shrinkage rate of Nepheline Syenite depends on the grade. The finest grades (A400 and 700) have the greatest melting power and shrinkage rate and are used in electrical porcelain and by manufacturers of ceramic wares (Taylor 1989). Grade A270 has a medium melting and shrinkage rate and is the most commonly used form of Nepheline Syenite in ceramic studios and schools. Grades A40–A200 (used by glass manufacturers) are the coarsest grades and therefore produce the lowest melting and shrinkage rates.

A technical data sheet contains an interesting photograph of three grade sizes, A400, A270, and A200, in their melted state ("Indusmin syenite in ceramic whitewares"). This photograph shows clearly the relative fusibility of three forms of Nepheline Syenite and explains the swing of the pyrometric cone equivalent from cone 3 (A400) to cone 5 (A270) to cone 6–7 (A200). A chart of comparative fluxing power of three grades of Nepheline Syenite and a potash feldspar showed that 70% A400, 75% A270, and 80% A200 were the equivalent of 100% potash feldspar. These three tests clearly demonstrate the significant differences in melting power produced by various grades of Nepheline Syenite.

Characteristics of Nepheline Syenite

When Nepheline Syenite is fired alone to high stoneware temperatures, it forms a lustrous gemstone surface. The only difficulty lies in its application to a ceramic form. The high-alumina content of Nepheline Syenite will cause crawling if a thick coat is applied. In addition, like the feldspars and Cornwall Stone, Nepheline Syenite does not remain evenly suspended in water. A small amount of Epsom Salts or soda ash will markedly improve suspension. And, lastly, an unfired, high-Nepheline Syenite glaze coat is extremely pow-

dery and easily flakes off the ware during handling. Small additions of bentonite clay or borax (1%–2%) will toughen the raw glaze coat. Despite these problems, Nepheline Syenite provides the stoneware potter with glaze surfaces that are both fascinating and unique.

1. Effect on Glaze Fusion

Nepheline Syenite will, alone and unaided, create a glossy white surface at both the lower and higher stoneware oxidation and reduction firing temperatures. The Nepheline component of Nepheline Syenite (22%) forms a eutectic bond with its soda feldspar (54%) to produce powerful melting action over a wide sintering range (Taylor 1989, #5, *Ceramic Industry* 1998, 138). This makes Nepheline Syenite an ideal glaze core and/or claybody melter for cones 4–5–6 firing temperatures. The substitution of Nepheline Syenite for a feldspar or Cornwall Stone will appreciably lower the firing temperature of the glaze or claybody. A comparison of the quantities of potash feldspar and Nepheline Syenite, which are necessary to produce an equivalent melt, showed that approximately 25% more potash feldspar is required to produce the same fluxing activity obtained with Nepheline Syenite (Indusimin Inc. 1967, fig. 2).

2. Effect on the Glaze Surface

Despite its increased melting action, the lower silica and higher alumina content of Nepheline Syenite can cause a gloss surface to become less shiny and more matt. This is clearly illustrated by comparative tests with small additions of calcium carbonate melter, fired to cone 9/10 reduction temperatures. The combination of 10% calcium carbonate melter and 90% feldspar, produced a gray-blue, smooth, glossy surface compared to the white, often-crawled surface of the feldspar fired alone. The same combination with Nepheline Syenite produced a nonglossy, gray-blue-green-yellow, matt surface sometimes interspersed with fluid puddles of green, crackled glass. This surface differed dramatically both from its white-orange, lustrous surface when fired alone and the comparative feldspar-calcium tests. The powerful Nepheline melter combined with the calcium carbonate melter to bring more of the alumina into the melt.

This result, together with the lower silica content of Nepheline Syenite, created a dense, stony surface, interspersed with puddles of glass. The same result occurs in more complex glaze formulas. *Figure 1.3, p80.*

3. Effects on Craze/Crackle Network

The craze/crackle network of a Nepheline Syenite glaze surface is greater than that of the feldspars or Cornwall Stone due to the high soda and low silica content of this unique glaze core. Soda has the highest expansion and contraction rate of all our ceramic materials. In addition, the lesser amount of nonexpanding silica in Nepheline Syenite enforces the high expansion and contraction rate of the glaze surface. The result is a strongly visible craze/crackle network.

4. Effect on the Color of the Claybody and the Glaze Surface

Unlike the fat, white, glossy surface of the feldspars, Nepheline Syenite produces a unique, orange-white, lustrous surface when fired alone to high stoneware temperatures in a reduction atmosphere. (See *Figure 1.4, p81.* for example of quintessential Nepheline Syenite surface.) The cause lies in the high soda content of Nepheline Syenite (10.6%). (Compare this soda content to the 6.8% soda content of the "soda" feldspar Kona F4. Our tests of an ancient supply of soda feldspar #56 [no longer available] bears a close resemblance to Nepheline Syenite with respect to fired color and surface texture, due to its high percentage of soda.) As described previously in the discussion of the soda ash tests, soda possesses an extraordinary ability to combine with the silica in the claybody and form a lustrous, orange-brown, glassy surface. In addition, small amounts of soda in a high-soda compound will volatilize after temperatures of 2192°F are reached. The escaping sodium vapor burns the edges of the glaze layers (Hamer and Hamer 1986, 250, 293–299). Both of these effects are clearly visible in the surface of Nepheline Syenite fired alone or in combination with other ceramic materials. When a glaze that contains a large amount of Nepheline Syenite melts at stoneware temperatures in both oxidation and reduction atmospheres, the claybody immediately below the glaze surface often appears darkened, reddened, and singed. If a faint wash of Nepheline Syenite is applied to a claybody, the same result occurs. Here again, the high soda content of Nepheline Syenite causes dramatic color changes in the claybody.

The high soda and alumina content of this unique glaze core also causes unique color effects in glazes that contain iron and copper. When the combination of 90% Nepheline Syenite, 10% calcium carbonate, and 1/2% iron oxide is fired to cone 9/10 reduction temperatures, the resulting surface is a stony-matt, gray-blue, often yellow-edged glaze surface, sometimes interspersed with flowing puddles of crazed-crackled, gray-blue-green glass. This surface differs significantly from the smoother, glossy, bluer surface of both potash and Kona F-4 feldspar tests. Different kiln firings of this same combination produced varied results, depending on kiln atmosphere, temperature, and the form of the test. Words are inadequate to describe the peculiar, elephant-gray, skinlike surface that resulted from one set of tests. The soft gray color reflected the underlying reduction claybody color (gray), as well as the carbon-trapping tendency of soda materials. Unlike the glossy, sparkling surface of the feldspar tests, this unique Nepheline Syenite surface was quiet, subdued, haunting, and evoked memories of ancient origins.

Conclusion

Comparison of the tests of Nepheline Syenite and the feldspars highlight the special qualities of Nepheline Syenite. The white-orange luster, the peculiar dark gray-blue-yellow-edged color, the stony matt-puddled glass texture, the increased flow and craze/crackle of the glaze surface, and, lastly, the burnt and singed claybody are all in marked contrast to the glossy, even-textured, white, gray, green, or blue feldspar tests. The surfaces produced by Nepheline Syenite show clearly that it is a highly individualistic and unique glaze core.

CORNWALL STONE[10]
Figure 1.3, p80; 1.8, p85.

Cornwall Stone is one of those rare materials that is manufactured exclusively for the ceramic industry as a glaze and claybody ingredient. This fact holds great importance for the potter because significant changes in the mineralogical and chemical structure of Cornwall Stone will undoubtedly affect its industrial use as a ceramic ingredient. Thus, the manufacturer of Cornwall Stone will take greater care to maintain a standardized product than a manufacturer of a foundry-plug fire clay. In the case of the fire clay, increased additions of calcium, iron, or manganese nodules would not be noted unless they affected the refractory function of the material. Yet such changes, even though small, could create disasters for a producer of ceramic ware. *It can never be stated too often that the predominate industrial function of a material determines its reliability for ceramic usage. This factor should always be considered by the potter when selecting a ceramic material for a glaze or claybody.*

Cornwall Stone is a light-colored, sometimes partially kaolinized granite, which intruded into folded sediments some 300 million years ago in the region of St. Stephen, St. Austell, Cornwall, England. Millions of years of chemical weathering and hydrothermal pressures began the kaolinization process—the decomposition of granitic rock into kaolin. Hence, we find deposits of Cornwall Stone next to large kaolin deposits. The line of demarcation between the Cornwall Stone deposits and kaolin deposits is not clear, but rather shows a gradual transition from granite to kaolin.

The history of Cornwall Stone is closely linked with the beginnings of the English porcelain industry in the eighteenth century. William Cookworthy discovered in the mid-1740s that the long-used, building stone deposits of Cornwall, in southwestern England, contained the secret ingredients of Chinese porcelain, namely kaolin and Petuntse (altered granite). From that time onward, these Cornwall Stone deposits have provided the stoneware and porcelain industry with a unique glaze core and claybody material.

The production of Cornwall Stone has fluctuated since its inception. Beginning with a modest 1,000 tons in 1807, it reached its zenith shortly before the first World War with the manufacture of 70,000 tons for foreign as well as domestic markets. After the first World War, the output of Cornwall Stone steadily declined due to the competitive impact of the United States and Scandinavian feldspar industries.

"Formerly Cornish China Stone was used extensively by many foreign potteries and china factories both in the body and in the glaze of wares, but today the British Ceramic Industry constituted the sole user."
(Presidential Address of C. V. Smale 1977)

Despite the loss of primary foreign markets, and the removal of Cornwall Stone as a separate listing in *Ceramic Industry Materials Handbook* (the feldspar section now contains a mere two sentence reference; *Ceramic Industry* 1998, 102), Hammill & Gillespie has continued to import Cornwall Stone for U.S. potters.

In 1989, a program to further improve and stabilize Cornwall Stone was underway.

"The traditional methods of extraction and processing of the stone have been very labor intensive and we are already well into a major capital expenditure program to further mechanize and improve these activities. Perhaps the most significant of the projects embarked upon is the installation of a new secondary crushing plant leading to grinding of the china stone prior to flotation in order to produce a low fluorine product. We expect the new material to be available in 1990 and it will enable us to supply not only to the potters' millers but also direct to the potters. We feel that with our involvement in processing the stone to a level at which the potter can obtain his requirement directly from us, we have better control over the product from extractive source to consumer use."
(C. V. Smale, letter to the author, 6 June 1989)

"Our Great Wheal Prosper Quarry is the sole U.K. producer of china stone, and with reserves in excess of a million tons we consider that we have a long and healthy future for the industry."
(C. V. Smale, letter to the author, 5 June 1989)

CORNWALL STONE*

TRADE NAME	Cornwall Stone, also known as Cornish Stone or China Stone
SOLE GEOGRAPHICAL SOURCE	St. Stephen, St. Austell, Cornwall, England. Goonvean Ltd.,
GEOLOGICAL SOURCE	Final granitic intrusion into folded sediments of lower Devonian period (380 million years) during the end of the Carboniferous Period (300 million years).

MINERALOGICAL STRUCTURE *

	Hard White	Purple Stone
Major		
Quartz	30%	29%
Soda Feldspar	13%	30%
Potash Feldspar	15%	10%
Muscovite (potash mica)	20%	23%
Kaolinite	10%	
Minor		
Fluorspar (fluorite)	Trace	1.5%
Topaz (alumina silicate with fluorine)		
Apatite (calcium phosphate)		
Amorphous Silica (opal)	12%	

CHEMICAL STRUCTURE

	Purple Stone*	Hard White*	D.F. Stone**	Cornish Stone-H&G***	Carolina Stone†	Average Granite††	Average Range of Feldspars†††
SiO_2	73.10	72.00	79.50	73.0	72.30	70.18	66–69
AlO_2O_3	15.00	18.20	12.00	16.0	16.23	13.49	17–19.6
FeO						1.78	
Fe_2O_3	0.15	0.10–0.16	0.06	0.14	0.07	1.57	0.07–0.15
CaO	2.05	0.20–0.50	0.20	2.0	0.62	1.99	0.14–1.6
MgO	0.08	0.15	Trace				
K_2O	3.60	5.0–6.0	3.80	3.7	4.42	4.11	3.8–1.6
Na_2O	3.70	0.15–0.35	3.90	3.7	4.14	3.48	3–7.0
Ti_2O	0.05	0.05		0.07			
F	1.25	0.25	0.08		$1.23(CaF_2)$		
P_2O_5	0.45	0.10					
L.O.I.	1.60	3.50	0.45	0.5	1.06	0.84	0.13–0.30

* Goonvean Ltd. Technical Data 1998.
** Hammill & Gillespie Technical Data; See also Grimshaw 1989, 323.
*** Hammill & Gillespie Technical Data 1989.
† *Ceramic Industry* 1966, 94 (January).
†† Bates 1969, 40.
†††Chemical Structure Chart of 5 commonly used soda and potash feldspars, p. 43.

Kinds of Cornwall Stone

The mineral content of Cornwall Stone consists primarily of feldspars, quartz, and muscovite mica. It may also contain minor amounts of fluorite,[11] kaolin, and amorphous silica (opal). Over the years, the mineralogical and chemical structure of Cornish Stone has undergone considerable change.

From 1960 to 1973, the removal of fluorite and mica from Cornish Stone produced a defluorinated quartz and feldspar material known as D.F. Stone.

The 1960 Clean Air Act in England, which restricted the emission of hydrofluoric gases, caused this material to undergo an extensive defluorination process.

The primary distinguishing characteristics of D.F. Stone's chemical structure were the reduction of fluorine, potash, calcia, iron, and alumina, and an increase in silica.

In 1973, due to the increased cost of the refining process, D.F. Stone was discontinued. Note that the chemical structure of D.F. Stone is different from the chemical structure of subsequent supplies of Cornwall Stone (see Chemical Structure chart, p. 35). Hence, dissimilar alumina-silica ratios would be present in different glazes, depending on the inclusion of either D.F. Stone or later supplies of Cornwall Stone.

Two kinds of Cornwall Stone or Cornish China Stone are manufactured for potters: the first is Purple Stone, which contains soda and potash feldspar, quartz, muscovite mica, and fluorite. A second kind is known as Hard White; in this stone some of the soda feldspar is replaced with kaolinite. In addition, Hard White contains less fluorine compared to Purple Stone. Hence, the fusion point of Hard White is higher than that of Purple Stone. A blend of both Purple and White Stone is also available.

One can expect variation in mineral and chemical structure, (particularly the alumina/silica ratio) depending on which supply of Cornish Stone is used. For this reason, if possible, it is important to obtain from your supplier the chemical data that corresponds to your particular supply of Cornwall Stone. The tests of Cornwall Stone that are discussed in this book were made with old supplies of white-colored Cornwall Stone (D.F. Stone ?) due to the fact that my studio and the 92nd Street Y had stockpiled large amounts of this material. Comparative tests performed with a later, bluish-green, fluorinated Cornwall Stone, (Purple Stone?) did not compare to the tests of the older, white Cornwall Stone in terms of their depth or luster. The Cornwall Stone subsequently supplied by Hammill & Gillespie was physically white in color and did not list fluorine as an ingredient in the chemical structure sheet. It is to be hoped that currently available supplies of this material (described by the Technical Manager as a "low fluorine product") are more like the older supplies used in our tests and will thus restore Cornwall Stone to its earlier greatness.

General Characteristics of Cornwall Stone

Despite the changes that have occurred in Cornwall Stone throughout the years, the following basic characteristics have remained constant.

1. Effect on Glaze Fusion

Cornwall Stone contains more silica and less melter oxides than do the feldspars. Silica has a high melting point and, unless combined with sufficient melters, will raise the firing temperature of the final result. For this reason, Cornwall Stone has a higher melting temperature than the feldspars. It appears stiffer and less melted when fired alone to stoneware temperatures. This is especially apparent at the lower stoneware temperatures. Even the potash feldspars show more fusion at the cone 5/6 oxidation firing temperatures than does Cornwall Stone. It is an obvious conclusion that Cornwall Stone would not be a first choice as a glaze core at these firing temperatures unless a stiffer surface is desired. On the other hand, Cornwall Stone becomes a natural choice as a glaze core for the higher stoneware firing temperatures. It is a particularly fine glaze core for a high-fired porcelain body. Cornwall Stone combines with 10% calcium melter to produce a silky, white surface, which bonds so harmoniously with the underlying porcelain body that glaze and claybody truly appear as one. *Figure 1.8, p85.* The white (defluorinated?) Cornwall Stone of some twenty-odd years ago in combination with low amounts of Whiting (calcium carbonate) and iron oxide produced fat, lustrous surfaces that were reminiscent of the ancient Chinese Song Dynasty celadons. Only further testing will determine whether or not the present and future supplies of Cornwall Stone possess this same gemstone quality.

When Cornwall Stone substitutes for the soda and potash feldspars in a glaze formula, it may drastically alter the glaze surface. It brings an increased amount of silica and less melter oxides to the glaze combination. This substitution can raise the firing temperatures of the glaze, with the result that a formerly shiny and glassy surface may appear more opaque and more matt. On the other hand, if the original surface of the glaze is already matt, or satin-matt, and if this result depends upon the low amount of glassmaker and a large amount of melter oxides, then the Cornwall Stone substitution may have just the opposite effect. Although the substitution of Cornwall Stone for the feldspar lowers the total melter content of the glaze formula, the large amount of auxiliary melters already present in the glaze can suffice to melt the additional silica brought in by the Cornwall Stone. In this way, the substitution of Cornwall Stone turns a satin-matt surface into a glassy transparency.

2. Craze/Crackle Pattern

The craze/crackle network of a glaze surface may lessen considerably when Cornwall Stone substitutes for a feldspar. Here again, the result is due to the higher silica content of Cornwall Stone. Silica inhibits the high expansion and contraction rate of the soda and potash melters. Comparative tests of feldspar and Cornwall Stone reflect their difference in silica content. In these tests both the feldspar and Cornwall Stone combine with 10% of the same calcium melter (Whiting). A strong craze/crackle network appears on the feldspar tests; the Cornwall Stone tests reveal a smooth and relatively uncrazed surface. (See Comparative Tests of Feldspathic Glaze Cores, pp. 42–43; *Figure 1.3, p80; 1.8, p85.*)

3. Color

The color of a glaze may change if Cornwall Stone is substituted for a feldspar. Cornwall Stone contains more iron oxide and other impurities than do the feldspars. These impurities, together with the increased silica content, can produce distinctive color differences, especially in high iron glazes.

We combined Cornwall Stone with 10% calcium melter and added varying amounts of red-iron oxide. The tests were fired to cone 9/10 temperatures in a reduction atmosphere. We performed the same tests with potash feldspar and compared the results. The test of Cornwall Stone with 12%–14% red-iron oxide produced a reddish purple shade that differed markedly from the deeper brown color produced by the potash feldspar. There were further differences of color when low amounts of iron were used. Cornwall Stone produced a milky, pale blue-green celadon in contrast to the more brilliant, gray-blue surface of the feldspar tests. Cornwall Stone has, at times, contained low amounts of titanium. Titanium has a strong effect on color and melt, and even low amounts will encourage yellow opacity. This variation in titanium content could explain the greener, milkier surface of our tests of a celadon glaze compared to the more brilliant blue surface of the same glaze with a potash feldspar (See Tests, p. 44).

4. Variability

There has been considerable variation over the years in the oxide structure of Cornwall Stone. Tests of Cornwall Stone celadon glazes at Parsons School of Design in New York consistently produced a green-blue color as compared to the bluer color of the same Cornwall Stone celadon glazes at the New York 92nd Street Y. The Parsons supply of Cornwall Stone used in these tests was a pale blue-green color in its raw state, which suggests the presence of fluorine. On the other hand, the supply of Cornwall Stone at the 92nd Street Y was white in color. Due to the fact that Cornwall Stone was not required for many of the glazes of the 92nd Street Y, new supplies of Cornwall Stone had not been bought for many years. Thus, the 92nd Street Y's supply of Cornwall Stone at the time of these tests may have been D.F. (defluorinated) Stone. More recent tests of the blue-green Cornwall Stone produced a shinier, glossier, and less-complex surface than comparative tests with the white 92nd Street Y Cornwall Stone. The difference is explained by the presence of fluorite in the blue-green Cornwall Stone. Fluorite is a highly reactive and volatile mineral (see p. 179) and could, even in very small doses, increase the fusion point of the glaze. Visible proof of the increased fusibility of the blue-green Cornwall Stone appeared when it was substituted for the white Cornwall Stone in a stony, opaque, cone 9/10 reduction glaze known as Pavelle (Figure 3.7, p229.). The usually smooth, matt surface of Pavelle glaze contained shiny, glassy spots.

CONCLUSION

In conclusion, there are two points to remember about Cornwall Stone:

1. The minerals in Cornwall Stone have varied in chemical composition throughout the years. Depending on the age and color of the supply, different supplies of Cornwall Stone will reflect this variation. Comparative testing of old and new, or white and blue supplies of Cornwall Stone is necessary in order to anticipate changes to a glaze surface that can occur when the supply of Cornwall Stone is changed.

2. Despite this variation, the chemical structure of all forms of Cornwall Stone consistently appears higher in silica and lower in melter oxides than feldspars. Cornwall Stone will, therefore, affect the glaze surface in accordance with the principles that govern an increase in silica content and a decrease in melter content (see Chapter 4, Silica). Thus, when fired alone, every type of Cornwall Stone will produce a less-melted and more refractory surface than what is produced by the feldspars. For this reason, Cornwall Stone becomes a natural glaze core at higher stoneware firing temperatures and, with a minimum of additional materials, creates unique glaze surfaces for stoneware and porcelain ware.

THE FUNCTION OF FELDSPARS AND ROCKS IN A CLAYBODY

Thus far, I have focused on the use of feldspars and rocks as glaze cores. This is, of course, their most important function. Feldspathic minerals also play an important, though smaller, role in claybodies.

The strong and gradual melting action of the soda and potash melters in the feldspars and rocks make them ideal materials for vitrification of stoneware and porcelain claybodies. In addition, because they transform the free silica in the claybody into silica glass, they help to determine the fit of the glaze on the claybody.

"With the possible exception of clay, feldspar is the most essential ceramic material in the whiteware industry. It is the universal flux used in all types of ceramic bodies. . . ."

(*Ceramic Industry* 1998, 102)

Less feldspar is added to a stoneware claybody than to a porcelain claybody. Stoneware claybodies are stronger, more open, and generally require less additional melter than porcelain claybodies. They are not as vulnerable to the slumping problem as are their finer-grained porcelain counterparts. In addition, stoneware clays themselves contain more melters in their mineralogical structure than do the porcelain (kaolin-ball) clays. Hence, the feldspar addition of a stoneware claybody does not usually exceed 10% of the total claybody formula. On the other hand, feldspathic minerals can constitute as much as 25%–30% of the porcelain claybodies. The pure, white-burning kaolin-ball clays that constitute the clay materials of the porcelain claybody formula require a high firing temperature (2600°F–3000°F) to reach their vitrification point. The addition of about 25% feldspar will lower this vitrification point to stoneware temperatures. The feldspar will also dissolve some of the silica and transform it into silica glass so that translucency results.

According to industrial research, potash feldspar is the preferred choice for claybodies because of its longer firing range, and general overall stability and strength (*Ceramic Industry* 1998, 102, 104). Potash feldspar dissolves silica more easily than soda feldspar and thus creates greater translucency in the porcelain claybody. Thermal expansion and slumping is said to be higher in claybodies with soda feldspars.

Despite the advantages of the potash feldspar as a claybody melter, feldspathic materials high in soda are often used in claybodies. As described below, Nepheline Syenite is a preferred melter in both porcelain and stoneware claybody formulas at the lower stoneware firing temperatures (Taylor 1989, #5; claybody formulas p.113). Stoneware bodies frequently use Kona F-4 feldspar as the source of the additional melter. However, note that the potash content of Kona F-4 is only 1.8% less than its soda content. Thus, despite its designation as a "soda" feldspar, it would probably not qualify as the kind of soda feldspar described in the

Ceramic Industry magazine report.

Both Cornwall Stone and Nepheline Syenite function as melters in a claybody in the same manner as feldspars. As stated on page 36, there are two kinds of Cornwall Stone distributed by the manufacturer, Purple Stone and Hard White. Hard White is the form of Cornwall Stone in which some of the soda feldspar has altered to kaolinite. Because Hard White Stone contains 10% kaolinite and is, therefore, more plastic than other feldspathic materials, it should function most successfully in a claybody. (Note that the extraordinary form and translucency of certain ancient Chinese porcelains has been attributed to the fact that they contained a rock material, not unlike Cornwall Stone in composition, that combined the properties of both plasticity and melter; hence, this material was a primary ingredient of both the claybody and the glaze [Wood 1978, 12; Hetherington 1948, 9–10]). Because of this rare combination of properties, it is a common practice in England to use Cornwall Stone in porcelain claybodies.

"Cornwall Stone is used extensively in England in whiteware bodies, replacing a portion of the spar and china clay which we, in this country, would use as such. Since it represents a stage in the transition from feldspar to china clay, bodies in which it is used are not subject to great strain in firing as when straight feldspar and china clay are used."

(*Ceramic Industry* 1966, 94)

Tests made by my students showed that Cornwall Stone is more plastic than feldspars or Nepheline Syenite. This effect was demonstrated by adding water to dry amounts of potash feldspars, soda feldspars, Nepheline Syenite, Flint, Cornwall Stone, and EPK kaolin. The students attempted to mold each of these materials into pinch pots. The point of this exercise was to actually feel the property of plasticity and to feel the consequences of its absence. Cornwall Stone was found to be more plastic than feldspars, Nepheline Syenite, or Flint. All of these materials fell apart and refused to take on any kind of shape. We found that it was almost possible to mold Cornwall Stone into a pinch pot. Here was visible evidence of the kaolinization process inherent in this mate-

rial. Cornwall Stone also contains feldspar, mica, and quartz in addition to kaolin; therefore, it would function as a melter in a claybody and fuse the pores of the clay structure with melted glass. Cornwall Stone contains more iron and titanium impurities than feldspars and may blur the pure whiteness of a porcelain body. Perhaps this is the reason Cornwall Stone does not often appear in a porcelain claybody in the United States. On the other hand, plasticity is an important goal for all claybodies, and for the functional potter this property is even more important than pure whiteness. Because a porcelain claybody usually contains 25%–30% nonplastic melter materials, it would make sense to utilize a more plastic melter such as Cornwall Stone.[12]

Despite the injunction against high soda melters, Nepheline Syenite frequently appears as a melter in the lower firing temperatures. According to the manufacturer, the advantages of using Nepheline Syenite in claybodies are as follows:

Lower firing temperatures, faster firing schedules, longer firing range and a stronger final product are said to result from the use of Nepheline Syenite compared to a feldspar. Prior to the use of Nepheline Syenite, fine commercial china had to be fired to the higher temperatures of cones 9–12. The result was a longer, more costly firing schedule and a narrower range of color. Nepheline Syenite achieves translucency in commercial chinaware fired to cones 3–6. The fast firing and low firing temperatures permits a wider and brighter color palette than previously achieved with feldspar. Increased shrinkage admittedly occurs with the use of Nepheline Syenite compared to the feldspars. An addition of 5%–10% Flint to the claybody is said to correct this problem (Taylor 1989, #5).

Note that industrial ceramic ware is usually slipcasted or molded into shapes that are more or less traditional forms. Handworked and/or wheel-thrown ware, on the other hand, especially those in which the form is unconventional and daring, impose greater requirements for working and firing strength. The problem of slumping during both the working and firing process of such ware is an ever-present danger. It seems likely that the high soda content of Nepheline Syenite could present a greater slumping risk than either a potash feldspar or Kona F-4 feldspar and is a factor to be considered in selecting a feldspathic melter for your claybody. However, any such possible disadvantage is more than balanced by the unique fusion powers of Nepheline Syenite. Refractory clays, such as the kaolin-ball clays of porcelain claybodies, require the presence of strong melters to ensure vitrification of the claybody at the lower firing temperature of cone 5/6. Then too, at this temperature, slumping is not as great a problem as it is at the higher temperatures of true porcelain.

CONCLUSION[9]

Feldspathic materials with a high potash content are more stable and contribute more strength and translucency to the porcelain claybodies than do the soda feldspathic minerals. Because Kona F-4 contains a fair quantity of potash, it should not greatly increase the risks of slumping in a claybody. Although Nepheline Syenite contains high amounts of sodium, the strong and gradual melting power of Nepheline makes it a useful melter in porcelain claybodies at the lower stoneware temperatures. Cornwall Stone is the most plastic of the feldspathic melters and would be a valuable melter in a porcelain claybody where the proportion of nonplastic materials is usually 25% of the total. A more detailed description of the role of the feldspathic materials in claybodies during the firing process will follow in Chapter 2, pp. 103–105.

TESTS OF FELDSPARS AND ROCKS

Theoretical analysis can never replace the practical experience of hands-on testing of ceramic materials. The personalities of the feldspars and rocks spring to life when they are tested alone and in various glaze and claybody combinations. It is only then that we can hope to understand their contribution to the ceramic surface. For this reason, an important part of this book describes comparative tests of basic ceramic materials and outlines further tests to be performed by the reader. Our comparative approach, which contrasts the effects of different feldspars and rocks, seeks to define the contribution that each material brings to a specific glaze magma. These tests also provided rich rewards by their production of far-ranging ceramic surfaces. Thus, we

achieved many different glaze surfaces from the simple substitution of a different feldspar or rock in the same glaze. We also obtained a wide range of ceramic surfaces by increasing or decreasing the proportion of each of the materials in the glaze formula. Finally, and most important, our tests showed that the *same* material could have profoundly different consequences in various ceramic recipes due to the unique oxide ratio that exists in each glaze and claybody formula. Thus, many of the conclusions describing the behavior of specific feldspars and rocks are limited to the particular kind of glaze in which they appear. General conclusions that hold true for all types of glazes are not essential. What is important is that you learn the contribution of each of the various materials to a specific glaze and see how the proportion of these materials can be altered to create an extraordinary spectrum of ceramic surfaces. Thus, much will be gained by staying within the framework of a few, simple glaze formulas.

The following considerations should be kept in mind when evaluating test results:

1. The testing procedure, in order to be successful, requires that you be familiar with the possible range of variations of the original glaze, including those that result from variable kiln temperatures. These variations in kiln temperature occur as a result of different firings as well as different positions of the glaze in the same kiln. The variations produced by different firing conditions can often obscure the changes produced by the substituted material itself. This is especially true in the case of the small test pot. *Different firings (changes in firing and cooling rates, position in kiln, etc.), different kilns, and different supplies of the same material may create more variations than any change in the glaze formula.*

2. Changes in the size and thickness of the test pieces and in the methods of glaze application (thick or thin glaze layers) may again create more variations in the final result than changes in the glaze formula.

3. The control of these variables has been attempted by the following procedures:

a. Two or more repetitions of the same test in different firings and at different studios with different kilns and different supplies. This practice was followed with the feldspathic glaze core and Whiting tests. The test conclusions were based on the results that reoccurred with the most frequency.

b. The placement of the original glaze on one-half of each test.

c. A uniform practice of test formation and glaze application for all test pieces.

4. The lack of glaze dissection tests for cone 9–10 oxidation is due to the fact that the primary firing temperatures available for my students were either cone 9–10 reduction or cone 5–6 oxidation.

5. Percentage oxide or molecular analyses of the original glaze and the glaze with the substituted material help to further explain changes produced by substituted materials. (By way of illustration, percentage oxide analyses appear for a cone 9–10 reduction glaze and its substitutions on pp. 45–47 and for a cone 5–6 oxidation glaze and its substitutions on p. 52).

6. A special word must be mentioned with respect to the reduction atmosphere tests. The surface and color of the reduction tests are necessarily subject to more variation than are the comparable oxidation tests. At stoneware temperatures there is considerable interaction between the claybody and the glaze. Thus, the color of a glaze is partly determined by the color of the underlying claybody. In a reduction firing, the color of the claybody is not necessarily constant, because it depends on the amount of atmospheric reduction that it receives. This amount can vary considerably, not just from firing to firing, but also within a single firing, depending on the position of the tests in the kiln. Thus, depending on these factors, the underlying color of the claybody can be pale, medium, or dark gray, and the color of the glaze surface will vary accordingly. We have attempted to compensate for this variation by the methods outlined in item number 3.

Before we proceed, a word of caution: The results of the following glaze core substitution tests are often obscured by the complex nature of the feldspathic glaze cores. The term glaze core by definition refers to an earth material that combines glassmaker oxide, adhesive glue oxide, and melter oxides in a prescribed ratio that is unique for that particular material. The changes to the glaze surface that occur in our feldspathic substitution tests result from the particular ratio of silica, alumina, and melter oxides in the substitute glaze core. For this reason, true understanding of these feldspathic glaze core tests must await subsequent tests that show how additions of separate melter, silica, or alumina minerals create a new oxide ratio and change the surface of the glaze. These subsequent tests will clarify the results of the more complex feldspathic substitution tests because they deal with separate additions of the three kinds of oxides that exist only in combined form in the feldspathic glaze cores. *Once again we see the interdependence of the ceramic materials within a glaze magma—an understanding of the dynamics of each material requires an equal understanding of all of the other materials within the ceramic surface. We are following in the path of a ceramic circle, and we must complete one entire revolution before we can untangle the separate strands of the richly woven ceramic surface.*

With these considerations in mind, we begin with the substitution tests of the feldspathic glaze cores and compare their contributions to the final glaze surface.

COMPARATIVE TESTS OF FELDSPATHIC GLAZE CORES
Figure 1.3, p80; 3.2, p224.

Cone 9–10 Reduction

G-200 GLAZE

Blue-gray color. Flower craze-crackle pattern where thick. Smaller, distinct craze-crackle where thin. High gloss. Transparent surface. Viscous flow.

G-200 Feldspar (potash)	90%
Whiting (calcium carbonate)	10%
Red Iron Oxide	0.5%

Test I The following materials were substituted for G-200 in 5 separate tests and fired at cone 9–10 reduction temperatures.

a. Custer Feldspar (potash)

Blue-gray color, bluer than G-200. Glassy bubbles in glaze layers. Test variable re craze-crackle network, but no flower pattern. High gloss. Transparent surface. Viscous flow.

b. Norfloat Feldspar[13]
(no longer available)

Grayer color than above. Glassy bubbles in glaze layers. Fine, small craze-crackle pattern. High gloss. Transparent surface. Viscous flow.

c. Kona-F4

Less brilliant blue-gray color. Fine, small, overall craze-crackle. High gloss. Transparent surface. Viscous flow.

d. Nepheline Syenite

Gray-green color. Fine, small craze-crackle pattern where thick. No visible pattern where thin. High gloss where thick; semi-matt where thin. Reduced transparency. Increased fluidity.

e. Cornwall Stone

Gray-blue color. No craze-crackle in majority of tests with medium to thin coat. Large craze-crackle network where thick. High gloss. Slightly reduced transparency. Viscous flow.

G-200 GLAZE

Test II Four of the above tests were repeated in the oxidation atmosphere.

a. G-200

Beige color where thin. Milky white where thick. Orange singe on edge of glaze. Flower craze-crackle pattern where thick. Smaller, distinct craze-crackle where thin. High gloss. Reduced transparency. Viscous flow.

b. Kona-F4

Beige color where thin. Beige-white where thick. Deep orange singe on edge. Craze-crackle network smaller than G-200. High gloss. Decreased transparency compared to reduction test. Viscous flow.

c. Nepheline Syenite

Beige color where thin. Beige-white where thick. Very deep orange singe on edge of glaze. Flower craze-crackle pattern where thick: where thin, small craze-crackle pattern. Highest in gloss and luster.

d. Cornwall Stone

White color. Pale orange singe. Craze-crackle network larger and more visible than on above tests. Surface pin-holed and opaque.

G-200 GLAZE

Test III Two of the above tests were repeated in the cone 5–6 oxidation atmosphere.

a. Kona F-4 — Soft white color. Small, craze-crackle pattern. Semi-opaque, semigloss.

b. Nepheline Syenite — Beige-white color. Pale orange singe. Small craze-crackle pattern. Semigloss surface, semi-opaque.

In order to evaluate the various changes described above, it is helpful to compare the percentage oxide analyses of these feldspathic glaze cores.

OXIDE ANALYSIS OF GLAZE CORES

	Custer[14]	G-200[14]	Kona F-4[14]	Nepheline[14] Syenite	Cornwall[14] Stone	DFStone[15]
SiO_2	69.20	66.24	67.2	60.8	72.9	80.0
Al_2O_3	17.0	18.91	19.5	23.3	14.93	11.0
Fe_2O_3	0.11	0.087	0.05	0.08	0.13	0.06
K_2O	10.0	10.94	4.8	4.8	3.81	3.8
Na_2O	3.1	2.67	6.6	10.4	4.00	3.9
CaO	0.3	0.90	1.6	0.2	2.06	2.2
Li_2O	—	—	—	—	0.02	
TiO_2	—	—	—	—	0.10	
F	—	—	—	—	0.90	0.08
P_2O_5	—	—	—	—	0.50	Trace
L.O.I.	0.2	0.15	0.3	0.3	0.61	0.45
A/S*	1:4	1:3.5	1:3.4	1:2.6	1:4.9	1:7.3

*Alumina/silica ratio. See also note 16.

CONCLUSIONS OF G-200 TESTS I–III

Test I: The tests revealed the following characteristic pattern for potash feldspars, soda feldspars, Cornwall Stone, and Nepheline Syenite:

a-b. The potash feldspar tests displayed the bluest color. Also in evidence is the different craze-crackle patterns displayed by the three potash feldspars. The flower pattern of G-200 repeated over and over again in various tests, as did the more glassy, bubble-trapped (trapped gas) surface of Custer Feldspar. The most brilliant blue color was obtained with Custer Feldspar, attributed to its higher silica and lower alumina content.

c. The Kona-F4 test appears the most similar to the potash feldspars with respect to color, and craze-crackle pattern.

d. The test of Nepheline Syenite (higher soda and higher alumina content) displays a greener color; a fine, small, craze-crackle pattern; a unique matt-shine surface; and increased flow. Soda has the highest expansion and contraction rate of all the melters, and this is reflected in the smaller, overall craze-crackle pattern. The increased alumina content of Nepheline Syenite is reflected in the spotty, matt-opaque surface that appears whenever Nepheline Syenite is combined with low amounts of Whiting. The increased fluidity of this test is due to the powerful fusion powers of Nepheline.

e. Cornwall Stone test reveals its increased silica content in the large, occasional craze-crackle pattern. The variable amounts of titanium and other impurities, combined with a lower alkali melter content, are evidenced by the grayer, less-brilliant color and slightly decreased transparency. Tests made in different studios with different supplies of Cornwall Stone produced variable results.

Test II: At the cone 9–10 oxidation temperature the feldspathic glaze cores follow a predictable pattern with respect to craze-crackle patterns and fusion. With the exception of their beige-white color, the surface of these tests do not differ remarkably from their reduction counterparts. However, Nepheline Syenite did not produce consistently the matt-shine surface so characteristic of the reduction tests. On the other hand, this surface did appear when this glaze was used on larger test forms on a porcelain claybody. The matt-opaque surface appeared on the outside of the pot. A glassy, crystalline, flower craze-crackle pattern appeared in the thick layers of glaze in the interior of the pot. The interior result was a gemstone surface reminiscent of the surface of ancient Chinese bowls.

The slightly less fused surface of the oxidation Cornwall Stone test compared to its reduction test bears witness to the increased fusion temperature of the reduction firing. This combination of Cornwall Stone and Whiting produced a superb glaze for a porcelain claybody, in both oxidation and reduction atmospheres. The milky-white, glossy surface blended so completely with the porcelain claybody that it was difficult to tell where the glaze left off and the claybody began.

The main difference between the Cornwall Stone oxidation and reduction tests was, of course, in color. The oxidation color was an ivory white, and the reduction color was a cooler, bluer white. Both surfaces were superb.

Test III: At the lower stoneware temperatures of cone 5/6, increased opacities and less-fused surfaces result from the simple combination of feldspathic glaze cores and Whiting. However, the combination of 80% Nepheline Syenite and 20% Whiting or Wollastonite (calcium-silicate) produced a silky, crackled, lustrous surface on a white claybody (see p. 160). Similarly, the combination of Kona F-4 (70%) with Whiting or Wollastonite (30%) produced a satin-matt, pearl-white surface, with as much depth and appeal as the higher-fired surface. Although I have also used this combination successfully on a stoneware claybody, once again, it appears most lustrous over a white clay.

Thus, at stoneware firing temperatures in both oxidation and reduction atmospheres, the simple combination of feldspathic materials with low amounts of calcium melter can produce a gemstone surface for the potter.

SANDERS CELADON
Figures 1.1, p78; 1.5, p82.

Gray-green.
Gloss. Transparent.
No visible craze.

Kona-F4 feldspar	44%
Whiting	18%
Flint	28%
EPK kaolin	10%
Barnard Clay	5%

Test I
Omit Feldspar — Dry, yellow-brown surface.

Test II
Replace Kona F-4 with:

a. G-200 Feldspar — Bluer color. No difference in surface.

b. Nepheline Syenite — Greener color with areas of black. Spotted coverage. Increased craze.

c. Cornwall Stone — Lighter color. Stiffer, less melted surface.

Conclusions

Test II

a. Note surface similarity of Kona F-4 and potash feldspar. Bluer color is the predictable result of potash feldspar in the celadon glaze.

b. Increased soda and alumina of Nepheline Syenite creates a greener color. Black spotting indicates carbon trapping, which is characteristic of high-sodium materials. Increased craze-crackle pattern due to lower silica and higher soda of Nepheline Syenite. (Higher silica materials reduce craze-crackle pattern.)

c. Lower melter and increased silica content of Cornwall Stone produces a less-melted surface. In the oxide ratio of this graze, the addition of Whiting (18%) is not powerful enough to withstand the reduction in soda and potash melters caused by the Cornwall Stone substitution. This melter reduction, together with the increased silica content of Cornwall Stone, raises the fusion point of the glaze and thus produces a stiffer and partially unmelted surface.

It is possible to adjust the glaze formula so as to compensate for the different oxide ratio introduced by the substituted glaze core and to achieve thereby, a surface as close to the original glaze as possible.[17] This was done successfully with Sanders Celadon (base) for each feldspathic substitution. The revised glaze formula for each substitution produced a surface that was closer to the original. The subtle color differences that resulted were due to the fact that we were not able to duplicate the exact soda and potash ratio of the original glaze core.

OXIDE ANALYSIS OF UNADJUSTED GLAZE WITH GLAZE CORE SUBSTITUTIONS[17]

	Kona F-4 Original Glaze	Custer	Nepheline Syenite	Cornwall Stone
SiO_2	68.96%	69.7%	65.6%	71.77%
Al_2O_3	13.96	12.6	15.7	11.62
Na_2O	3.2	1.5	5.1	1.95
K_2O	2.3	4.9	2.4	1.86
CaO	11.96	11.3	11.2	12.10
F	—	—	—	0.40
P_2O_5	—	—	—	0.24
A/S Ratio	1:4.9	1:5.5	1:4	1:6

BATCH RECIPES OF ADJUSTED GLAZE WITH GLAZE CORE SUBSTITUTIONS

	Kona F-4 Original Glaze	Custer	Nepheline Syenite	Cornwall Stone
Glaze Core	44%	41.8%	35.2%	73.7%
Whiting	18	19.36	21.7	17.8
Flint	28	25.32	36.2	6.2
EPK Kaolin	10	13.5	6.8	2.3
Barnard	5	5	5	5

MIRROR BLACK I

High gloss; dense black.

Custer feldspar	56.0%
Whiting	16.0
Flint (silica)	20.5
Ball Clay	7.5
Red Iron Oxide	5.0

The following three materials were each substituted for Custer Feldspar in three separate tests:

Test I. Kona F-4 — Pitted surface; increased fluidity; green-black-brown color.

Test II. Nepheline Syenite — Increased pitting. Increased fluidity. Matt-gloss surface. Deeper green color.

Test III. Cornwall Stone — Closest to Custer. Decreased fluidity. Fat rolls at edge of pot. Still pitted. Green-black color.

Conclusions

Test I. Note color change from black to green-black-brown with Kona F-4 substitution. The Kona F-4 glaze core lowers potash and silica content and increases the alumina content. Increased flow is an additional response to the increased sodium content of Kona F-4.

Test II. Increased green color and fluidity result from the substitution of a glaze core with higher sodium and alumina content and lower silica content.

Test III. The substitution of Cornwall Stone, which contains low soda and alumina and high silica, produces a surface most like the original glaze surface. The color remains black, although it contains a greenish cast.

OXIDE PERCENTAGE ANALYSES OF MIRROR BLACK (BASE) WITH GLAZE CORE SUBSTITUTIONS

	Custer*	Kona F-4	Nepheline Syenite	Cornwall Stone
SiO_2	69.0%	68.0%	64.2%	72.5%
Al_2O_3	13.0	14.6	16.9	12.0
K_2O	6.2	3.0	3.1	2.3
Na_2O	2.0	4.1	6.4	2.4
CaO	9.5	10.3	9.4	10.7
A/S Ratio	1:5.3	1:4.66	1:3.8	1:6

*Original glaze core. See also note 18.

RON'S SATURATED IRON

	Dense black-brown gloss; Thin: red-brown.
Kingman Feldspar (potash)	43%
Whiting	13
Flint (silica)	19
EPK kaolin	15
Red Iron Oxide	10

The following three materials were each substituted for Kingman feldspar in three separate tests.

Test I. Kona F-4 — Oranger color.

Test II. Nepheline Syenite — Increased orange color.

Test III. Cornwall Stone — Blacker color. Satin surface. Reduced gloss.

Conclusions

Tests I–II. Note color change toward orange, which increases even more with the substitution of Nepheline Syenite. In both cases, the substitutions provided more soda and alumina, which could be responsible for the increased orange color.

Test III. The increased silica and lower melter content of the Cornwall Stone substitution has lowered the fusion point of the glaze and produced the blacker color of the surface. Note the similar black surface color of the Cornwall Stone substitution in the Mirror Black Test. The higher the firing temperature (increased fusion), the redder and more mahogany is the color of a high-iron glaze surface. (For a dramatic surface transformation from black to red, in an oxidation atmosphere, see Rotten Stone, pp. 69–70; *Figures 1.14, p91; 3.8, p230*).

TEMPLE WHITE

Figure 4.8, p272.	Semi-gloss. Semi-transparent. Gray-white color.
G-200 Feldspar (potash)	34.7%
Dolomite	19.6
Whiting	3.1
Flint	18.9
EPK kaolin	23.6

The following four materials were each substituted for G-200 feldspar in four separate tests.

Test I. Kona F-4 — No discernible difference but test is slightly more opaque and less glossy than original.

Test II. Nepheline Syenite — Opaque, non-gloss, stony surface. Yellow-brown color similar to Rhodes 32.

Test III. Cornwall Stone — Much increased gloss and transparent gray color.

Test IV. Volcanic Ash — Increased gloss; grayer color; more transparency.

Conclusions

Test I. Note similarity of Kona-F4 and G-200.

Test II. Lowered silica and increased alumina create opacity, mattness and yellowness of color.[19]

Test III. Conversely, the substitution of a glaze core with increased silica produces more gloss, transparency, and a grayer color. (The grayer color reflects the gray color of the reduced claybody.)

Test IV. Note the similarity of Volcanic Ash test to Cornwall Stone. (See Volcanic Ash, pp. 64–68.)

RHODES 32

Figure 4.2, p266.	Opaque, stony, matt. Yellow-brown-white; orange flecks. White when thick.

Custer Feldspar (potash)	48.9%
Dolomite	22.4
Whiting	3.5
EPK kaolin	25.1

Test I: Omit Custer Feldspar: Yellow-brown-white. Dry surface.

Test II: The following three feldspathic glaze cores were substituted for Custer feldspar in three separate tests.

a. Kona F-4	Similar surface but oranger and redder than control.
b. Nepheline Syenite	Stonier, drier surface. Greener-yellower-brown color.[20]
c. Cornwall Stone	Shinier surface; Yellower-orange color. Test repeated with similar color result. Test repeated at different studio resulted in color change to gray-green plus surface change of increased gloss. Possibly due to different kind of Cornwall Stone.

Conclusions
Test II

a. The smallest surface difference occurred with the substitution of Kona F-4. But note the oranger color cast, possibly due to the increased soda content.

b. Once again, it is possible that the increased sodium content of Nepheline Syenite caused the oranger color of porcelain test. However, the stoneware test of Nepheline Syenite resulted in a less-orange color. In this case, the reduced silica and increased alumina content, together with the iron in the claybody, could account for the yellow-brown color. Clearly, the lowered silica content is responsible for the less-glassy surface.

c. The shinier, glassier surface is an expected result with

the substitution of a higher-silica glaze core. But the color change to increased yellow-orange (which repeated in a further test) instead of gray-green is a surprise because the increased silica of the Cornwall Stone substitution would produce an increased transparency of the surface; this in turn would result in a grayer color due to the gray color of the reduced claybody beneath the glaze. Note the color change to the expected gray-green when the test was performed in a different studio, with a different supply of Cornwall Stone. *Never forget that different firings in different studios with different supplies of the same materials may create more differences than changes in the glaze formula test itself.*

Comparative Oxide Analyses of Temple White and Rhodes 32 with Kingman Feldspar (Potash), Nepheline Syenite, and Cornwall Stone Glaze Cores					
	Temple White		**Rhodes 32**		
	Kingman[21]	Nepheline Syenite	Kingman[21]	Cornwall Stone	D.F. Stone[22]
SiO_2	63.0%	60.0%	54.3%	58.3%	60.5%
Al_2O_3	18.0	20.5	23.0	20.8	18.5
Na_2O	1.0	4.0	1.5	2.4	2.3
K_2O	4.0	2.4	7.0	2.3	2.3
CaO	10.0	10.0	11.6	12.9	13.0
MgO	3.0	3.0	3.3	4.0	3.0
A/S	1:3.5	1:2.9	1:2.36	1:2.8	1:3.3

SUMMARY OF CONCLUSIONS OF CONE 9–10 REDUCTION TESTS

These tests underscore the importance of the oxide ratio of the glassmaker silica, the adhesive alumina, and the melter oxides of soda, potash, and calcia in the production of the following specific color and surface effects.

1. Kona F-4 and the potash feldspars appear to cause the

least change in the surface when substituted for one another in the glazes that we tested. However, there were color differences. The differences that occurred in the low-iron celadon glazes were not dramatic. However, the high-potash, high-silica, lower-alumina feldspars produced the bluest color. Increases in soda and alumina resulted in a grayer and/or a greener shade of celadon. Greater differences occurred with respect to the high-iron glazes. These color and surface differences were caused as much by the changes in silica and alumina content as by the changes in soda and potash content and indicate the sensitivity of the iron-black-brown color to changes in the alumina-silica ratio. In some cases, the difference in the ratio was considerable, as indicated by the percentage oxide analysis. The black color appears to depend on high silica and relatively low alumina. Whenever a lower silica and higher alumina potash feldspar was substituted, a greener, browner, and generally less-black color appeared. Changes in the soda and potash content may also cause some subtle color changes. A saturated-iron glaze with a mahogany surface color changed to a redder, browner, less-mahogany color with the substitution of a higher-soda, higher-alumina feldspathic glaze core (Ron's Saturated Iron glaze).

2. Nepheline Syenite lowers silica content and increases the soda and alumina content of a glaze. Thus, a green-brown-yellow color increased opacity, flow, and craze-crackle often resulted from the substitution of this glaze core in the low- and high-iron glazes that were tested. The results in this case were fairly consistent and made it easy to identify the Nepheline Syenite substitution in each case. In certain glazes, the high soda and alumina content of Nepheline Syenite will encourage a red-orange color. The mahogany surface of the above-mentioned saturated-iron glaze changed to a redder, more orange color with the substitution of Nepheline Syenite for the potash feldspar glaze core. Similarly, Nepheline Syenite produced a more orange color when it substituted for a potash feldspar in a high-alumina, opaque, stony-matt glaze on a porcelain claybody (Rhodes 32). The stoneware tests produced less-consistent results.

3. Cornwall Stone increases the silica content and lowers the alkali melter and alumina content. In addition, it adds a variable amount of titanium, fluorine, and iron, depending on the particular supply of Cornwall Stone that is used. Thus, the substitution of this glaze core can disturb the balance of the oxide ratio and cause the surface to become (a) more opaque, less shiny, muted, and a less-brilliant color if substituted in a gloss, transparent glaze; or (b) more transparent and shinier if substituted in a matt, opaque glaze.

On the other hand, if a glaze depends on a high silica ratio, the substitution of Cornwall Stone can cause less disturbance to the resulting color and surface than a lower-silica feldspar. We found this to be the case in a high-iron, high-silica glaze where Cornwall Stone replaced the potash feldspar glaze core (Mirror Black).

4. Glaze surfaces can change from stony-matt to semigloss translucent and, conversely, from semigloss translucent to stony-matt, by means of glaze core substitutions. Thus, the low-silica, high-alumina Nepheline Syenite transformed the semigloss, semitransparent, gray-white surface of Temple White into the yellow-brown opacity of Rhodes 32. Similarly, Cornwall Stone with its higher silica content caused Rhodes 32 to resemble Temple White when it substituted for the lower silica potash feldspar glaze core.

5. No absolute rule will hold true for all glazes, other than the fact that changes in the color and surface may occur with the substitution of different feldspathic glaze cores. In each case, the nature of the changed surface (whether it becomes more or less transparent, more or less shiny, or greener or blacker in color), as well as the degree of change, will depend on the ratio of the glassmaker oxide to the adhesive oxide in both the original glaze and the new surface with the substituted glaze core. In order to ascertain the differences more precisely, a percentage oxide analysis and/or molecular analysis should be performed. (See Appendix A1 for methods of glaze calculation, and Chapter 4 for exploration of silica-alumina ratio and its effect on glaze surfaces.) However, the percentage oxide analysis does not tell the whole story. Many complex variables, in addition to chemical structure, determine the final result. The color and surface of both a glaze and an earth mineral depend not solely on the oxide ratio, but also on the size, refractive indices, and number of crystalline formations that they contain. This in turn depends upon the firing and cooling cycle. In the final analysis, the reasons why something happens in the tests being

considered are not as important as the happening itself. Irrespective of cause, what is important to us is the fact that new surfaces appear when we substitute different feldspathic glaze core materials.

Cone 5–6 Oxidation

The following tests (unless otherwise specified) were performed with the stoneware claybodies in use at that time at the Crafts Students League, New York and Clay Arts Center, Portchester, N.Y. See page 139 for claybody formulas.

ALEXANDRA'S GLAZE [23]		
Cornwall Stone	40%	White, edged in orange.
Nepheline Syenite	40	Opaque, satin-matt.
Whiting	20	

Note the satin-matt surface produced by the calcium melter, Whiting. As we shall see in later tests this satin opacity is more characteristic of the calcium melter at higher oxidation and reduction temperatures. Recent tests of this glaze on cone 6 red and white claybodies consistently produced a stony-matt surface at the cone 6 firing temperature. However, the substitution of Nepheline Syenite 40 for Cornwall Stone 40 and Wollastonite 20 for Whiting 20 produced the satin-matt surface. (See Nepheline Syenite Glaze, p. 160; *Figure 4.7, p271.*) The substitution of potash feldspar (Custer) for Nepheline Syenite produced the following predictable results.

1. The potash feldspar, which is more refractory than Nepheline Syenite, raised the melting temperature of the glaze and produced a less-melted surface.

2. The decrease of the soda content caused the disappearance of the orange singe at the edge of the glaze, which is characteristic of sodium materials. (See pp. 25–26.)

TRANSPARENT BLUE (CRAFTS STUDENTS LEAGUE)		
Custer Feldspar (potash)	44 grams	Glossy blue.
Colemanite	20.0	No craze-crackle.
Whiting	1.0	
Zinc	3.0	
Flint	24.0	
EPK kaolin	1.0	
Cobalt Carbonate	0.5%	
Copper Carbonate	4.0%	

Test I Omit Custer: Dry surface, pale turquoise color.

Test II The following four glaze core materials were each substituted for Custer Feldspar.

a. Nepheline Syenite	Greener color. Craze.
b. Kona F-4	Greener color.
c. Cornwall Stone	Greener color. Higher gloss.
d. Petalite	Blue-green-gray.
(Lithium glaze core)	Satin-matt surface.

Conclusions
Test II

a–b. The greener color could be caused by the increased alumina and soda content of Kona F-4 and Nepheline Syenite. Note the craze-crackle of the Nepheline Syenite test due to the lower silica and higher soda content, which produces a higher contraction rate in the glaze.

c. The increase in surface gloss is probably due to the increased silica content of Cornwall Stone. The power of the colemanite melter is sufficient to melt the additional silica despite the reduction in alkali content brought about by the Cornwall Stone substitution. The reason for the increased green color is not clear to me. Different supplies of Cornwall Stone contain variable amounts of soda and potash, and other materials, such as iron, titanium, fluorine, and phosphorous pentoxide. All of these materials could influence the resulting color and account for the color change. Whatever the cause, the substitution of Cornwall Stone in this glaze encourages the green color and inhibits the blue.

d. See lithium glaze cores, pages 53–64.

RANDY RED

Figure 3.11, p233.		Red-brown-green mottled. Gloss.

Kona F-4	19.8%
Gerstley Borate	31.7
Talc	13.9
Flint	29.7
EPK kaolin	4.9
Red Iron Oxide	15.0

Test I. Omit Kona F-4	Redder color. Matt surface.
Repeat test	Milky-grey color. Matt surface.
Repeat of test in different studios:	Deep purple-red-black. Matt surface. Red-mahogany. Matt surface.
Test II. Omit Gerstley Borate	Dry, red-brown.

Test III. Substitute for Kona F-4:

a. Nepheline Syenite	Red-yellow flecked surface. Lower gloss.
Repeat test.	Duller red, greener. Bright green where thick.
b. G-200 (potash)	Duller red. Greener. Slightly less fusion.
c. Custer (potash)	Redder color.
d. Cornwall Stone	Even mahogany color. Satin-matt surface.
e. Lepidolite (Lithium)	Yellow color. Increased fluidity.
Repeat test.	Blue-red.
Repeat test.	Greener. More flow.
f. Spodumene (Lithium)	Yellow cast. Fewer red streaks.
Repeat.	Yellow-red-blue
Repeat	Purple-green-burgundy.
g. Petalite (Lithium)	Brown.
Repeat test.	Mahogany. Satin-matt.

Conclusions

Test III.

a. The lowered gloss results from the lower silica and higher alumina content of Nepheline Syenite.

b–c. Potash feldspars are more refractory than the soda glaze cores; similarly, the reddening power of potash is not as strong as soda. These two properties would account for the reduced red color and the slightly lower gloss of the test surface.

d. The change from gloss to a satin-matt surface could result from the higher melting temperature of Cornwall Stone. The titanium and other impurities known to occur in Cornwall Stone could account for the mahogany color of the test surface.

e–f–g. See lithium glaze cores, pages 53–64.

All of the glaze-core substitution tests produced color and surface changes and indicate the tremendous sensitivity of this glaze to any change in either its composition or firing environment.

Note: For additional tests of glaze core substitutions, see Appendixes B1 and B2.

CORNWALL STONE GLAZE (Crafts Students League)

Custer Feldspar	43.6%'	White, semigloss;
Cornwall Stone	21.8	Semitransparent.
Whiting	17.8	
Zinc Oxide	7.9	
Titanium Dioxide	4.0	
EPK kaolin	4.9	

Test I. Omit Custer — Yellow-white; pinholed; stony.

Test II. Omit Cornwall Stone — White with snowflake spots. Grainy surface.

Test III.
For Custer Feldspar substitute:

a. Kona F-4 — Stony, white, pinholed.

b. Nepheline Syenite — Stony, white, yellow. Grainy texture.

c. Lepidolite — Yellow, dry, stony surface.

d. Spodumene — Lighter yellow color. Drier surface.

e. Petalite — Pale yellow-white flecked with white specks. Stony surface, but not as dry as Lepidolite or Spodumene. More flow than either.

Conclusions

Test III.

a. Note the surprising result of a stonier surface with Kona F-4 substitution. Because the fusion point of Kona F-4 is slightly lower than that of the potash feldspar, I expected this substitution to produce a slightly more rather than less-melted surface. All the tests performed at Crafts Students League with their supplies of Kona F-4 and Custer feldspar produced the same atypical results (see Katherine Choy Test, p. 52). These results do not correspond with the comparative tests of Kona F-4 and Custer Feldspars performed in other studios.

b. The lower glass and increased alumina content of Nepheline Syenite has predictably changed the semigloss, semitransparent, white surface to a stonier, yellower texture. Note the similar yellow color that appears when the Custer

glaze core is omitted in Test I. The omission of Custer glaze core reduces both the glass and melter content of the glaze.

c–d–e. See Lithium glaze cores on pages 53–64.

PORTCHESTER COPPER

Figure 1.6, p83.

Nepheline Syenite	55%	Black-green color.
Whiting	25	Satin-matt surface.
Flint	10	
EPK kaolin	7	
Tin oxide	3	
Copper oxide	2% addition	

Test I. Omit Nepheline Syenite — White, unmelted, stiff.

Test II.
Substitute for Nepheline Syenite:

a. Kona F-4 — Greener color; shinier surface

b. Custer — Similar to above test.

c. Cornwall Stone — Lighter, bluer color. Stiffer surface. Lower gloss.

d. Volcanic Ash — Similar to Cornwall Test, but greener and yellower color.

e. Lepidolite — Brilliant blue-green color. Pitted surface.

f. Spodumene — Bluer color. Increased fluidity. Matt-shine surface.

Conclusions
Test II.

a–b. The increased green color and shinier surface could result from the increased glass-silica content contributed by Kona F-4 and Custer feldspars compared to the original glaze core, Nepheline Syenite.

c. The more refractory Cornwall Stone glaze core has raised the fusion temperature of the glaze. Note that at the cone 5–6 temperature, Whiting is not a powerful melter.

d. The Volcanic Ash has a similar chemical composition to Cornwall Stone, and this is reflected in the similarity of the tests.

e.–f. See lithium glaze cores pp. 53–64.

COMPARATIVE OXIDE PERCENTAGE ANALYSES OF PORTCHESTER COPPER

	Nepheline Syenite *	Custer	Kona F-4	Cornwall Stone	DF Stone
SiO_2	53.2	58.5	57.2	61.9	66.1
Al_2O_3	17.6	13.65	15.2	12.4	9.1
Na_2O	6.5	1.9	4.1	2.5	2.4
K_2O	3.1	6.3	3.0	2.4	2.4
CaO	16.1	16.2	17.0	17.3	16.4
SnO_2	3.4	3.4	3.4	3.4	3.5
A/S	1:3	1:4.3	1:3.76	1:5	1:7

*Original glaze core. See also note 24.

KATHERINE CHOY

Figure 4.6, p270.

Nepheline Syenite	53.9%	White-yellow color;
Whiting	11.7	Satin-matt surface
Lithium carbonate	4.7	interspersed with
Zinc oxide	10.3	matt spots.
Flint	1.5	(Recent test on
EPK kaolin	17.9	porcelain claybody
		produced
		a gloss surface.)

Test I. Omit Nepheline Syenite — Overall yellow color. Satin-matt surface.

Test II. Substitute for Nepheline Syenite:

a. Kona F-4 — Drier surface. Stony white.

b. Custer — Shinier surface.

c. Cornwall Stone — Less-yellow color. Increased fluidity.

d. Lepidolite — Yellow cast. Lower gloss.

e. Spodumene — Brown yellow color; Cracked rim.

f. Petalite — High gloss; pinholes.

Conclusions
Test II.

a. Kona F-4 has a higher fusion point than Nepheline Syenite, and its substitution for Nepheline Syenite in this glaze has resulted in a more-refractory, less-fusible glaze surface. However, note the opposite result with an even more refractory feldspar, Custer feldspar (test b below).

b. The substitution of the potash spar, which has an even higher fusion point than Kona F-4, has surprisingly changed the surface into a shinier higher gloss. Admittedly, this feldspar has a high-silica content that could have changed the alumina-silica ratio of this glaze. However, this does not explain why the surface has increased in dryness with the substitution of Kona F-4, which also would have increased the glass content of this glaze. The test of the Kona F-4 substitution was repeated, with identical results. The fact that the Crafts Students League tests of the Custer and Kona F-4 feldspars fired alone consistently showed the Kona F-4 feldspar to be drier than the Custer spar has led me to believe that these surprising results are perhaps only indicative of the particular supplies in use at the Crafts Students League at the time these tests were performed. These tests are atypical and do not correspond with the preponderance of comparative tests of Custer and Kona F-4 that have been made at other ceramic schools and studios.

c. The stronger yellow and reduced whiteness of the surface suggests increased transparency. (Underlying color of the claybody is buff.) The increased silica content produced by Cornwall Stone could possibly account for this result.

d–e–f. See lithium glaze cores on pages 53–64.

SUMMARY OF CONCLUSIONS OF CONE 5–6 OXIDATION TESTS

The most obvious and admittedly least satisfying conclusion to be drawn from the five cone 5–6 tests is that each glaze appears to have its own unique balance of glassmaker, adhesive, and melter oxides. Substitutions of glaze cores caused considerable disturbances in the surface texture and color, especially in the case of the opaque, satin-matt glazes where the alumina-silica ratio appears most sensitive. In certain transparent, high-gloss glazes, the balance of the oxide did not appear to be as delicate. In these glazes, the potash and soda feldspars substituted for each other without causing drastic surface changes. Thus, a transparent, high-gloss glaze with a glaze core of potash feldspar and colemanite as an auxiliary melter accepted all of the core substitutions without discernible surface changes except for color. As in the case of the cone 9–10 reduction tests, the substitution of the soda feldspars for the potash feldspar caused subtle color changes from blue to green. (Transparent Blue, Test II) The changes in surface from gloss to matt and opaque to transparent were highly variable and again depended on the particular oxide balance that existed in each glaze. Thus, in a satin-matt glaze, a substitution of Cornwall Stone for Nepheline Syenite caused a lower gloss (Portchester Copper, Test IIc). On the other hand, in a transparent, high-gloss surface, the substitution caused no change. These different results with Cornwall Stone can be explained by the fact that Cornwall Stone has both a lower melter content and an increased silica content. In the satin-matt glaze, the fusion of the glaze surface required the presence of the total amount of the soda and potash melters provided by the glaze core. Because the calcium melter, Whiting, is not a powerful melter at the lower temperatures of cone 5–6, any reduction of the glaze core melters would cause a less-melted surface to appear. However, the powerful melter Colemanite, which appears in the Transparent Blue glaze, is able to absorb the reduction in soda and potash melters caused by the substitution of Cornwall Stone for the potash feldspar. Only the greener color evidences the glaze core substitution. For all of these reasons, hard and fast rules of behavior are not possible. The consequences of the substitutions are predictable only within the framework of each glaze formula. The accuracy of predictions is increased by percentage oxide analyses or molecular calculations of the glaze with and without the substituted material.

ADDITIONAL GLAZE CORES

LITHIUM GLAZE CORES
Figure 1.9, p86.

The Hall of Gems of the American Museum of Natural History in New York City displays a glittering panorama of 2,000 minerals that have formed in the diverse environments of the earth. This extraordinary mineral spectrum of every conceivable shape, color, and form is the inspired handiwork of natural forces. Nowhere is the awesome creativity of nature more clearly demonstrated than in the formation of the lithium minerals.

In a series of mineral depositions that took place over millions of years, silica and alumina were bonded with the melter oxides of lithium, soda, and potash (Bates 1969, 256–261; Hall of Gems, American Museum of Natural History, New York). A rhythmic cycle of slow crystallization, followed by injections of lithium-rich fluids, and ending with another slow recrystallization period, gave birth to the lithium glaze core minerals of Spodumene ($Li_2O \cdot Al_2O_3 \cdot 4SiO_2$; 8% Li_2O), Lepidolite ($2H_2O \cdot Li_2O \cdot K_2O \cdot 2Al_2O_3 \cdot 6SiO_2$; 4.1% Li_2O), and Petalite ($Li_2O \cdot Al_2O_3 \cdot 8SiO_2$; 4.9% Li_2O).

Violent earth upheavals cracked and sheared off portions of existing rock formations and replaced them with new rocks crystallized from fresh siliceous magma that had flowed into the cracks and crevices of the older rock formations. Protected by the outer layers of the older rock, the magmatic solution cooled slowly into large-grained, high-silica rock formations known as pegmatites. Pegmatites are the exclusive source of the lithium glaze core minerals just listed. Their large crystal size, which is characteristic of all pegmatitic formations, facilitates the extraction of comparatively pure minerals. Thus, pegmatites are also an important commercial source of large-grained feldspar, quartz, and mica.

During the later stages of pegmatitic formations, a residual watery magma with high concentrations of lithium flowed into the cracks and fissures of some preexisting pegmatitic rock layers and slowly cooled into coarse-grained lithium pegmatites. The lithium minerals of Spodumene, Lepidolite, Amblygonite, Eucryptite, and Petalite formed during the last stages of lithium pegmatitic formations when additional residual magmatic fluids bearing high concentrations of soluble soda and potash replaced some of the lithium in the lithium pegmatites. Hence, they are known as lithium replacement minerals. This complex birth process accounts for the presence of soda, potash, fluorine, and rubidium in their chemical structure.

The lithium minerals provide the potter with incomparable glaze cores whose presence in a glaze is unmistakable. The fiery touch of the volatile lithium melter, combined with complex, accessory minerals, creates hard, durable, acid-resistant, craze-crackle-free surfaces of gemstone quality. Of all the glaze cores used by the potter, the lithium glaze cores are the most distinctive and unique—they are also the most treacherous. If their special properties are not fully understood, they can wreak havoc and destruction.

THE LITHIUM MELTER:
LITHIUM CARBONATE

The pure element lithium and the pure compound lithia (lithium oxide) do not exist separately in nature. Their intense chemical reactivity drives them into varied combinations with other elements and minerals. The purest form of lithium is Lithium carbonate (Li_2CO_3), which produces 40.4% Lithia when the carbonate is changed into lithium oxide (Li_2O) at the end of the kiln firing. Although lithium carbonate can be processed from the lithium mineral Spodumene, a more economical source results from the evaporation of lithium brines in a dry arid climate. These lithium carbonate deposits result from solar evaporation of brackish pond waters containing high concentrations of lithium salts (Foote Mineral Company, Bulletin 312). Major deposits of this mineral are found in Silver Peak, Nevada, Searles Lake, California, and in Great Salt Lake, Utah.

LITHIUM CARBONATE

Li_2CO_3
40.4% Li_2O

TRADE NAME   Lithium Carbonate
Li_2CO_3 40.4% Li_2O

GEOGRAPHICAL SOURCE*   Salar del Hombre Muerto, Argentina; FMC Corp. Perth, Australia; Gwalia Consolidated LTD, F&S International, Inc. (Rep) Silver Peak, Nevada; Antofagosta, Chile; Cyprus Foote Mineral Co. (Chemetall GMBH parent company).Salar de Atacoma, Chile; SQM Chemicals (MINSAL)

GEOLOGICAL SOURCE   Solar Evaporation Ponds. Processed from Spodumene.

CHEMICAL STRUCTURE**

Lithium Carbonate (Typical)	Technical Grade
Li_2CO_3	99.200
Na	100 ppm
K	20 ppm
Mg	20 ppm
Ca	100 ppm
Cl	50 ppm
SO_4	30 ppm
Fe_2O_3	20 ppm
H_2O	200 ppm
L.O.I.	5,500 ppm

*Derek J. McCracken, Mike Haigh, "Lithium minerals in a state of flux," *Industrial Minerals* 1998; *Lithium Minerals Review*, 1997.
**Gwalia Consolidated Ltd. Product Catalogue of F&S International Inc. 1997

General Characteristics

In order to understand the unique properties of the lithium glaze cores, it is helpful to compare the general properties of the lithium melter, lithia (lithium oxide), with those of the soda and potash melters. *Figure 1.2, p79.*

Lithia has powerful fusion, hardness and thermal properties that are superior to those of soda and potash. (Glaze core tests, pp. 49–53 Appendices B1, B2; Quality Manager Report, Bikita Minerals (Pvt) LTD, 1998)

1. Although pure lithia has a high melting point of 3092°F, lithium carbonate is a powerful melter—it melts alone at temperatures of 1202°F–1327°F[26] well before soda (Na_2O, 1652°F), sodium carbonate (Na_2CO_3, 1580°F) and potassium carbonate (K_2CO_3, 1643°F). Note that potash (K_2O) has a melting point of 1291°F (Hamer and Hamer 1986, 361–363). Industrial tests have shown that as little as .5%–1% Lithium carbonate increases the fluidity and gloss of a glaze and heals the pinholes left by escaping gases (Foote Mineral Company, Bulletin 319; *Ceramic Industry* 1993, 83). Our tests produced similar results. The replacement of 5% Whiting with 5% lithium carbonate increased transparency and gloss of a test with 90% potash feldspar and 10% Whiting ($CaCO_3$) compared to comparative tests with 5% Soda Ash (Na_2CO_3) at cone 5/6 oxidation and cone 9–10 oxidation and reduction firing temperatures. Similarly, the substitution of a lithium glaze core for a soda and potash feldspar in three different glazes increased fluidity in all glazes (see Appendix B1).

2. Although Lithium carbonate has been described as "the most water-soluble of all of the alkali metal carbonates," its water solubility decreases with increase of temperature (Foote Mineral Company, Bulletin 312). As a glaze and claybody material, unlike its soda and potash counterparts, Lithium carbonate is said to be "only slightly soluble in water" (Cyprus Specialty Metals, *Application of Lithium in Ceramics*, 5:76).

3. Lithia has a strong effect on the color and strength of the claybody. A thin wash of lithium carbonate and water will burn and singe the stoneware clay and turn it yellow-orange-brown in an oxidation atmosphere and a deep, mahogany purple in a reduction atmosphere. Although this effect is clearly visible on an iron-bearing stoneware claybody, even the iron-free porcelain claybody reveals at times the touch of lithium by turning a pale, yellow-orange in an oxidation atmosphere and a smoky-gray streaked with orange and pink in a reduction atmosphere at cone 9/10 firing temperatures. Hence, lithium carbonate can be used most effectively as a thin wash on the claybody in the same manner as soda and potash materials, but with even more dramatic results. Note that even though you may wash an unfired, high-lithium glaze off a bisque pot, the fired claybody will still reveal the fiery traces of lithium. During the initial glazing process, the lithium materials flow into the open pores of the bisque pot. A subsequent washing of the pot will not remove them, and the hidden presence of lithium in the pores of the pot will burn the exposed skin of the claybody a dark-reddish brown in a reduction fired kiln.

A heavy coat of high-lithium glaze may cause problems for the claybody. When a glaze containing the powerful lithium melter flows into the open clay pores of a bisque pot, the volatile presence of the lithium melter can lower the firing temperature of the claybody and make it brittle and weak. Hence, when applying lithium-based glazes, a thin to medium thickness of coat is best.

4. The presence of lithium in a glaze or claybody will affect the glaze color, especially if iron or copper minerals are the primary colorants in the glaze. The reduction-fired, blue-green celadon glaze will display a greener and yellower cast when lithium replaces some of the potash and soda melters of the glaze core (see Appendix B2). Lithium tends to darken and blacken iron spots that break through the glaze surface from the underlying reduction-fired stoneware claybody. Small amounts of Lithium carbonate will blacken, gray, and redden the milky-white surface produced by soda and potash glaze cores. A cone 9/10 reduction test of 90% Cornwall Stone and 10% Whiting (smooth, gray-blue surface) was compared to 90% Cornwall Stone, 5% Lithium carbonate, and 5% Whiting. The test with 5% lithium produced a mottled, brown-black-orange-brown, lustrous, glossier surface.

In an oxidation atmosphere, equally striking color changes occurred. A soft, satin-matt, copper-green surface became a brilliant, blue green when lithium glaze cores replaced Nepheline Syenite. (See Portchester Copper, Test II, p. 51, and Appendix B2.)

5. Lithium-melter materials possess a rare thermal property that cause them to expand on heating without subsequent contraction on cooling. This unique thermal property makes high lithium-melter materials indispensable

ingredients for the creation of pyroceramics (i.e., claybodies and glazes with increased resistance to thermal shock, such as oven-to-table ware and flameware). However, unless the claybody itself is balanced with equivalent amounts of expanding lithium materials, this thermal expansion property may interfere with the fit of the glaze on the claybody with disastrous consequences for the fired ware. The high expansion of a lithium-melter glaze can create a glaze coat that is too large for the claybody. The pressure of the oversized glaze in its attempt to remain on the claybody can cause the fired ware to crack and shatter. Even more serious is the fact that portions of an oversized glaze may loosen and fall off edges and rims in splinters of glass. The compatibility of claybody and glaze movements is especially crucial for the successful production of ovenware. Ovenware reaches the temperatures of cristobalite inversions,[27] at which point, if the claybody is without the benefit of thermal shock-resistant materials, it will expand and contract once again. Such movements of the claybody beneath an immobile glaze coat can crack the claybody at its weakest point. In addition, the presence of the high-lithium glaze in the pores of the claybody may cause the ware to become slightly overfired and brittle. This will further weaken the claybody so that it is unable to withstand unequal pressures of any kind. (A casserole of ovenware stoneware, which had been glazed with a cone 9/10 reduction glaze containing 5% lithium carbonate, cracked after baking 30 minutes at 400°F oven temperatures.) Glaze splintering, or claybody cracks, need not happen immediately—it sometimes takes a period of days, weeks, or even months for these disastrous consequences to occur. Note that this problem of glaze fit known as "shivering" is the opposite of "crazing." Shivering is the cracking of the claybody from the pressure of an oversized glaze; crazing is the cracking of the glaze from the pressure of an oversized claybody, caused by high-contracting glaze materials such as soda and potash feldspathic melters or by the lack of contracting quartz materials in the claybody. The pressure of the tight glaze on the claybody is relieved by the glaze breaking up into a network of tiny cracks known as craze-lines. Although not a desirable result for functional dinnerware, it is a less serious condition than an oversize glaze, because, at the very least, the clayform remains intact, and except in the most extreme cases, the glaze coat is still bonded with the pot.

Tests of high lithium-melter glaze cores cracked, shivered, or shattered within a short period after the firing. The substitution of Spodumene for Kona F-4 feldspar in Sanders Celadon caused the glaze to shiver off the rim of the test pot. The same result occurred with Petalite, another high lithium-melter glaze core. Shivering resulted in both tests, despite the readjustment of the formula in accordance with a recalculated molecular analysis, based on the substituted lithium glaze core.[28] In this connection, it is important to note that this celadon glaze already contained large quantities of silica. Silica has a very low contraction rate, and its inclusion in a lithium-based glaze increases the possibility of shivering. In such a case, it would be possible to minimize the shivering problem by lowering the amount of silica. Conversely, the substitution of a lithium-melter glaze core for the high-crazing soda and potash feldspars would reduce the crazing of the glaze surface. In this way, what is a problem for one glaze surface may become the solution for another. The important thing is to understand the special properties of a ceramic material and to use them to produce the desired surface.

SPODUMENE
Figure 1.9, p86.

The lithium glaze core Spodumene combines 5%–7% lithium melter with a high alumina content (up to 25%); low amounts of soda, potash, manganese, and phosphorous; variable amounts of silica (62%–75%); and iron oxide (.07%–3%).

The name Spodumene comes from the Greek meaning "reduced to ashes," and describes the typical gray-white color of the mineral (Frye 1981, 717). Spodumene belongs to the complex group of minerals, known as "Pyroxenes," or "Strangers to the Fire." This romantic designation distinguishes them from granites and feldspars on the basis of their crystalline and chemical structure. The pyroxene group is typically lower in silica and higher in magnesium, calcium, and iron. Nature often obscures rigid classifications; thus, the chemical analyses of Spodumene do not clearly reflect these differences.

SPODUMENE

$Li_2O \cdot Al_2O_3 \cdot 4SiO_2$
8% Li_2O (theoretical)

TRADE NAME Spodumene

GEOGRAPHICAL* SOURCE Perth, Australia; Gwalia Consolidated LTD. Manitoba, Canada; Tantalum Mining Corp. TANCO. Zimbabwe: Bikita Minerals (Pvt) LTD. Domestic Deposits: Mines Closed.

GEOLOGICAL SOURCE* Pegmatites, (coarse-grained granites) that have intruded deeply weathered pre-Cambrian gneiss and schist. Over 225 million years old. Mineralization of some deposits may be of Archean age.

CHEMICAL STRUCTURE**

| | Fine Spodumene | | Concentrates |
	Gwalia Con. (Australia)	TANCO (Canada)	Bikita Minerals (Zimbabwe)
SiO_2	64.00	65.00	65.00
Al_2O_3	26.50	25.00	26.50
Fe_2O_3	0.10	0.12	0.07
Li_2O	7.50	7.10	7.30
Na_2O	0.20	0.30	0.15
K_2O	0.20	0.20	0.10
CaO	0.10		
P_2O_5	0.25	0.35	
MnO_2	0.05	0.06	

P.C.E.*** 2552°F Cone 14–15 (Hamer and Hamer 1986, 363).
Note: 70 Feldspar + 30 Spodumene = P.C.E. 2088°F Cone 2
60 Neph.Syn. + 40 Spodumene = P.C.E 2043°F Cone 01.
(*Ceramic Industry* 1989, 112; 1998, 166)

*Bates, 1969, 235, 245, 256–259; Bikita Minerals (Pvt) LTD, 1998, Quality Manager Report; McCracken and Haigh, "Lithium Minerals in a State of Flux," *Industrial Minerals*, Aug. 1998.
**F&S. International Inc. Product Catalogue 1997; Haigh and Kingsworth 1990, Table 6, *Glass Magazine*; Bikita Minerals (Pvt) LTD., Product Specifications, 1998.
***Pyrometric Cone Equivalent. See pp. 9–10.

At the Kings Mountain, North Carolina, site, over 25 million years ago (late Paleozoic period), highly reactive lithium-enriched magmatic solutions flowed into the cracks and fissures of the quartz pegmatite to form lithium-enriched pegmatite. During the middle stages of this formation, potassium-enriched magma replaced some of the lithium to form large deposits of the mineral Spodumene (Bates 1969, 258–259). However, these mines are now closed due to the cheaper deposits available in Chile and Argentina (Derek J. McCracken, F&S International, August 1998).

Characteristics of Spodumene

Spodumene has a fiery personality; its outstanding features are as follows:

1. A thin coat of Spodumene produces orange, brown, and purple colors in both oxidation and reduction atmospheres. A thick coat of Spodumene alone will not totally melt at high stoneware temperatures of cone 9/10. The top layers consist of an orange-pinkish-white powder that flakes off when touched. True to this pure state, when combined in a glaze, Spodumene helps to achieve fiery orange-white-black surfaces in both oxidation and reduction atmospheres. In combination with Nepheline Syenite, Kona F-4, and soda ash, Spodumene produces a black-rimmed, reddish, orange-white, semigloss surface at cone 9/10 reduction temperatures (Shino Carbon Trap glaze, *Figure 1.10, p87*). In the oxidation atmosphere, 60% Spodumene combines with mixed wood ash and Gerstley Borate to create unique wood-fired effects (p. 171; *Figure 1.12, p89*).

2. Spodumene has a high melting point of 2552°F (cone 14–15). Compare this to the melting point of potash feldspar, which is 2192°F (cone 6). The refractory nature of Spodumene is explained by its low melter content, which is less than 8%; compare this to a feldspar, which has a melter content of almost double that amount. The lithium mica mineral, Lepidolite, which by itself melts into an opaque, semigloss, orange-white surface at high stoneware temperatures, has a melter content in excess of 20%. However, the lithium melter in Spodumene will form a powerful eutectic relationship with any soda and potash melters that are present in the rest of the glaze materials, and this eutectic bond will drastically increase the fusion of the glaze. Thus, when

30%–40% Spodumene is combined with feldspar or Nepheline Syenite respectively, the fusion point of the feldspar combination drops to cone 2; with Nepheline Syenite, fusion plunges to cone 01 (*Ceramic Industry* 1989, 112; 1993, 106; see also Foote Mineral Company, Bulletin #317).

3. A glaze in which Spodumene is the primary glaze core will rarely display a craze-crackle network. On the contrary, Spodumene may create the more dangerous condition of glaze-shivering and/or claybody cracking. This condition occurred in many of the tests in which Spodumene replaced the soda and potash feldspar glaze cores. The reasons for this lie in the peculiar thermal expansion of lithium materials, which occurs in the case of Spodumene at about 1976°F. At this temperature, Spodumene expands in volume more than 30%, as it converts irrevocably to beta Spodumene. At the conclusion of the firing process, during the cooling period (400°F–500°F), the cristobalite formed in the stoneware claybody undergoes its final contraction. However, the Spodumene materials remain in their expanded, swollen state, and the clinging pressure of the now oversized glaze can crack the claybody. Even worse, sections of glaze may flake off the clayform. If Spodumene materials in the glaze are balanced with higher-expansion glaze materials or if the claybody itself contains sufficient beta Spodumene to offset the expansion of the beta Spodumene in the glaze coat, Spodumene can create smooth, craze-free glaze surfaces that strengthen the fired ware. The beta phase of Spodumene, which increases resistance to thermal shock, is one of the primary ingredients in the production of industrial heatproof flameware and oven-to-table dinnerware (*Ceramic Industry* 1998, 166).

4. Spodumene changes the color, surface texture, and flow of a glaze, when it replaces the feldspar as a glaze core. The gray, blue-green of Sander's Celadon changed to a yellow-brown-green when 44% Spodumene replaced 44% Kona F-4. This change occurred even in the Molecular Revision Test, in which the batch formula of Sander's Celadon was recalculated to provide for the different oxide ratio contained in Spodumene. Spodumene changed the semiopaque, gray-white surface of the Temple White glaze (cone 9/10 reduction) to a yellow, orange-brown (see Tests, Appendix B2).

Surface and texture changes occurred in all of the tests in which Spodumene substituted for the feldspar glaze core, in both oxidation and reduction atmospheres, and in low and high stoneware firing temperatures. The surface became more opaque, without any craze-crackle network, and flowed very strongly. Here again, as in the case of Lepidolite, the lower silica and higher alumina of Spodumene opacifies the surface; the combined power of the lithium and calcium melters increases the flow of the glaze; the low contraction rate of the lithium melter produces the uncrazed, crackle-free surface. When the silica and alumina were adjusted in the molecular revision test, so as to restore the balance disturbed by the substitution of the low-silica, high-alumina Spodumene glaze core, the opacity and flow of the surface was less than in the trial-and-error test, but still greater than in the original glaze with a feldspathic glaze core. As in the Lepidolite tests, the yellow-green color, the craze-free surface, and the occasional shivering of the glaze on the rims of the test pots bear witness to the unique presence of a lithium glaze core.

5. Spodumene has a strong effect on the color of a claybody. When it substitutes for a feldspar in a low-iron, oxidation claybody, it changes the pale, yellow-white color into a warm, orange-brown. Thus, Spodumene breathes life into the often dead color of oxidation-fired stoneware. Note that the lithium melter in the claybody may affect the color of the covering glaze. Light, transparent glazes will become muddier and browner in shade. Dark, iron-bearing glazes will show less change and would be a better choice for Spodumene claybodies.

6. Spodumene will affect the contraction rate of the claybody, and thus will determine the fit of the glaze and the strength of the fired claybody. Spodumene in a fired claybody will no longer contract or expand when subjected later to oven temperatures. For this reason, additions of 30%–50% Spodumene are included in flameware bodies at cones 9 to 11 so as to enable them to withstand the thermal shock of open flames. According to data contained in materials from Val Cushing's glaze class (1986, Alfred University), a safe flameware claybody at cones 9 to 11 firing temperatures should contain 50%–60% shock-resistant materials. At cone 9/10 temperatures, 60% was the preferred amount.

Spodumene alone was found to be too refractory to produce vitrification at cone 11. According to Weltner (Alfred University, Val Cushing Claybody Book, 1986), combinations of 20% Spodumene and 30% Petalite proved most successful at cone 11 firing temperatures. It is advisable to include an equivalent or even greater amount of Spodumene in the glaze in order to equalize the expansion and contraction of a high-Spodumene claybody with the glaze (See Lawrence 1972, 194; Joanides 1976, 38–40).

Kinds of Spodumene

The potter should be aware that the supplies of Spodumene have changed significantly in the last 15 years. Prior to 1984, the Foote Mineral Company, currently known as Cyprus Foote Mineral Company (owner of a predominant domestic Spodumene source located in Kings Mountain, North Carolina) once produced a ceramic grade of Spodumene with a low iron content and the expanded, beta-phase form of Spodumene known as Thermal Grain (Foote Mineral Company, Bulletin 313A). After 1984, the supply of low-iron Spodumene became exhausted and was replaced with a chemical grade containing 1.45% to 3% iron oxide (*Chemical Data* of 7/1/88, courtesy of Cyprus Specialty Metals). As of 1990, Cyprus Foote closed the Kings Mountain operation because of the "low cost production of the Sociedad Chilena de Litio brines of which Cyprus Foote have an 80% ownership" (Haigh and Kingsnorth 1990, 2). According to a letter of January 1990 from Charles E. Larson, Market Development Engineer, Cyprus Foote Mineral has now "discontinued the Spodumene business." This ended our onetime, readily available, primary domestic source of Spodumene (the basis for the tests and glazes described in the prior section) and illustrates, once again, the ephemeral nature of our ceramic materials.

Although domestic sources of low-iron Spodumene are no longer available, Australia, Zimbabwe, and Canada produce significant supplies of this material.[29]

Low-iron Spodumene is mined in Manitoba, Canada, by TANCO, Tantalum Mining Corporation of Canada LTD, in the Bernic Lake pegmatite region. Gwalia Consolidated in Perth, Western Australia, is the largest producer of Spodumene. It produces a ceramic grade of low-iron Spodumene with a lithium content of 6.5% and an iron content of 0.10%. Although the minimum order requirement is a full truckload of 22 short tons, there are "stocking distributors" in Hadley, Massachusetts, Pittsburgh, Dallas, and Los Angeles who supply the smaller ceramic markets (Derek J. McCracken, F & S International Inc., August, 1998). Thus, although the current market for low-iron Spodumene is still primarily industrial, it has now become available once again to the nonindustrial ceramic market.

For those potters who still have domestic supplies of Spodumene, note that the substitution of higher-iron, lower-lithium Spodumene in a glaze or claybody that originally contained lower-iron Spodumene can cause drastic changes in the surface color. A glaze that consisted of 35% Spodumene changed from a warm orange-brown into a dull, gray-green brown when high-iron Spodumene was introduced into the glaze. Because ceramic suppliers in the recent past sold only domestic supplies of high-iron Spodumene, carefully test older supplies of Spodumene for iron content before plunging into a large-scale operation with a Spodumene glaze or claybody.

LEPIDOLITE
Figure 1.9, p86.

Lepidolite is a lithium mica, which is found with Spodumene, Amblygonite, and Petalite minerals. The lithium mica known as Lepidolite forms when high-potash fluids flow into lithium pegmatites and partially replace the lithium contained in the pegmatite. The fact that it is found in scaly aggregates caused it to be named after the Greek word for scale (Fyre 1981, 656).

Lepidolite is an important ingredient in the manufacture of industrial glasses because it combines powerful fluxing action, high alumina content, and lowered crazing (*Ceramic Industry* 1998, 125). In claybodies, the combination of Lepidolite with Nepheline Syenite and feldspars in low-temperature bodies is said to produce "semivitreous and vitreous bodies having good strength and firing range" (Ibid.). In view of this industrial research, it would be interesting to test Lepidolite as a melter in our claybodies.

In prior years, Tantalum Mining Corporation of Canada (TANCO) reported that it was the world's significant supplier of rubidium-rich Lepidolite. However, the *1997 Lithium Minerals Review* by Derek J. McCracken of F&S International Inc. and Mike Haigh of Gwalia Consolidated Ltd. lists only Portugal (Sociedad Mineria de Pegmatites) as a producer of Lepidolite. "I am not aware of Lepidolite being available in the U.S." (Derek McCracken, August 1998). This situation is unfortunate for potters because, as the following tests will show, Lepidolite creates fascinating and unusual glaze surfaces at stoneware firing temperatures. Even in its raw powdery state, its presence is unmistakable. The sparkling, silvery grains of Lepidolite distinguish it from the dull, floury powders of feldspar and other glaze-core minerals. Potters who are lucky enough to have stockpiled supplies of Lepidolite should use them sparingly. As was once the case of low-iron Spodumene supplies, current supplies of Lepidolite will be irreplaceable until Lepidolite becomes available once again to potters. In this connection, note that Bikita Minerals in Zimbabwe reports that "small reserves of high-grade Lepidolite are available with no market" (Lithium and Allied Minerals at Bikita Minerals, Report of Quality Manager, Bikita Minerals

---

LEPIDOLITE*

$2H_2O \cdot Li_2O \cdot K_2O \cdot 2Al_2O_3 \cdot 6SiO_2$
4.1% Lithia (theoretical)

TRADE NAME   Lepidolite

GEOGRAPHICAL SOURCE
Portugal: Sociedad Mineria de Pegmatites.
Zimbabwe: Bikita Minerals (Pvt) LTD.

GEOLOGICAL SOURCE   Lithium pegmatites partially altered by intrusions of potash-enriched magma. Crystalline formation is that of a mica.

CHEMICAL STRUCTURE**

(Typical)   Range of Maturation*
                         2012°F–2372°F

	% OXIDE
Li_2O	03.85
K_2O	09.00
Na_2O	00.55
Fe_2O_3	00.07
Al_2O_3	28.60
SiO_2	50.40
Cs_2O	00.50
Rb_2O	03.56
CaO	00.05
MgO	00.03
P_2O_5	00.05
F	3.50

*Source: *Ceramic Industry* 1998, 120; Haigh and Kingsnorth 1990 (January); Frye 1981, 656-657; Bates 1969, 256-260; Hamer and Hamer 1986, 194, 361; *The Lithium Minerals Review* 1997, F&S International Inc.
**Ceramic Industry* 1998, 125; Bikita Minerals (Pvt) LTD Product Specifications 1998.

(Pvt) LTD, 1998). Whether or not this supply could become available to the nonindustrial ceramic industry depends on further investigation by an adventuresome supplier.

Glaze Function

Lepidolite is a lithium mica that performs a glaze core function in ceramics similar to that of a feldspar. Despite this functional similarity, both the chemical structure and melted surface of Lepidolite differ markedly from that of a feldspar. Compared to a feldspar, Lepidolite has a high alumina content (25%) and a correspondingly low silica content (55%). Compare this to the 17%–20% alumina, and 65%–70% silica of the feldspars, or even to the 23.6% alumina, and 60% silica of Nepheline Syenite. Most important is the highly complex structure of Lepidolite. In addition to containing an average of 4% lithium, it contains sizable amounts of potassium, fluorine, and rubidium. Hence, it is not surprising that Lepidolite has a unique visual appearance when fired to stoneware temperatures. A thin layer of Lepidolite in a high-stoneware, reduction firing displays a warm, orange-red, semigloss surface; this surface becomes a milky-white, semigloss when thick. In an oxidation firing of both high and low stoneware, Lepidolite fires to a stiff, white opaque surface where thick and a faint, brown-orange surface when thin. A thick coat will tend to crawl. The low contraction rate of the lithium melter is reflected in the absence of a visible craze network. The fired surface of the Lepidolite tests appeared less glossy than did comparable feldspar tests due to Lepidolite's higher alumina and lower silica content. The pitted surface of the Lepidolite tests reflects the presence of volatile fluorine gas that has broken through the layers of the nonfluid glaze coat and escaped into the atmosphere.

Interesting color and surface changes occur when Lepidolite substitutes for the feldspar glaze core, Kona F-4, in Sanders Celadon glaze. In the trial-and-error tests, the total amount of Kona F-4 was replaced by the same amount of Lepidolite. In the molecular revision tests, the recipe batch was recalculated from the molecular formula of the glaze. Lepidolite replaced Kona F-4 and the amounts of flint and kaolin were adjusted to compensate for the changed percentages of silica and alumina caused by the substitution of Lepidolite for Kona F-4. The original blue-gray-green, transparent surface of Sanders Celadon glaze changed in the trial-and-error test to a milky, opaque-blue where thick and a yellow-green where thin. The surface of the molecular revision test was slightly yellower, milkier, and less bright than the original glaze.

The semiopaque, gray-white, satin-matt glaze known as Temple White (see p. 46) changed to an opaque, stony, yellow-white surface when Lepidolite substituted for the potash feldspar in a trial-and-error test. (Note that in all tests, the original glaze appeared on one-half of each test so that an accurate comparison could be made.)

These changes from a gray, transparent gloss to a yellowish stony opacity, which occurred in the trial-and-error tests of both Sanders Celadon and Temple White, is partly the result of Lepidolite's increased alumina and lowered silica content. As we will see in Chapter 4, increased alumina has a drastic effect on both color and opacity of a glaze surface. However, as was the case in the molecular revision tests, the lithium, fluorine, and rubidium minerals in Lepidolite would also cause color changes even after adjustments for alumina and silica differences have been made. Our tests show that the lithium melter shifts the original blue-green color of a celadon, high-feldspar glaze toward a less brilliant, yellow-green-brown shade.

In an oxidation atmosphere, different color results occur. When Lepidolite replaced the high-soda, low-silica, high-alumina Nepheline Syenite in a low stoneware glaze in which the colorant was copper, (see p. 51 for formula), the dull, soft, black-green surface changed to a brilliant blue-green.

Lepidolite will also cause textural changes in the surface of the glaze when it replaces the feldspathic glaze core. The low contraction rate of the lithium melter in Lepidolite will reduce the craze network of the glaze surface caused by high-shrinking melters of soda and potash. Thus, there were no visible craze marks on the surface of the Sanders Celadon glaze test with the Lepidolite substitution. The original glaze with Kona F-4 feldspar showed distinctly visible craze lines. Although the substitution of Lepidolite erased the craze network, it did not cause the glaze surface to shiver. None of the tests in which Lepidolite substituted for the feldspathic glaze cores, shivered or cracked, unlike the tests of the other lithium glaze cores. Lepidolite combines lithium (4%) with high-shrinking potassium and rubidium (9%) and fluorine (5%). Consequently, even high amounts of Lepidolite in a glaze would not ordinarily produce the glaze fit problems of shivering and cracking that occur with the purer lithium glaze cores of Spodumene and Petalite.

Lepidolite can alter the opacity and gloss of the surface. In both the Sanders Celadon and Temple White tests, Lepidolite caused a shift toward a more opaque, milkier surface. This change was, of course, more noticeable in the trial-and-error tests where we did not compensate for Lepidolite's different weight and lower silica, higher alumina content. The surface of the Sanders Celadon trial-and-error test, although opaque, flowed very strongly, and appeared to be a "running matt"—similar to the surface, which often results when 10% calcium carbonate is added to 90% Nepheline Syenite. Although the surface of the molecular revision test was less opaque, less matt, and less fluid than the trial-and-error test, it still showed distinct differences from the original glaze surface. The color changed from blue-green to a pale, milky, yellow-gray-green; the surface appeared slightly less glossy and more opaque; there were no craze marks on the surface compared to the several distinct craze lines on the control. And finally, the surface of the test was highly pitted, unlike the smooth, glassy surface of the original glaze. Thus, the distinct personalities of the potent materials contained in Lepidolite (which include, in addition to lithium, even larger quantities of potassium, fluorine, and rubidium) will leave their unique mark on any glaze surface in which they appear. Because of Lepidolite's distinctive touch, it becomes a very special glaze core. Hoard and cherish your existing supplies, because more may not be forthcoming. Never forget that the disappearance of a particular ceramic material from the ceramic scene is the only constant we have. Hence, it is best not to become dependent on any one glaze material, but rather to find permanence only in your ability to successfully adapt and change your glaze surfaces in accordance with the ever-changing state of available earth materials.

AMBLYGONITE
$2LiF \cdot Al_2O_3 \cdot P_2O_5 \cdot nH_2O$[25]

Amblygonite typically contains 9.50% lithia, and next to lithium carbonate, has the highest lithium content of the lithium materials described in this book. In addition, its substantial alumina (29%) and phosphorus (50%) content make it a useful source of alumina-phosphate. Domestic sources are in South Dakota and Wyoming. Bikita Minerals produces Amblygonite on a "minor scale." Its cone 4 PCE would make it a useful, high-lithium material for the lower firing temperatures. However, it is said to have "limited availability" (Haigh and Kingsnorth 1990 [January] Table 1; *Ceramic Industry* 1998, 64; Report of Quality Manager, Product Specifications, Bikita Minerals (Pvt) LTD, 1998).

PETALITE
Figure 1.9, p86.

Petalite contains the highest silica content and the lowest coefficient of expansion of any of the lithium minerals, including Spodumene, whose coefficient of expansion in a 9–10 reduction firing test was about five times higher than that of Petalite (Cushing, 1986). The reaction of Petalite to high heat is extraordinary! When heated to temperatures above 1900°F, Petalite converts irreversibly "into a solid solution of silica in beta spodumene" with no thermal expansion rate (*Ceramic Industry* 1998, 142).

The largest producer of Petalite is Bikita Minerals, in Zimbabwe. Hammill & Gillespie is the sole supplier of Petalite in North America for the ceramic industry.

Glaze Function

With the exception of its lithium content, Petalite has a skeletal structure similar to Cornwall Stone. Petalite, like Cornwall Stone, has a high silica content and a relatively low amount of melter and iron oxide—its iron content is even lower than that of Cornwall Stone. All of this is in dramatic contrast to the low-silica, high-alumina, and high-iron content of both Lepidolite and Spodumene. The primary difference, of course, lies in the significant amount of both silica and lithium contained in Petalite, and it is the combination of these two minerals in Petalite that gives it such valuable thermal shock properties.

Petalite fired alone at high stoneware temperatures displays a white, semigloss, opaque surface where thick. A thin layer is semiopaque with a high gloss. Despite the lower total melter content compared to Spodumene and Lepidolite, the higher silica content in Petalite results in a glossier and more fused surface when fired alone. This higher silica content could also account for Petalite's minimal thermal expansion rate when introduced in a claybody, which makes it the indispensable ingredient of the flameware claybody and glaze. And finally, Petalite's low iron content makes it an important ingredient for white or light-colored glazes that require lithium without additional iron minerals.

At low stoneware temperatures it remains stiff and unmelted when fired alone. However, the addition of but 10% Whiting produces an astonishing change and turns the unmelted surface into a milky-white, satin-matt glaze.

At both high and low stoneware firing temperatures, all of our tests of Petalite shivered at the rims and eventually cracked or broke apart under the stress of this rigid, expanded lithium material. Thus, the use of large amounts of Petalite in a glaze, without a corresponding amount in the claybody, could present a serious glaze-fit problem. This problem occurred when Petalite substituted for a potash feldspar in the semiopaque, milky-white, high-stoneware, reduction Temple White glaze (see p. 46). The lithium melter in Petalite changed the gray-white color into a greener, yellower shade. The rim of the test pot showed distinct evidence of shivering. Comparable tests with Spodumene did not produce shivering, which bears witness to the low contraction rate of Petalite, as compared to the other lithium glaze cores. The same result occurred when Petalite substituted for Kona F-4 in Sanders Celadon. The pure blue-green-gray color turned greener and browner. The glaze flowed more strongly, despite the fact that the alumina and silica had been readjusted in accordance with the molecular requirements of the formula. Again, there was evidence of shivering on the rim of the Petalite test pot. However, no evidence of either shivering or cracking appeared when Petalite substituted for Nepheline Syenite in the oxidation, low-stoneware, highly crazed, Portchester Copper glaze (see p. 51). The craze-network disappeared with the substitution of Petalite. The dull green color of the surface, produced by copper oxide, changed to a more brilliant blue-green. The Petalite substitution also caused the glaze to flow more strongly.

Despite the glaze-fit problems produced with Petalite, the silky, satin surfaces that it displays at both low and high stoneware temperatures in both oxidation and reduction atmospheres should tempt us to use more of this material.

Our tests point to the general direction in which to go. One can lower the silica content of the glaze or substitute high-contraction melters to temper the high expansion rate of Petalite.

PETALITE	
$(Li_2O \cdot Al_2O_3 \cdot 8SiO_2)$ 4.9% Li_2O	
TRADE NAME	Petalite
GEOGRAPHICAL SOURCE	Zimbabwe: Bikita Minerals (Pvt) LTD; North American Supplier: Hammill & Gillespie, Inc.
GEOLOGICAL SOURCE	Lithium pegmatites partially altered by intrusions of soda and potash enriched magma.
CHEMICAL STRUCTURE*	(Typical)
SiO_2	77.25
Al_2O_3	16.90
Fe_2O_3	0.04
Li_2O	4.20
K_2O	0.26
Rb_2O	0.07
Na_2O	0.70
MgO	0.04
CaO	0.11
Cs_2O	0.04
P_2O_5	0.04
P.C.E.**	2600°F (cone 15)

*Bikita Minerals (Pvt) 1998
**Ceramic Industry 1998, 143.

CLAYBODY FUNCTION

As a last resort, in dealing with glaze-fit problems caused by the use of Petalite in a glaze, one can always introduce Petalite into the claybody as well, and thus equalize the expansion and contraction rates of the claybody and the glaze; this is, of course, Petalite's major industrial use (*Ceramic Industry* 1998, 142–143). Because of its high melting point (2600°F, cone 15), Petalite functions as a primary melter only in high-temperature, cones 9 to 14 claybodies. In this connection, it is interesting to note that Spodumene was not found to be as successful as Petalite in creating flameware claybodies even at cone 11 because it proved to be more refractory than Petalite and, therefore, did not produce as vitrified a claybody. On the other hand, Petalite alone created too short a firing range. The combination of 20% Spodumene and 30% Petalite at cone 11 and 30% Spodumene and 30% Petalite at cone 10 produced the desired results (Weltner 1986).

The ease with which Petalite forms eutectic bonds with feldspars and Nepheline Syenite causes it to perform well as an auxiliary flux in low-stoneware claybodies. According to industrial tests, a combination of 45% Petalite and 55% F-4 feldspar lowered the melting point of Petalite from cone 15 to cone 4. And when 55% Nepheline Syenite combined with 45% Petalite, the PCE plummeted to cone 2 (Foote Mineral Company, Bulletin 301). Hence, in combination with feldspathic materials, it has an extraordinary firing range. Combinations with Nepheline Syenite, Talc, and Whiting created fusion at cone 2; a combination with Kona F-4 in the same formula created a melt at cone 4 (Cushing 1986). The presence of Petalite as an auxiliary flux in a cone 5–6 claybody is said to increase thermal shock resistance and lower the maturing temperature without decreasing the firing range of the claybody (*Ceramic Industry* 1998, 143). On the other hand, different tests showed that the use of Petalite in a cone 11 flameware claybody shortened the firing range of the claybody; it was necessary to replace 20% Petalite with Spodumene in order to reduce this effect (Weltner 1986). Here again, one's own testing experience provides the most important guide in determining how much of Petalite to use in a claybody.

At the time our tests were made, both Lepidolite and Spodumene were readily available and were included as primary lithium glaze cores for our stoneware glazes. We have not tested Petalite to the same extent either as a glaze core or as claybody melter; hence, the personality of this lithium glaze-core remains mysterious and its potential unknown. The cool, snowy gray surfaces that Petalite revealed in our preliminary tests, together with its paramount availability, compels further exploration. Petalite is a lithium glaze core of the future for the curious and adventurous potter.

Grateful acknowledgment is made to F&S International Inc., Gwalia Consolidated LTD, TANCO (Canada), Cyprus Specialty Metals, Foote Mineral Company, and Bikita Minerals (Pvt) LTD for Lithium Glaze Cores source material.

VOLCANIC ASH GLAZE CORES
(PUMICITE) *Figure 1.16, p93; 3.10, p232.*

The ceramic materials discussed thus far derive, for the most part, from igneous magma that cooled slowly beneath the layers of the earth's crust. We come now to a material that is an immediate result of fast-cooled volcanic eruptions—namely Volcanic Ash, or Pumicite. Volcanic Ash forms from frothy, siliceous lava, which is hurled from the erupting volcano into the cooler atmosphere of the earth. If the rate of ejection is very fast, the heated, molten liquid swells with gaseous vapors that expand on cooling. As these gases escape into the cooler earth's atmosphere, they will leave behind empty holes. The result is a porous, rough-textured material known as Pumice (Bates 1969, pp. 39–40). The finer, powdery form of this same material is Pumicite, or Volcanic Ash. Should the rate of ejection slow down, the magma will lose its contained gases before it reaches the earth's surface; in this condition, the magma cools immediately into the amorphous, noncrystalline, natural glass known as Obsidian. Should this same magma cool slowly beneath the upper layers of the earth's crust, the slow, pressurized cooling would create the coarser-grained rock known as Granite. Faster cooling (which would occur closer to the earth's surface) produces the fine-grained rock known as Rhyolite. (*Figure Intro.1, p11.*) If it is possible to create such totally different earth materials as Granite, Rhyolite, Obsidian, or Pumice simply by altering the time and manner of solidification, then, by the same token, the cooling cycle of each kiln firing must play an equally crucial role in determining the final glaze surfaces. Thus, when a potter fires a pot glazed with Volcanic Ash, the kiln's cooling cycle may recreate the conditions for the formation of Obsidian; the glaze coat becomes, in fact, a coat of Obsidian (see McDowell 1981, 34–37). Once again, an understanding of the principles of geology—in this instance, the importance of the cooling cycle in the formation of different minerals and rocks—can help explain the mysterious surface variations which occur all too often in the ceramic process.

KINDS OF VOLCANIC ASH[30]

The possibilities for variation in the case of Volcanic Ash are endless. Volcanic Ash, or Pumicite, is simply the solid phase of fast-ejected lava. Because every eruption may have a different chemical composition from prior and subsequent lava flows, there can be as many different kinds of Pumice and Pumicite as there are volcanic eruptions. Volcanic erup-

tions are generally of three kinds. They can be basaltic in nature—low in silica and high in iron and magnesium. The black Volcanic Ash of the Hawaiian Islands is an example of this type. Secondly, volcanic eruptions can be andesitic. This kind of lava flow contains a medium amount of silica, iron, and magnesium. The chemical analysis of Mt. St. Helens Volcanic Ash suggests that it originated from this type of lava flow. The third, and most common, type of eruption is granitic in composition and is high in silica and low in iron. The Great Plains Volcanic Ash sold by ceramic suppliers prior to the eruption of Mt. St. Helens was of this type.

Even within a single lava flow, the resulting minerals may have considerable variation of structure. Thus, Pumice or

---

VOLCANIC ASH

(Pumicite)

TRADE NAME Volcanic Ash

GEOGRAPHICAL SOURCE Great Plains Ash: Great Plains region of central United States
Mt. St. Helens Ash: Washington.

GEOLOGICAL SOURCE Volcanic lava (Pumicite)
Great Plains Ash: Granitic magma. Pleistocene (2.5 million–5000 yrs.)
Mt. St. Helens: Andesitic(?) magma. May 1980.

CHEMICAL STRUCTURE

	Great Plains[1]	Mt. St. Helens[2]	Cornwall Stone[3]	Average Granite[4]
SiO_2	72.51%	61%–66%	72.00%	71.00%
AlO	11.55	16.3–19.7	18.20	14.50
FeO	1.20	4.5–7	0.10–0.16	3.80
TiO_2	0.54	0.50–0.75	0.05	
CaO	0.68	4–6	0.20–0.50	2.70
MgO	0.07	1.75–3.75	0.15	1.09
K_2O	7.84	1–2	5.00–6.00	3.89
NaO	1.79	4.2–4.6	0.15–0.35	2.99
P_2O_5	0.10			
F			0.25	
P.C.E	Cone 04–4		Cones 3–11	Cone 11–12

[1]*Ceramic Industry* 1998, 177.
[2]McDowell 1981 (January), 36.
[3]Goonvean LTD. 1998.
[4]Englund 1984, 13.

Pumicite from a single lava flow can consist of variable amounts of frothy silica glass and/or amorphous glass, together with minute grains of quartz, feldspars, or other minerals blown out of the underlying layers of the earth's crust by the explosive force of the volcanic eruption. Fortunately for potters, materials blown from a single eruption usually contain a fairly similar chemical composition; each single eruption can blanket hundreds of square miles with Pumice and Pumicite of generally uniform composition.

Great Plains Volcanic Ash

The Pumicite carried by ceramic suppliers prior to the Mt. St. Helens eruption was mined in the Great Plains section of the United States. Its original source is believed to be the eruptions of the Valles Mountain in central New Mexico, containing the largest caldera in the world, with a diameter of 15 miles, a circumference of more than 50 miles, and caldera walls that stretch 2,000 feet high. The powdery blanket of ash that resulted from the various eruptions of this giant volcano during the Pleistocene period of earth history (2.5 million–5,000 years ago) was carried by winds and water to the Great Plains section of the United States. These Pumicite deposits have been the main ceramic source of volcanic ash prior to the eruption of Mt. St. Helens on May 18, 1980.

Note that the chemical structure of Great Plains Volcanic Ash is similar to the chemical structure of Cornwall Stone, and that Cornwall Stone, in turn, is a partially kaolinized state of granite. Consider the fact that the clays known as Bentonites are direct products of the kaolinization of Volcanic Ash and that the natural glass Obsidian results from the cooling of this same lava under slightly more pressure than exists in the case of Pumicite or Volcanic Ash. Thus, the material of Great Plains Volcanic Ash, Obsidian, Cornwall Stone, Granite, Rhyolite, and Bentonite clays may have a single parental source. It is the atmospheric conditions—the rate of cooling and further weathering processes—that reshape the same volcanic magma into each of these different earth materials.

Great Plains Volcanic Ash functions very effectively as a high-silica glaze core at the stoneware firing temperature. When fired alone, this Volcanic Ash reveals its siliceous,

frothy, gaseous origins at both low and high stoneware temperatures in oxidation and reduction atmospheres. The textures of the tests resemble frozen froth. Their color reflects the iron impurities that this Volcanic Ash contains.

100% VOLCANIC ASH		
9/10 Oxidation	**9/10 Reduction**	**5/6 Oxidation**
Gray, greenish color.	White-blue color.	Pale, purply-gray-brown color.
Bubbled, frothy, pitted.	Bubbled, frothy.	Bubbled, frothy, pitted surface.
Glassy.	Glassy.	Glassy.
Nonfluid.	Nonfluid.	Nonfluid.

Great Plains Ash Combined with Calcium Melter Cone 9/10 Reduction

Volcanic Ash combined with 10%–40% Whiting show results similar to comparative tests with Cornwall Stone and Whiting. As more and more Whiting is added to Volcanic Ash, the pale, blue-green-gray, glassy surface changes to a mottled, yellow-gray-green, matt-shine surface, flecked with yellow-brown crystals.

The substitution of Volcanic Ash for glaze-core Kona F-4 (44%) in Sanders Celadon transformed the pale, transparent, gray-green, glassy surface into a more opaque, dark-brown surface, mottled with streaks of dark green-black. The increased iron content of Volcanic Ash, compared to Kona F-4, was reflected in this color change. On the other hand, when the customary iron colorant in Sanders Celadon was omitted from the Volcanic Ash substitution test on a porcelain claybody, the substitution of Volcanic Ash for Kona F-4 created a glossy, fatter, uncrazed, gray-blue-green surface. The higher silica content of Volcanic Ash compared to Kona F-4 resulted in the craze-free surface. The gray-green color was contributed by the iron-impurities in Volcanic Ash. A similar surface occurred with the substitution of Cornwall Stone for Kona F-4, although the color of the Cornwall Stone test was paler than the Volcanic Ash test, due to Cornwall Stone's lower iron content.

Note: Volcanic Ash tends to settle out quickly and, like the feldspars, does not remain evenly in suspension in a liquid glaze mixture. For this reason, additions of 1%–3% Bentonite, or some stronger suspending agent, is recommended whenever any kind of Volcanic Ash is a major constituent of a glaze.

The following tests show the potential of Volcanic Ash for providing beautiful color and opalescent surfaces at the oxidation temperatures. Note that no additional colorants were added.

VOLCANIC ASH GLAZE
Cone 9/l0 Oxidation

Volcanic Ash	68%	Pale blue-gray where thin.
Boron Frit 3134	32	Pale green where thick.
		Glassy. Crackled.
		Opalescent.
		Beautiful!

Cone 5/6 Oxidation (Porcelain claybody)
Figure 1.16, p93.

Volcanic Ash	68%	Pale yellow.
Colemanite	32	Glassy.
		Crackled.

Volcanic Ash	68%	Blue-gray-white-yellow.
Boron Frit 3134	32	Opalescent.
		Glassy. Crackled.
		Beautiful!

Volcanic Ash	68%	Gray-green color.
Boron Frit 3124	32	Pitted, bubbled surface.
or		
Boron Frit 3195		Glossy.

SEED BUBBLES

See Figure 3.10, p232.		Purple-blue, glossy surface when applied over iron slip.
Volcanic Ash	62.3%	
Boron Frit 3134	28.3	
Grolleg China Clay	9.4	

PORTCHESTER COPPER

Figure 1.6, p83.		Black-green color.
Cone 5-6 Oxidation		Satin-matt surface.
Nepheline Syenite	55.0%	
Whiting	25.0	
Flint	10.0	
EPK kaolin	7.0	
Tin Oxide	3.0	
Copper Oxide	2.0	

Test I. Cornwall Stone substituted for Nepheline Syenite: Result: Stiffer surface, lighter, bluer color.

Test II. Volcanic Ash substituted for Nepheline Syenite: Result: Similar to Cornwall Stone surface, but green-yellow color instead of bluer color, due to higher iron content.

Mt. St. Helens Volcanic Ash[31]

The eruption of Mt. St. Helens on May 18, 1980, provided the ceramic industry with a new and different source of Volcanic Ash. This Volcanic Ash contains less silica, more magnesia, and more iron than the Great Plains Volcanic Ash, which would suggest that its lava flow was andesitic rather than granitic in origin. The iron content of Mt. St. Helens ranges from 4.5%–7%, and for this reason it would make a good glaze core for a brown-black glaze.

Tests of this Volcanic Ash fired alone reflect the sensitivity of iron-enriched materials to the atmosphere of the fire. A student fired this ash to cone 9/l0 temperatures with heavy reduction and obtained a dense, brown, metallic, satin surface that was pitted, but displayed more fluidity and even coverage than did a comparable test with Great Plains Volcanic Ash. A potter reported in *Ceramics Monthly* that moderate reduction produced "red-brown speckles on a black field," and that heavier reduction resulted in a uniform, red-brown metallic surface (McDowell 1981, 34, 36). This potter also reported good results from combinations of this Volcanic Ash with low amounts of lithium carbonate, Gerstley Borate, dolomite, and other melters. At cone 5/6 oxidation temperatures, our tests of Mt. St. Helens Ash remained unmelted with uneven coverage. Its color was purply-brown. The surface was glossy, pitted, and crawled. However, additions of powerful auxiliary melters such as lithium carbonate or Gerstley Borate would create usable glaze surfaces at the lower stoneware temperatures (see Paul Lewing, "Mt St Helens Ash Glaze," *Ceramics Monthly*, 1998 [June–August] 42).

CONCLUSION

No other ceramic material connects us more closely with the beginnings of our earth. The source in this case comes directly out of the fire of the earth's creation. The high silica content of Volcanic Ash combined with its soda, potash, and calcia melters make for a superb glaze core. In addition, the iron and other impurities contained in Volcanic Ash create exciting color and surface effects that cannot be duplicated with the feldspathic glaze cores. As a substitute for Cornwall Stone, Great Plains Volcanic Ash provides increased color and surface depth. Mt. St. Helens iron-rich Volcanic Ash proves to be an exciting glaze core for a dark, iron-brown-black surface and is a timely addition in view of the much lamented loss of Albany Slip clay. Both kinds of Volcanic Ash enable the potter to achieve unique glaze surfaces with a minimum of glaze materials.

ROTTEN STONE

Figure 1.13–15, pp90–92.

Rotten Stone has been my pride and joy for more than 30 years—a never-ending source of extraordinary glazes and claybodies. In dramatic contrast to its ceramic value is its name—a synonym of Rotten Stone is terra cariosa, meaning rotten earth.

Rotten Stone is described as any "friable, lightweight, earthy residue consisting of fine-grained silica and resulting from the decomposition of siliceous limestone (or of a shelly sandstone) whose calcareous material has been removed by the dissolving action of water" (*Glossary of Geology* 1974, 619).

Keystone Filler & Mfg. Co., the manufacturer of Rotten Stone, describes it simply as a "decomposed siliceous sandstone from natural sediment" (Product Data Sheet, 3/97).

Rotten Stone is not manufactured for ceramic use, and ceramic suppliers usually do not include it in their extensive lists of available ceramic supplies. It is manufactured for industrial use as a soft abrasive and is used in the manufacture of brake linings. Despite this unglamorous background and name, I have found that Rotten Stone, more than any other ceramic material, produces incomparable, sensuously satin glaze and claybody surfaces.

Rotten Stone's chemical structure resembles a potash feldspar with a high iron content (7.65%). This structure, which includes almost 5% magnesium and calcium oxides, provides the silky, smooth chocolate surface that is Rotten Stone's unique characteristic. The high iron content together with various other unlisted impurities, such as Vanadium (1989 Emission Spec analysis), cause bubbles and eruptions to appear whenever the glaze layers are thick. These eruptions will create spectacular, lustrous, silvery oil spots in the oxidation atmosphere if the heat of the fire, the cooling period, and glaze thickness are just right; if the temperature is too low, the cooling period too fast, or the glaze too thinly applied, these same eruptions will fracture the otherwise smooth glaze surface. Admittedly, Rotten Stone is a most variable and unpredictable material, holding the potential for incredible gemstone surfaces at one end of its spectrum, and disastrously blistered and wildly deranged surfaces at the other. Not every potter will be willing to take on this risk. Yet to some, the seductive, smooth, satin surfaces and silvery oil spotting that occur some of the time are worth the many intervening failures, and it drives one on to pursue the reason WHY. Let the reader be the judge as we describe the tests of Rotten Stone in stoneware claybodies and glazes.

ROTTEN STONE

TRADE NAME	Rotten Stone
GEOGRAPHICAL SOURCE	Muncy, PA; Keystone Filler & Mfg. Co. Empire Products Co., Bayville, NJ. (supplier)
GEOLOGICAL SOURCE	Soft, crumbly stone, resulting from the decomposition of siliceous limestone by the action of moving waters.

CHEMICAL STRUCTURE*

SiO_2	62.28%
Al_2O_3	20.71
Fe_2O_3	7.65
CaO	2.52
MgO	2.02
Na_2O	1.21
K_2O	3.61

EMISSION SPEC. ANALYSIS (PERFORMED BY CHEMIST IN 1989)

Al_2O_3, Fe_2O_3, TiO_2, SiO_2	Major constituents.
MgO	10.00%
CaO	0.10%
V	0.08%
Na_2O	0.50%

*Keystone Filler & Mfg. Co. Product Data Sheet 1997.

CLAYBODY FUNCTION
Figure 2.2, p145.

Rotten Stone in a claybody performs the double function of colorant and melter. This dual role is particularly important for the cone 5/6 oxidation firing temperature, where melter and colorant are necessary to compensate for the lower fire and the still, oxidizing atmosphere. The cone 5/6 claybodies described below contain 20% Rotten Stone, which contributes to both the brown and black color and to the silken surface of the fired claybodies. Unlike the iron-bearing clay, Cedar Heights Red Art (see p. 129), the addition of Rotten Stone appears to improve workability. More extensive testing over a longer period of time would be necessary to determine the effect of Rotten Stone's variable impurities on the stability of these claybodies. Thus far, no adverse effect has been observed.

CONE 5/6 OXIDATION

	Rotten Stone Brown	Rotten Stone Black
Jordan Clay	40	40
Ball Clay	15	15
Custer Feldspar	25	25
Rotten Stone	20	20
Barnard Clay		10
Cobalt Oxide		2

Rotten Stone Brown	Rotten Stone Black
Smooth-grained, pale red-brown color, very plastic. Needs more testing to determine if impurities in Rotten Stone would eventually cause bloating.	Silky, smooth surface; deep black color with greenish cast. Bubbles appear at cone 6, indicating limited range of cone 4–5. Reducing Custer feldspar to 20% might increase its range to include a hot 6. Throws well, highly plastic.

CONE 9/10 OXIDATION

Rotten Stone I

Pine Lake Fire Clay	5.0 lb.
Jordan Clay	2.5 lb.
Ball Clay	1.5 lb
Rotten Stone	1.0 lb.

Pale brown fired color; smooth, fine-grained fired surface; highly plastic.

Note: All three claybodies would now require the substitution of another clay for Jordan clay.

GLAZE FUNCTION
Figures 1.13–15, pp90–92.

Cone 9/10 Oxidation

Rotten Stone is a material that needs the fire of high stoneware temperatures, melter assistance, and an oxidation atmosphere to bring out its magic touch. At the higher stoneware temperatures of cone 9/10 oxidation, Rotten Stone fired alone produces a dark brown, stony-matt surface. The excitement begins with the addition of a small amount of melter. The combination of but 10% of calcium carbonate, Whiting, with 90% Rotten Stone creates a silky, soft, brown-black coat with lustrous oil spots at the cone 9/10 temperatures. If the fire climbs to cone 11, the color and surface of this Rotten Stone glaze changes to a smooth, silky, reddish-brown, satin, gloss surface that almost resembles its appearance in the reduction atmosphere. Although the glaze surface is completely opaque, it closely hugs the form of the underlying claybody and thus becomes an integral part of the clay structure.

When this same Rotten Stone glaze is applied over a white, magnesium-based glaze known as "Charlie D" (see p. 158 for formula) the result is a satiny, chocolate-colored surface with occasional oil spotting. This combination is particularly effective on a porcelain claybody; the satiny, sensuous feel of the porcelain surface at both cone 9/10 and 11/12 is indescribable—it is much like stroking the velvety coat of a cat or caressing a baby's skin. Although the glaze is opaque, it merges with the clayform in such a way that they become indivisible. The numerous mugs and bowls produced with this combination have been a delight to touch and hold.

In the reduction atmosphere, this glaze combination produces a dark red color, flecked with black streaks, which somehow does not possess the darkly sensuous quality of the oxidation surface. *Thus, Rotten Stone is one of those rare materials that performs best at cone 9–10 oxidation temperatures. Figures 1.13–14, pp90–91.*

There are, however, certain problems that arise with this combination of Rotten Stone and Whiting. Much of the beauty of the surface depends on lustrous oil spots that result from the explosions of gases contained in Rotten Stone. If the glaze layers are too thick or if the cooling and firing period are too short, the craters and blisters will not heal over to form oil spots, and the surface will be marred. Then again, if the fire is too hot, if the cooling period is too long, or if the glaze layers are too thin, the surface will even out and oil spots will never develop or will disappear.

In order to effect a more complete conversion of bubbles into oil spots, half of the calcium melter was replaced by more powerful melters. When 5% of Whiting was replaced with 5% Borax, the result was an extraordinary reddish-

black surface, complete with large, lustrous oil spots. *Figure 1.14, p91.* Another method of dissolving bubbles and craters in the thicker layers of an oil-spotted Rotten Stone-Whiting surface is to apply a more fluid glaze over the thicker sections of Rotten Stone-Whiting glaze. The glaze we used for this purpose was a shiny, black, cone 5–6 glaze that contained powerful lithium and boron melters.[32] A thin coat of this shiny, high-gloss glaze over the more matt, and less-fluid Rotten Stone-Whiting glaze did in fact accomplish an overall, satin, brown-black surface, without the flaws of craters and bubbles. *Figure 1.13, p90.* Although the resulting surface was handsome and unflawed, it did not show evidence of oil spots. The solution to the problem of how to reduce bloats without destroying oil spots remains basically one of application, temperature and cooling period—the glaze must be thick enough to contain the explosive gases that will form oil spots; the temperature must be high enough and remain long enough to heal craters and allow the formation of oil spots—too much heat will even out the glaze surface and cause oil spots to disappear; and finally, the rate of cooling must be sufficient to retain oil spots once formed. Thus, one can expect only a 50% rate of success in the achievement of an unblemished, oil-spotted surface.

Cone 9/10 Reduction
Figure 1.14, p91.

Combinations of Rotten Stone, calcium carbonate, and calcium silicate materials in the reduction atmosphere naturally result in different colors and surface textures than occur in the oxidation atmosphere. The combination of 90% Rotten Stone and 10% Whiting (which in the oxidation atmosphere is a dramatic black-brown surface sprinkled with silvery black oil spots), produces a mahogany-red or brown surface with only a rare occurrence of oil spots. The surface is satin-matt and hugs the underlying claybody so closely that throwing rings and faint depressions in the clay form are clearly revealed as an integral part of the glaze surface.

The tests described on page 74 show that increases of calcium materials in both oxidation and reduction atmospheres cause the original brown-black or mahogany red-brown color to change to yellow-green. This change of color reflects the characteristic bleaching effect of calcium on iron-bearing materials. The combination of 70% Rotten Stone and 30% Whiting, fired in a reduction atmosphere, produced a handsome, yellow-green, silken surface. A similar surface was obtained in the oxidation atmosphere; however, the color was more mustardy and the surface somewhat less satin. *Figure 1.15, p92.*

Combinations of feldspathic materials and Rotten Stone produce some exciting results that resemble early Korean glazes of the Yi Dynasty. In particular, the combination of 85% Nepheline Syenite and 15% Rotten Stone creates a brilliant, orange-gold-white, pearly, lustrous, gemstone surface on a porcelain claybody. *Figure 1.4, p81.* Increased additions of Rotten Stone yield different but equally spectacular surfaces. For example, 60% Nepheline Syenite and 40% Rotten Stone produced a black, glossy, lustrous surface sprinkled with silvery, iron spots. More recent tests in a different kiln, using different supplies, resulted in an equally lustrous surface; however, the color was mahogany and there were no oil spots.

Cone 5–6 Oxidation
Figure 1.15, p92.

Although Rotten Stone's performance at the lower stoneware temperatures hardly equals its performance at the higher stoneware temperatures, our tests show that it can be used as a glaze core at these temperatures in combination with the right proportion of more powerful melters.

Rotten Stone fired alone at cone 5–6 temperatures produced a stiff, unmelted, reddish brown surface. However, when two-thirds Rotten Stone was combined with one-third colemanite and fired to about cone 7/8, a brown-black, lustrous, oil-spotted surface was the result. This suggested that it might be possible to use Rotten Stone at the lower temperatures. In pursuit of this aim, at cone 5/6, we combined 50%–70% Rotten Stone with 30%–50% colemanite. Although these combinations did not produce oil spots, they created a glossy, to satin-matt, mustard-yellow-brown surface. Similar results occurred with combinations of Rotten Stone and wood ash. Thus, although the potential of Rotten Stone as a cone 5/6 material needs more extensive testing, these tests show that it could function effectively at the low stoneware temperature in combination with the right proportion of powerful melters.

GLAZE TESTS

FELDSPAR AND ROTTEN STONE			
Cone 9–10 Reduction			
	Cornwall Stone	**Nepheline Syenite**	**Kona F-4 Mix***
25% Feldspar 75% Rotten Stone	Mahogany red-brown. High gloss. Opaque. No craze. Luster.	Dark brown. Lower gloss. Opaque. No craze. Crawled. Pitted.	Medium brown reddish cast. Medium gloss. Opaque. No craze.
50% Feldspar 50% Rotten Stone	Deeper red. High gloss. Luster. Opaque. No craze.	Blacker brown. Lower gloss. Satin surface. Opaque. No craze. Pitting.	Richer color. Medium gloss. Satin surface. Opaque. No craze.
60% Feldspar 40% Rotten Stone		Rich purple-red-brown. High luster. Satin surface. *Different Studio:* Black-brown. Silver oil spots.	
75% Feldspar 25% Rotten Stone		Lighter, redder brown. Black streaks. High gloss. High luster. Opaque. No craze. Faint pitting.	Black-gray (thick). Red brown (thin). Black streaks. Lower gloss. More luster. Mottled. Opaque. No craze. Satin surface. Beautiful Test!
80% Feldspar 20% Rotten Stone		Pearly-white-gray (thick). Reddish-orange brown (thin). High gloss. Pearly luster. Craze.	Grayer, whiter (thick). Lower luster. More satin surface. Opaque. No craze.
85% Feldspar 15% Rotten Stone		Pearly gray-white. Orange (thin). High craze Gloss.	Blue-gray. Brown (thin). Satin surface. Opaque.

*Kona F-4 80% / Whiting 20%; gray-blue; transparent; gloss.

ROTTEN STONE—MELTER TESTS

100% Rotten Stone
0% Melter

Stoneware:Cone 9/10 Reduction
Variable results due to variation in firing temperature and atmosphere.
Black, red-brown or mahogany color.
Dry unmelted surface, or stony opaque surface.
Occasional pitting.

Stoneware: Cone 9/10 Oxidation
Dark-brown color.
Dry, pitted, crawled surface.

Stoneware: Cone 5/6 Oxidation
Dark red-brown or brown color.
Dry, unmelted, bubbled surface.
Stony, smooth surface if thin coat.

	Whiting	**Wollastonite**	**Colemanite**
90% Rotten Stone **10% Melter**	*Stoneware:Cone 9/10 Reduction* Variable results due to variation in firing temperature and atmosphere. Black, brown, or red-brown color. Smooth, silky, satin texture. Occasional bubbles or pitting. Rare oilspotting.	*Stoneware:Cone 9/10 Reduction* Red-brown, brown-black color speckled with deeper brown iron spots arising from claybody. Satin matt surface. Slightly higher gloss than in Whiting test.	*Stoneware:Cone 9/10 Reduction* Reddish brown. Speckled, mottled surface. Satin surface.
	Porcelain: Cone 9/10 Reduction Similar to stoneware test. Usually red-brown in color. Smoother, more satin texture. Rare oilspotting.		
	Stoneware: Cone 9/10 Oxidation Variable due to firing temperature. Dark black-brown or red-brown. Silky, smooth satin texture. Occasionally bubbled and pitted where very thick. Oilspotting where medium thick.	*Stoneware: Cone 9/10 Oxidation* Dark black-brown color. Higher gloss than Whiting test. Less oilspotting than Whiting test.	*Stoneware: Cone 9/10 Oxidation* Blacker color. Same satin surface. Some oilspotting.
	Porcelain: Cone 9/10 Oxidation Similar to stoneware test. Black-brown color (unless cone 10/11 is reached, and then mahogany color similar to reduction color, with no oil spots). Smoother, more satin texture. Frequent oilspotting in medium coat.	*Stoneware: Cone 5/6 Oxidation* Red-brown color. Dry, stony surface.	*Stoneware: Cone 5/6 Oxidation* Red-brown color. Unmelted, bubbled surface.

Chart continued on following page.

ROTTEN STONE—MELTER TESTS (cont.)

	Whiting	Wollastonite	Colemanite
80% Rotten Stone 20% Melter	*Stoneware:Cone 9/10 Reduction* Greenish, yellow-brown color, streaked with black. Opaque, satin surface. Occasional pitting.	*Stoneware:Cone 9/10 Reduction* Increased brown, lesser red color. Increased black speckling. Increased gloss and flow. *Stoneware: Cone 5/6 Oxidation* Brown color, flecked with yellow. Silkier surface than 90-10 test.	*Stoneware:Cone 9/10 Reduction* Redder color speckled with darker brown spots. Higher gloss. *Stoneware: Cone 5/6 Oxidation* Black color. High gloss.
70% Rotten Stone 30% Melter	*Stoneware:Cone 9/10 Reduction* Stronger yellow, mottled black-brown color. Increased satin surface. Similar to Ash glaze surface, especially on porcelain. Can be runny and streaked as in Ash glaze. *Stoneware: Cone 9/10 Oxidation* Yellow-brown, ash-like surface.	*Stoneware:Cone 9/10 Reduction* Matt yellow ringed with green-gray glass. Ash-like surface. Increased fluidity. Similar in oxidation atmosphere. *Stoneware: Cone 5/6 Oxidation* Mustard yellow color. Dense, opaque, satin surface.	*Stoneware:Cone 9/10 Reduction* Similar to test above. *Stoneware: Cone 5/6 Oxidation* Black where thick; brown where thinner. Mottled, uneven coverage. High gloss.
60% Rotten Stone 40% Melter	*Stoneware:Cone 9/10 Reduction* Increased yellow color. Green-gray color where thick. Brown color where thin. Increased fluidity. Pinholing.	*Stoneware:Cone 9/10 Reduction* Brown spotted with black, lustrous iron spots and flecked with yellow. Mottled effect. Satin surface. *Stoneware: Cone 5/6 Oxidation* Increased yellow color. Increased gloss. Satin surface.	*Stoneware:Cone 9/10 Reduction* Brown color ringed with glassy black where coat is thicker. Satin surface where brown; black rings have high gloss. *Stoneware: Cone 5/6 Oxidation* Similar to 70-30 test. Increased yellow brown color, less black. Very high gloss.
50% Rotten Stone 50% Melter			*Porcelain: Cone 5/6 Oxidation* Yellow-brown color. Increased high gloss.

CONCLUSIONS

Rotten Stone and Feldspar Tests

1. The color provided by the iron in Rotten Stone will vary from mahogany to brown to black, depending on the particular feldspar that is used, together with the firing temperature. Note the redder cast with Cornwall Stone and the blacker cast with Nepheline Syenite in the 50–50 combination.

2. The soft, satin surface that is so characteristic of Rotten Stone reaches its peak in the test of 50% Kona F-4 mix (80% feldspar, 20% Whiting) and 50% Rotten Stone. We found that a mere 10% addition of Rotten Stone to the Kona F-4 mix, changed this high-gloss, transparent surface to a silky, satin-matt!

3. Lower percentages of Rotten Stone produced the more dramatic gemstone surfaces; in particular, the combination of 15% Rotten Stone and 85% Nepheline Syenite, fired at cone 9–10 reduction temperatures, consistently on both stoneware and porcelain, produced a pearly, gray-white, gemstone surface, edged in orange, where the glaze layers were thinner.

Rotten Stone and Melter Tests

1. The firing temperature and atmosphere will produce more variations in surface color than will different melters of Whiting, Wollastonite, Colemanite, or even lithium carbonate. A test of 5% lithium carbonate and 5% Wollastonite did not appear markedly different from a test of either 10% Whiting or 10% Wollastonite. On the other hand, a test of 5% Borax and 5% Whiting resulted in a more intense, reddish-black surface with larger, and more lustrous oil spotting than did the 10% Whiting test at cone 9/10 oxidation temperatures. Note that the range of surface color from brown to black to red is closely linked to the temperature of the firing. The color of a high Rotten Stone surface changed from brown to black and finally to red as the firing temperature rose from cone 8 to cone 11, and these color changes occurred irrespective of which melter was substituted or whether the atmospheric conditions were oxidation or reduction. Because the same results occurred in the oxidation atmosphere, we concluded that temperature and not atmosphere is the determining factor. *Figure 1.14, p91.*

2. Colemanite is once again the effective melter for lower firing temperature. Note the shiny, black surface that 20% colemanite produces with 80% Rotten Stone at cone 5/6 oxidation and how the surface moves toward a yellower, browner cast as the colemanite is increased. The same shift from darker brown to yellow occurs in the cone 9/10 reduction tests of increased amounts of Whiting and Wollastonite and bears witness to the yellowing, bleaching effect of increased calcium materials on the iron-bearing Rotten Stone.

GLAZE APPLICATION

The physical qualities of Rotten Stone in a glaze mixture are both contradictory and extraordinary. For some potters, it remains perfectly in suspension for long periods of time, impervious to changes in environmental temperature that solidify so many of our glaze mixtures. Yet other potters report just the opposite experience and complain that it does not remain in suspension for more than a short period of time. We found that a glaze mixture containing 90% Rotten Stone and 10% Whiting thickened drastically after remaining in a closed bucket over a period of time (a few weeks) and required subsequent additions of more and more water in order to thin it to a proper consistency for glaze application. In this connection, the thickness of the glaze coat is crucial for successful results. If Rotten Stone coats the pot too thickly, it will crack and peel off—particularly if it is a second coat or if a different glaze underlies the coat of Rotten Stone. Hence, second coats, thick coats, and fast drying should be avoided. Rotten Stone glazes perform best with but one medium coat, which should dry slowly on the pot in a relatively humid atmosphere.

These difficulties in application, combined with the variability of different supplies of Rotten Stone, make Rotten Stone a high-risk material. Yet, although there are, admittedly, more risks in Rotten Stone than exist in standardized, refined ceramic materials, there are infinitely more rewards. The allure of that satin, glowing, black-brown surface shining with dark, lustrous oil spots, can be so great, that it must be pursued over and over again, even in the face of many failures. It is precisely this kind of challenge that gives excitement and purpose to the exploration of ceramic surfaces.

Notes

[1] Note Wood's statement that "feldspar is by no means an indispensable material for high-fired glazes and true feldspathic glazes are actually rather rare in Chinese ceramics", 1978, 7. For us the significant point is the fact that a granitic rock material provided the core of the incomparable Song Dynasty celadon glazes. Fine geological distinctions between true feldspar, mica, granite and petuntse are not relevant here.

[2] Epsom salts are a hydrated magnesium sulfate ($MgSO_4 \cdot 7H_2O$), first discovered in Epsom, England (Frye 1981, 64). For additional suspension materials see pp.135–138.

[3] Hammill & Gillespie 1985, Chemical Data Sheet.

[4] Melting points of materials in this work derived from Hamer and Hamer 1986, 358–365.

[5] This is particularly true of soda, which has the highest rate of shrinkage; the thermal expansion and contraction rate of potash is less than that of soda.

[6] "Health Hazards for Raw Materials", Cerritos College, Art 56.

[7] Potash feldspars are insoluble materials; high-soda feldspathic materials such as Nepheline Syenite are said to be slightly soluble (Hamer and Hamer 1986, 299).

[8] Sources: Taylor 1989 #5; Indusmin, Inc. 1967, "Mining and Milling Nepheline Syenite"; Indusmin Syenite in ceramic whitewares; Unimin Corporation Technical Data 1998; Bates 1962, 230–231; *Ceramic Industry* 1998, 138.

[9] Note: Materials such as Cornwall Stone and soda and potash feldspars contain free silica (quartz) as an accessory mineral and, therefore, pose the risk of silicosis from inhalation of dust. Because Nepheline Syenite does not contain free silica, it does not present this risk. See Material Safety Data Sheet, U.S. Dept. of Labor, Occupational Safety and Health Administration, (September) 1988.

[10] Source: Presidential Address of C.V. Smale 1977. Letters to the author, C.V. Smale, June, July 1989. March 1993.

[11] CaF_2 Calcium fluoride.

[12] Porcelain claybody Tiso B contains 25% Cornwall Stone, p. 113.

[13] Norfloat Feldspar, a Norwegian feldspar, is no longer available for United States potters.

[14] Oxide analyses based on Hammill & Gillespie Data Sheet, 10-25-85. Current sheet at time the tests were made.

[15] DF (defluorinated) Cornwall Stone. Oxide analysis based on C. V. Smale 1977, 7. Not clear which kind of Cornwall Stone was the source of the 92nd St. Y's supply, where tests were made. Note difference in A/S ratio compared to Cornwall Stone.

[16] Alumina/Silica Ratio. See p. 241.

[17] See Appendix A1 for percentage oxide analysis method. Slight variation of figures due to use of KF-4 chemical data, which was current at time of tests.

[18] Original glaze core.

[19] Oxidation tests show orange singe on edge of glaze layer, which is typical of high-soda material.

[20] On porcelain claybody and on test in different kiln, color is oranger and redder.

[21] Potash Feldspar with 67.3% SiO_2; 17.8% Al_2O_3, 11.9%, K_2O, and 2.5% Na_2O. No longer available.

[22] DF Cornwall Stone was probably the material used in these tests. Note the different A/S ratio with the other Cornwall Stone.

[23] Claybody: Stoneware claybody of Parsons School of Design.

[24] Original glaze core.

[25] Amblygonite ($2LiF \cdot Al_2O_3 \cdot P_2O_5 \cdot nH_2O$) and Eucryptite ($Li_2O \cdot Al_2O_3 \cdot 2SiO_2$) contain a higher lithia content of 10% and 11.9% respectively. Because they were not available in our ceramic studios, they are not included in our tests.

[26] Foote Mineral Company, Bulletin 312 gives 1327°F as its melting point; Hamer and Hamer 1986, 361 gives 1202°F as its melting point.

[27] 400°F–500°F, see pp. 102–103, 240.

[28] For comparative tests of the lithium glaze cores Spodumene, Lepidolite and Petalite with the feldspathic glaze cores in stoneware glazes see Appendix B2.

[29] Tantalum Mining Corporation of Canada Ltd., TANCO. Tantalum Mining Corporation — An overview, courtesy of Amalgamet Canada; Interviews with Charles A. Merivale, General Manager 1990; Kusnik and Terry.

[30] Source: Tarbuck and Lutgens 1987, 59–61. Bates 1969, 44–48. *Ceramic Industry* 1998, 177. McDowell 1981, 34–37.

[31] Source: McDowell 1981, 34–37.

32 Rotten Stone Black   Cone 5–6 Oxidation

	grams
Lepidolite	20
Rotten Stone	50
Nepheline Syenite	30
Gerstley Borate	10
Talc	8
Bone Ash	2
Cobalt Carbonate	3

Figure 1.1 Stoneware Tests.

Cone 9–10 reduction.
Left test: left side—Sanders Celadon without 44% feldspar; *right side*—Sanders Celadon with 44% feldspar.

Center test: Rhodes 32 glaze with potash feldspar 48.9%.

Right test: Rhodes 32 glaze with chemical equivalent of potash feldspar 48.9%.

Figure 1.2 Porcelain melter test pins by Timna Neuman. (Parsons School of Design, 1984.)

Cone 9–10 reduction.

Vertical Rows:
Left:
    Top—Soda ash (thick).
    Bottom—Cryolite.

Middle:
    Top—Bone ash.
    Center—Soda ash (thin).
    Bottom—Fluorspar.

Right:
    Top—Lithium Carbonate.
    Bottom—Potassium Carbonate.

Figure 1.3 Feldspars and Rocks; Test pots by Barbara Beck.

Stoneware. Cone 9–10 reduction. Glaze: Feldspar 90%, Whiting 10%, Red iron oxide, 1/2%.

Pots, *left to right:* Potash feldspar, Cornwall Stone, Soda feldspar.

Pot, rear, center: Nepheline Syenite.

Rocks, left to right: Soda feldspar, Potash feldspar, Nepheline Syenite, Cornwall Stone.

80

Figure 1.4 Porcelain Bowl by author.

Cone 9–10 reduction.
Glaze: Nepheline Syenite 85%, Whiting 15%.
Rocks:
  left—Nepheline Syenite. (1.3 billion years old.)
  right—Rotten Stone,

81

Figure 1.5 (top) Stoneware Teapot by Sue Browdy.

Cone 9–10 reduction. Sanders Celadon glaze with brush of Rotten Stone 90%, Whiting 10%.

Figure 1.5 (bottom) Porcelain House by Rosemary Lax Stoller.
(Photo by Chris Dube.)

Cone 9–10 reduction. Glaze: Sanders Celadon.

Figure 1.6 Animal sculptures by Sue Browdy.

Portchester red-brown claybody. Portchester copper glaze. Cone 5–6 oxidation.

Figure 1.7 Jar with Saturated Iron Glaze by Sue Browdy.

Stoneware. Cone 9–10 reduction.

Figure 1.8 Porcelain Shell by Sue Browdy.

Cone 9–10 reduction. Glaze: Cornwall Stone 90%, Whiting 10%, Red iron oxide 1/2%.

Figure 1.9 Lithium Glaze Cores and Rocks.

Tests, *left to right:* Glaze of 100% Spodumene, Lepidolite, Petalite, cone 9–10 reduction, stoneware.

Rocks, *left to right:* Spodumene, Lepidolite, Petalite. (Collection of Department of Earth and Environmental Sciences, Columbia University, New York.)

Figure 1.10 Stoneware Jar by Sue Browdy.

Cone 9–10 reduction. Wale's Shino. Soda ash 4%; Spodumene 16.20%. Primary differences from Shino
Carbon Trap (p.158): higher Kona F-4 content (19.5%; Carbon Trap 10.8%), lower clay content (Ball clay 17%;
Carbon Trap: 15% Ball Clay plus 10% EPK).

Figure 1.11 Covered Jar by Sue Browdy.

Cone 9–10 reduction. Stoneware. Lau Lustre glaze (5% Lithium Carbonate).

Figure 1.12 Strip Basket by author.

Cone 9–10 oxidation, stoneware. Spodumene-wood ash glaze (Spodumene 60%, mixed wood ash 40%, Gerstley Borate 10%).

Figure 1.13 Stoneware Jar by Sue Browdy.

Cone 9–10 oxidation. Glaze: Rotten Stone Black, thinly applied over Rotten Stone 90%, Whiting 10%.

Figure 1.14 (top)

Rotten Stone 90%,
Whiting 10%

Clockwise from left:
Shell, Cone 9–10 reduction;
Plate, Cone 11 oxidation;
Bottle, Cone 9–10 oxidation.

Figure 1.14 (bottom)
Four Porcelain Shells
by Sue Browdy.

Left row: Rotten Stone 90%,
Whiting 10%
  Top—Cone 9–10
  oxidation.
  Bottom—Cone 9–10
  reduction.

Right row: Rotten Stone
90%, Whiting 5%, Borax 5%.
  Top—Cone 9–10
  reduction.
  Bottom—Cone 9–10
  oxidation.

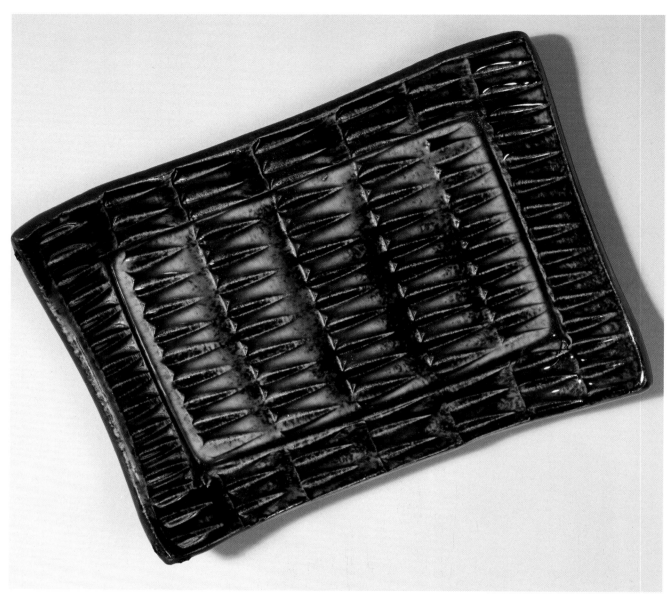

Figure 1.15 Rotten Stone plate by Jacqueline Wilder.

Grolleg porcelain. Cone 5–6 oxidation. Glaze: Rotten Stone 70%, Gerstley Borate 30%.

Figure 1.16 Volcanic Ash Plate by Jacqueline Wilder with three burned nuts covered with volcanic ash from past magma flow near Hana, Hawaii.

Grolleg porcelain. Cone 5–6 oxidation. Glaze: Volcanic Ash 68%, Frit 3134 32%.

LAB II GLAZE CORES

Potash and Soda Feldspars, Nepheline Syenite, and Cornwall Stone provide the basis for the bulk of stoneware and porcelain glaze surfaces. It is possible to understand their unique personalities and to recognize the contribution each makes to the glaze surface by testing them alone and in simple combinations with calcium and other melters. Though first seen as drab white powders, they reappear at the end of our testing process in their rightful form—ancient earth materials with unlimited potential for the creation of rich and unique ceramic surfaces. These tests are set forth in the hope of accomplishing this result.

1. Glaze four identical test pots, each with one of the four glaze cores above. As you apply each glaze core, concentrate on how each flows onto the test pot. Record any difficulties with any of these four materials during the physical application period.

2. You should become familiar with the surfaces created alone by the soda and potash melters. Not only will this enable you to understand the various feldspars and rocks in which they are the primary melters, but it will also show you their decorative potential when used alone. The contrasting personalities of these melters become very clear when placed side by side on the same test.

Take two to four identical test pots. Glaze half of each with soda ash or baking soda, and half with pearl ash or potassium carbonate. They are melters and can flow rapidly; only glaze the inside of each pot. *Wear rubber gloves and a mask when handling soda ash, pearl ash, and potassium carbonate. They are caustic, and potassium carbonate is highly toxic.*

3. Select two glazes with which you are familiar. One should produce a matt or satin-matt surface, the other a glossy, transparent surface. Each formula should contain a feldspar or feldspathic rock. Mix each glaze without the feldspar or feldspathic rock. Place this glaze (without the feldspar or rock) on the second half of the test pot.

4. Substitute each of the three other feldspars and/or rocks for the particular feldspar or rock in each of your glazes. For example, if your glaze contains 50% potash feldspar, mix it with 50% soda feldspar; next, mix it with

50% Cornwall Stone, and finally, mix it with 50% Nepheline Syenite. Again, each test must have a control. The original glaze with its potash feldspar should appear on half of each pot.

5. Repeat Tests #1 and #4 using one of the lithium glaze cores and/or Volcanic Ash and/or Rotten Stone in place of soda and potash feldspars or rock.

6. Fire each test pot at as many firing temperatures as are available to you.

GLAZE APPLICATION

1. Add enough water to the dry powders to achieve the consistency of light cream.

2. Dip one side of the pot, then redip the top. This will produce a thin and thick application.

3. Each test must contain a control so the results of changing the materials in the glaze can be clearly assessed. The control will consist of the original glaze and must appear on one-half of each test that is fired. In this connection, remember that the half of the test pot that is glazed first (whether it be the control or substituted glaze) will obtain the thicker coat of glaze. The open pores of the bisque pot will absorb the water from the glaze liquid and dampen the side of the pot as yet unglazed. One can compensate for this by making the second glaze thicker than the first or, ideally, by glazing one-half of the pot with the original glaze on a different day, thus giving the pot a chance to dry before applying the second glaze.

4. Mark on the bottom of the test which half is the control and which half is the test substitution.

5. Mix at least 200 grams of each test. It is very difficult to glaze with small amounts of glaze liquid. A test that has been poorly glazed because of insufficient glaze is a waste of time. The cost of materials discarded is minuscule compared to the time and labor invested in making the glaze. Keep leftover liquids in tightly closed, dated, and clearly labeled jars so you can determine the changes, if any, that take place over time. It is useful to keep extra amounts of the glaze liquid in case you need to repeat a test.

6. For your safety, always wear an OSHA approved industrial mask before mixing glaze powders and mix glazes in a well-ventilated room.

Chapter 2 | CLAYS AND CLAYBODIES

INTRODUCTION

A true craftsperson understands the material of his or her craft. The Song Dynasty potters knew the properties of the clay that formed their pots and the pots of their ancestors through generations of hands-on experience. This knowledge was handed down to their children along with the clay deposits, and they in turn, passed it on to their children. Today, some 800 years later, there are still some potters who experience this kind of physical and spiritual union with their clay, particularly among the fast-disappearing indigenous cultures of the world.

> "The pueblo potter has a true understanding, a harmony with his materials. Europe may have had this once, but no longer does. The Indian people are still making pottery in their own way—their intuition for their materials is still alive."
>
> (Peterson 1977, 92)

Many of us, however, are not so blessed. We neither quarry our clay nor live near its source and are thus cut off from a concrete, visible connection with the primary material of our craft. Ceramic clays are shipped to us from unknown regions, mined by unknown hands and machines, and tested by strangers for but a brief period of time. We fill abstract formulas with these unfamiliar clays in order to obtain our claybodies. These modern conditions are bemoaned by the great Japanese potter Hamada in his extraordinary conversations with Bernard Leach (Leach 1975, 100–101).

> "Today potting is like a department store; clay is bought by sending a postcard to any area of the world and asking for a particular clay and getting it in a few days."
>
> (Leach 1975, 101)

If our craft is to thrive under these modern conditions, we must fill the void of distance with a different kind of knowledge. Recent findings of clay mineralogists tell us much about the nature of clays. Once we open our minds to the ever-changing formations of the clay minerals as they leave the earth to enter the fire, we find a new and exciting world that stretches the boundaries of our imagination and restores our connection with clay.

WHAT IS CLAY?

Clay is an astonishingly simple four-letter word that describes the most complex and significant material known to humans. It is an essential mineral for the ceramics, refractory, petroleum, paper, adhesives, radioactive waste disposal, cement, fabrics, fertilizer, food, wine, beer, soil, medicine, pharmaceuticals, cosmetics, paint, leather, plastics, rubber, soap, polishing compounds, and water clarification industries (Fyre 1981, 78–79). In the last 10 years, clay minerals have played an increasingly important role in advanced space and computer technologies that have revolutionized our modern age. Thus, clay supports and nourishes all life on the earth's surface.

The dictionary describes clay as a mineral, a rock, a particle size, and lastly, a "symbol of the material of the human body. . ." (*Webster's New Universal Unabridged Dictionary* 1979, 335). This last usage appears over and over again in the great literature and poetry of the ages, including biblical writings, and the poetry of Rudyard Kipling, Lord Byron, Thomas Love Peacock, and John Dryden, to name but a few.

"Lo, I have wrought in common clay
Rude figures of a rough-hewn race . . ."

(R. Kipling, *Soldiers Three*
1888 Dedication Stanza 2.)

"This is the porcelain clay of humankind"

(John Dryden, *Don Sebastian*
1690 Act I, sc. i.)

Scientific data supports the connection between life and clay.

". . .[t]he kind of relatively mild, watery conditions that were presumed required for the origin of life on Earth would inevitably have generated clay minerals in abundance. In this respect at least, life and clay go together."

(Cairns-Smith and Hartman 1986, 80)

THE CLAY-MAKING PROCESS[1]

All forms of matter on our earth must adapt to their ever-changing environment or be doomed. They search for the most stable form of existence that will put them into a temporary state of equilibrium with their environment. Both the living and nonliving share this goal. Their mutual strivings bear witness to the driving force of perpetual change that underlies our universe, and shapes the destinies of everything contained within it (see Stevens, Wm. K. "New eye on nature: the real constant is eternal turmoil." *New York Times,* 31 July 1990, 1–2(C)).

Sedimentary clay minerals are transformations of igneous rocks existing at or near the earth's surface. Some of these igneous rocks are formed quickly when deeply buried magma flows upwards close to the earth's surface; other igneous rocks, such as granite, originate from magma that slowly solidifies beneath the protecting top layers of the earth's crust under conditions of low oxygen, high heat, and pressure. Movements within and upon the earth force magma and igneous rocks upwards into the cooler, less-pressurized environment of the earth's surface. Here they are first exposed to air, rain, snow, ice, winds, rising magmatic gases,

fire, climatic changes, plant growth, and animal behavior—the forces of perpetual change and disruption that restlessly shape and reshape the earth's surface. In their search for stability in this new, ever-changing environment, igneous rocks change into clay minerals. The geological term for this transformation process is weathering. The weathering process is highly complex, and some aspects are still being debated by geologists. The end result of the weathering process, the clay minerals, may require thousands or even millions of years for completion; the sheer length of this process makes it impossible to duplicate in a laboratory. Fortunately, many clay minerals (particularly kaolinite) form more quickly in reaction to high-temperature gases and water vapor. Thus, chemists have created certain clay minerals in the laboratory by subjecting finely ground feldspars to simulated hydrothermal weathering conditions. In this way, much has been learned about the formation of clays.

Physical and chemical weathering forces of the earth's environment change igneous rocks into clay minerals. Physical weathering breaks up the igneous rock into its mineral grains of feldspar, mica, and (if a granitic rock) quartz. The mineral grains are then pulverized into smaller and smaller particles and eventually become a primary ingredient of soils, rivers, lake beds, and ocean basins. The increased surface area of these fine-grained particles makes them highly susceptible to chemical weathering.

In comparison, chemical weathering changes the chemical composition and the crystalline structure of the mineral grains. This change is activated by the powerful, dissolving force of water. Water is one of the most powerful solvents on earth and is the essential medium for the birth of all clay minerals. Fourteen percent of a clay mineral is estimated to be water. One hundred pounds of dry clay releases 11 pints of water (U.K. measure), which, in turn, becomes 370 cubic feet of steam during the firing process (Hamer and Hamer 1986, 62). Free water molecules, polarized with negatively charged oxygen and positively charged hydrogen atoms, attract molecules of silica, potash, soda, and other elements contained in the feldspar and mica grains. Water molecules draw these molecules away from their three-dimensional feldspathic and micaceous bonds. Other water molecules, attracted by the remaining silica and alumina molecules, slip into the spaces formerly occupied by the departing silica and potash molecules and realign themselves to form the two-dimensional, layered structure of the clay mineral.

Clay minerals are found in three kinds of watery environments: hydrothermal, continental (soils, rivers, and lakes), and oceans. Mention has already been made of the hydrothermal environment, which creates the most ordered form of the kaolinite clay minerals. Heated water and steam, together with other gases, accompany rising magma and create pure kaolin deposits, such as are found in Cornwall, England.

Waters of the earth's atmosphere shape clay minerals found in soils. Atmospheric water, in the form of rain or snow, flows downward into the soil layers. As described previously, the dissolving action of the water removes silica and various metallic ions from the feldspathic mineral grains of the topsoil layers and weaves the remaining molecules into the tightly bonded structure of kaolinite. The watery solution enriched with the highly charged molecules drawn from the feldspathic mineral grains seeps downward toward deeper layers of soil to form yet different kinds of clay minerals. Among the clay minerals so formed are the extraordinary montmorillonites. The key to life, the transition between the living and the nonliving, is said to begin with this type of clay mineral (see Cairns-Smith and Hartman 1986).

Our earth is mostly water—ocean waters cover 70.8% of the earth's surface and are the final burial grounds for fine-grained rock particles, volcanic ash, and cosmic dust, which are carried by winds, melted glaciers, and rivers to their oceanic grave. Rivers flowing into the ocean enrich ocean waters with various kinds of dissolved elements and minerals, such as potassium, sodium, and calcium carbonate. Volcanic eruptions under the ocean floor spew forth a rich assortment of magma and minerals, accompanied by hot reactive gases. Thus, ocean waters teem with myriad assortments of electrically charged atoms (ions), which react with the drifting particles to form various mineral deposits. These particles, together with indigestible remains of plants and animals, drift about ocean waters in accordance with their natural cycle—*beginning, ending, sinking, settling.* As part of the sinking process, they may float in deeper and deeper layers of ocean waters for periods of weeks, months, or even

years, depending upon their size and weight. Eventually, they will settle on the ocean floor and, after millions of years, build up deep layers of sediments, some of which are deep-sea muds containing rich deposits of clay minerals. These minerals will surface on the earth's crust whenever ocean waters recede or when "mountain building" tectonics raises the seafloor above sea level, as has happened so often in the earth's history.

It is believed that the actual alteration of igneous rock particles to clay minerals occurred prior to ocean deposition due to the inhibition of the weathering process in the ocean. The absence of frost, currents, and variations of temperature in the ocean retards physical weathering processes. Chemical weathering is reduced because the high concentration of potassium, magnesium, and sodium ions in seawater prevents these elements from leaving the rock particles. Hence, the majority of clay deposits of deep ocean muds are composed of previously weathered clay particles. On the other hand, substantial montmorillonite clay deposits form near deep ocean trenches as a result of heated seawater interacting with oceanic basalt and gabbro.

Among the clay minerals found in deep-ocean sediments are the complex, unstable, high-iron clays known as illites, which appear throughout the deep-ocean sediments of mid-latitude Pacific and Atlantic waters. In contrast, the stronger-bonded, stable kaolinite clay minerals, which require heat for formation, are found in marine sediments of warm, tropical waters (Gross 1987, 89–91).

In addition to water and temperature, the composition of the parent rock is an important factor in the clay-making process. Ash from volcanic eruptions alters to form bentonite clay minerals (see pp. 136–138). Granites of the continental crust, low in iron and magnesium and rich in quartz and potassium feldspar, tend to form kaolinites. Rocks high in sodium feldspar, magnesium, and iron and low in quartz, which occur in both the oceanic and continental crust, decompose into the complex, unstable montmorillonite and illite clay minerals. However, nature refuses to be packed into neat, tidy categories, and we find granitic rocks altering to montmorillonite and other kinds of unstable clay minerals, depending on the particular environmental conditions that surround the granitic rocks during the decomposition process.

In addition to specific clay minerals, most clays contain various amounts of other minerals, such as quartz, mica, iron, magnesium, and calcium, which were present in the parent rock. They remain in the clay mineral because the process of clay-making is on going and is never completely finished. The continuous, ever-changing nature of this process accounts for the infinite number of mineral variations that can exist in specific clays at any one period of time.

OXIDE STRUCTURE AND CERAMIC FUNCTION OF CLAY MINERALS

The oxide structures of most clays, because of their feldspathic parentage, reveal the familiar trinity of glassmaker (silica), adhesive (alumina), and melter oxides. The specific ratio of these oxides in a particular clay mineral will ultimately determine the ceramic function of the clay and depends on the extent of the clay-making process. For example, the kaolinite clay mineral has undergone a more complete clay-making process and, therefore, contains larger amounts of alumina and lower amounts of silica, iron, potash, soda, and/or calcium than exist in either the parent feldspar rock or in other kinds of clay. This in turn means that kaolinite clays will function as an economical source of alumina in glazes, and as an important ingredient of white, or light-burning, claybodies.

On the other hand, if the kaolinization process is not completed, the oxide structure of the resulting clay mineral will more closely resemble the parent rock. If, in addition to feldspar, the parent rock contains iron and/or magnesium minerals, iron and/or magnesium oxides will also appear in the oxide structure of the clay mineral. Clays that contain predominant amounts of such clay minerals (such as Albany Slip and Red Art clays) function as dark-colored glaze cores, or claybody colorants at the higher firing temperatures and as iron-bearing claybodies at the lower firing temperatures.

INSTABILITY OF CLAY MINERALS

Over a period of time, the driving forces of nature transform one kind of clay mineral into another. Watery, acidic solutions, enriched with highly charged hydrogen ions and metallic ions drawn from prior clay minerals, percolate through the soil layers and change pre-existing clay minerals into different kinds of clay minerals. This fact has important consequences for the potter, because it means that no matter how stable they were in the past, all clay sites are highly vulnerable to change, as deeper and deeper layers of the earth's crust are mined.

"Even when taken from two closely-separated areas, a particular clay may show considerable variation. The difference between a deep-mined clay and the same seam under shallow overburden is so well marked that it is often impossible to use the two materials for the same purpose."

(Grimshaw 1980, 311).

The more stable kaolinitic clay minerals can also change into the unstable montmorillonites by means of this watery process. Thus, although clay minerals remain stable for a long period of time, the transforming powers of seeping, mineral-enriched waters may eventually change all of them into completely different clay minerals. In addition, climatic changes of increased humidity together with land erosion can force deeply buried clay layers upwards toward the surface, where once again they will change in response to the demands of a different environment. And finally, cultivation of the soil will change the kinds of clay minerals that exist in the soil layers of the field.

"A striking change in the plastic behavior of a clay often occurs when it is taken from the land on which there has been extensive cultivation. The fertilizers used in the soil are sufficient to alter the exchangeable cations even in clays at a considerable depth and so change their character."

(Ibid., 1980, 311).

Thus, clay, the foundation of the potters' craft, is as ephemeral and transient as human life itself.

CHARACTERISTICS OF CLAY MINERALS

PLASTICITY

The first and most important characteristic of clays is plasticity. The word *clay* comes from the German verb *kleben*, to stick to, and that is, of course, what immediately comes to mind when we think of clay. It is a material that "sticks to" the hand and "sticks to" itself in response to the touch of the hand. In other words, the identifying characteristic of plasticity is intrinsic to the meaning of the word *clay*.

Plasticity refers to the ability of a material to form and retain the shape directed by an outside force. "This is one of the most important of the properties of clay and one which is least understood" (Parmalee 1937, VIII–16). The plasticity of most clay minerals derives from the unique crystal structure of their molecules, which are minute in size and platelike in shape. These crystals form flat, two-dimensional sheets that touch each other on only two sides. There is a disproportionately large ratio of surface area to mass in these platy crystals. When water floods these two-dimensional sheets, it creates a strong bond between them in much the same way that a wet, flat plate bonds to a table surface. The water also acts as a lubricant and causes the platy crystals to slide over one another. The strongly bonded, sliding sheets will take on whatever shape they are directed toward by an outside force.

In contrast, the molecules of a nonplastic material, such as the parent feldspar, for example, are larger and touch each other on three points. Water does not create the same kind of bond or lubricating force between the molecules. Add water to a feldspar and try to form a shape with it. Its only response to the hands' pressure will be to fall apart into a sodden mass. This is the best way to understand the true meaning of the word *plasticity*, which for the potter is the identifying characteristic of clay.

PARTICLE SIZE

Particle size is an all-important characteristic of clay minerals and is of crucial importance in the identification of clay minerals. All clay minerals possess a fine-grained, minute particle size. *The Encyclopedia of Mineralogy* describes this trait as the most significant factor in the identification of a material as a clay.

"Clays are characterized primarily by their small particle size, which is usually taken as less than 2 μm.[2] Coarse, medium and fine clays have ranges about 2–0.5, 0.5–0.2, and below 0.2 μm respectively. . . .On this basis any material ground to less than a 2 μm particle size becomes a clay."

(Frye 1980, 69)

According to the above definition, a material could be classified as a clay even though it does not possess the property of plasticity. (Note Frye's reference on p. 69 to Flint clays, which are fine-grained but not plastic. In this clay, the bonds between the crystals are "firmly cemented" so that they do not slide past each other as they do in most other clays.) Although plasticity may not figure in a geological definition, it is everything to a potter. If a material is not plastic, it is not clay, insofar as a potter is concerned. Because this is a book for potters, plasticity is always the first and most important characteristic of a clay mineral.

Subject to the exception noted above, the fineness of the particle size determines the plasticity of the clay mineral. The minute clay crystal contains more surface area and therefore greater water-bonding capacity than the larger particle size crystals of other materials, such as feldspar. Consequently, the bonds between the clay particles and the sliding power produced by water increase in strength as the particles become smaller and more minute. Hence, clays with smaller particle size are more plastic and take on more water than clays with a larger particle size. This has important consequences for the potter, as we shall see when we consider the characteristics of various ceramic clays.

CHEMICAL AND CRYSTALLINE STRUCTURE

A third identifying characteristic of clays is their chemical and crystalline structure. All clays contain significant amounts of silica (SiO_2), alumina (Al_2O_3), and water (H_2O). More important is the fact that the atoms are bonded together in flat, two-dimensional sheets of two or more layers. Sheets of silica, alumina, and other metallic oxides are interspersed with sheets of hydrogen-oxygen molecules (HOH). The top layer consists of silicon and oxygen atoms; the second layer is alumina and (in the case of clay minerals other than perfectly ordered kaolinite) other metallic ions such as magnesium, calcium, iron, sodium, and potassium. The third layer contains the hydrogen and oxygen atoms. It is the second layer of aluminum and other metallic atoms that holds the most interest. This is the layer in which exchanges of aluminum are made with other metallic atoms. The exchange property of the aluminum layer creates the complexity and disorder of some clay molecules, for here, many different atoms can appear. This fascinating exchange property includes the ability to store, exchange, and transfer energy, and it is the reason certain clays are thought to be a possible link between inorganic and organic matter (see Cairns-Smith and Hartman 1986).

METAMORPHOSIS BY FIRE

A fourth unique feature of clays is their metamorphosis by fire into a stronger material. Most earth materials are weakened and broken by the heat of the fire. Clays, on the other hand, exchange their fragile and perishable existence for a hard, durable form that is capable of lasting for thousands of years. They become hard, water-impermeable materials with a new crystalline structure and different physical properties. Though clays lose plasticity in the heat of the fire, they gain permanence in its stead.

CONCLUSION

A layered, hydrous, fine-grained silicate of alumina with the properties of plasticity in the raw state and hard strength in the final, fired state constitutes the identifying characteristics of clay minerals; this is what is meant here by the term *clay*.

KINDS OF CLAY MINERALS

Clay minerals are classified by geologists according to the symmetry of their crystalline structure, and although this approach means little to most potters, there are in fact certain relevant factors that contribute to an understanding of clays.

There are six general groups of clay minerals, and numerous subgroups within each group. (These groups are worth mentioning, if only because of their exotic titles!)

Classification of Clay Minerals[3]

Kaolinite-Serpentine Group
  Kaolinite minerals
  kaolinite, halloysite
  (kaolins-ball clays, fireclays)
  Serpentine minerals
  lizardite, chrysolite (fibrous)

Clay Micas—Illites
  Similar to muscovite and biotite in composition
  (Albany Slip clay, Cedar Heights Red Art clay)

Smectites
  Montmorillonite
  Fine-grained, important swelling properties
  (bentonites)

Vermiculites

Chlorites
  (Sheffield Slip Clay)

Allophane
  Amorphous clay
  Imogolite—may be an early stage of crystallization
  of allophane.

The first group is known as the Kaolinite-Serpentine group and contains mostly kaolinite. Kaolinite is the major clay mineral in the kaolins, or china clays, which are the basis of the porcelain claybodies, and the main source of alumina for glazes. The purest kaolinites are formed by the hydrothermal weathering of granitic rocks and, consequently, have a well-formed, large crystalline structure, low capacity of ion exchange, and lower plasticity and iron impurities than the other clay minerals. Also included in this first group is the clay mineral livesite, which has an identical chemical composition to kaolinite, but is finer grained, and consequently more plastic. Livesite is a major clay mineral in the white-burning kaolin-ball clays, which make up the essential core of the porcelain claybodies. Livesite appears in many other sedimentary clays, such as ball clays and fireclays.

The Serpentine portion of this group of clay minerals results from the weathering of basaltic rocks that are high in magnesium and iron. The clay minerals lizardite and chrysolite are formed by the same hydrothermal processes as kaolinite.

A second group of clay minerals bears the name of Smectites. Smectite comes from the German word *smektis* to wipe off, or clean, and refers to the cleansing property of removing oils and grease, which is typical of these clay minerals (*Webster's New Universal Unabridged Dictionary* 1979, 1713). This group contains the extraordinary montmorillonite clay minerals, which are said to possess the secret of life. Their name derives from the place of discovery, Montmorillon, France (*Webster's New Universal Unabridged Dictionary* 1979, 1166). Unlike kaolinite, their crystalline structure is disordered and subject to innumerable variations. They freely exchange their molecules for those of other materials and are, therefore, the most variable and exciting of all the clay minerals.

A startling characteristic of some clays in this second group is their tendency to swell and increase in size when immersed in water. Bentonite, which frequently appears in low percentages in both glazes and claybodies as a suspending agent, provides a prime example of this feat. I will never forget the day I mixed up some 900 grams of Bentonite with 100 grams of Whiting, added water to the mixture, and watched in amazement as one container of liquid swelled to a volume of almost three containers!

A third group of clay minerals consists of clay micas and are known as Illites and Glauconites. These clay minerals, like the Smectites, contain a disordered crystalline structure and are a complex mixture of minerals. They are often found in combination with the Montmorillonite clay minerals. Although related to muscovite and biotite micas, they are finer grained, and contain less potassium and more water. Albany Slip clay, Red Art clay, and other similar iron-bearing lower-fire clays contain major amounts of Illite clay mineral.

A fourth group of clay minerals includes the Vermiculites, which again are similar to Smectites. Vermiculites possess an even higher exchange property, and also swell in water, although their rate of swelling is lower than the Smectites. Potters use Vermiculite to achieve a lighter-weight claybody.

Chlorites, fibrous Palygorskite, and gel-like Allophane, which appears in some sedimentary clays, make up the fifth and sixth groups. Sheffield clay, which is an iron-bearing, lower-firing clay, contains chlorite minerals.

This classification of clay minerals is impressive in its awesome complexity and creative nomenclature; yet, for a potter, it is more important to know *how* a particular clay will behave in a glaze or claybody. Thus, the only meaningful classification of clay minerals for a potter is one based on the working properties of the clays. There can be many differences, depending upon the kinds of clay minerals contained in the clays. That is the reason why most claybodies are a balanced formula of different clays. Each clay contributes one or more of the following qualities needed to make up a good claybody:

I. Plasticity:	How does the clay hand-build and throw on the wheel?
2. Wet strength:	Does the clay slump when it takes on water?
3. Dry strength:	Does the clay crack during the handling or drying process?
4. Firing strength:	Does the clay shrink excessively, slump, or crack when subjected to the heat of the fire?
5. Firing range:	Is the clay mature or immature at the desired firing temperature? Will water seep through its pores? Can the clay tolerate a range of firing temperatures?
6. Color/Texture:	How does the clay contribute to the desired color and texture of a glazed or unglazed piece after firing?
7. Thermal shock:	Does the clay crack or dent during the firing or cooling cycle?
8. Glaze fit:	Does the glaze crack or shiver off the fired clay form?

In contrast to the requirements for glaze surfaces, it is preferable for a claybody to include many different kinds of clays. The workability and resulting surface of the claybody is usually better if this approach is followed. Bear in mind that if one of the clays used in the claybody is highly unplastic, it may interfere with the general plasticity of the claybody, despite the presence of other plastic clays. Red Art clay is not as plastic or workable as A. P. Green fireclay, Jordan clay, or Gold Art fireclay. Thus, we suspected that the workability and plasticity problems of the Portchester 5–6 claybody were due to the presence of 25% Red Art clay, notwithstanding the fact that the balance of the claybody was made up of plastic clays, such as A. P. Green fireclay and Jordan or Gold Art Clay.

The fact that the claybody does not depend solely on one or two clays is a good thing in view of the ephemeral nature of clays. East Coast potters have recently had a bitter dose of the perils in overdependence, when Jordan clay suddenly became unavailable. Suppliers had difficulty finding an acceptable substitute, and as a result, many high-Jordan claybodies were put on hold. A number of years ago, deposits of A. P. Green fireclay contained sizable amounts of limestone (calcium carbonate), which caused bloats in the fired ware. Pine Lake fireclay was substituted until it too became contaminated with limestone nodules, and once again, bloated ware was the result. In both of these cases, the fireclay constituted a sizable proportion of the claybody. Changes in the mineral content can easily occur in any kind of clay, because all clay deposits are continually subjected to ongoing weathering forces. The use of different clays in a single claybody is one way to minimize the consequences of unavailability and changed mineral content.

THE FIRING PROCESS AND ITS EFFECT ON A CLAYBODY[4]

The question "*What is a stoneware glaze?*" is answered as follows: A stoneware glaze is a prescribed ratio of glassmaker (silica), adhesive-glue (alumina), and melters. When sufficient heat is applied to this trinity, a layer of melted glass is bonded to the walls of the clay form.

The primary mineral source for a stoneware glaze is feldspar, whose chemical oxide structure reflects a similar trinity. The question now to be asked is "*What is a stoneware claybody?*" A claybody, once again, is a trinity of glassmaker, adhesive-glue, and melter oxides. The ratio of these materials is now altered, because the focus here is on structural form and workability, rather than on surface effects. Alumina—a primary ingredient of the mullite crystal, which provides the strength of the fired clay form—increases in proportion to the melter content in the ratio.

The high-alumina clay minerals are the primary mineral source for a claybody, because workability and fired strength are now the paramount considerations, and only the clay minerals provide these properties. Stoneware claybodies usually contain about 75%–80% clay. The remainder is made up of small quantities of nonplastic quartz, feldspar, mica, and other melter minerals that help the claybody achieve proper fusion and glaze fit at the desired firing temperature. Thus, we are in fact replacing some of the potash and silica that have been removed from the clay minerals by the clay-making process. And because the kiln fire removes the water produced by the clay-making process, it could well seem as though we have recreated the feldspar that gave birth to our clay minerals. However, it takes millions of years of slow cooling at high temperatures to recreate the feldspathic crystalline structure. Although the fired claybody is in fact stony and rocklike, it has now become a manufactured creation of mullite, partially dissolved quartz particles, and glass. Most important, although the kind of oxides and minerals in the claybody is the same as exist in the feldspar and glaze, the proportion of each has altered because of the changed needs of the potter.

A functional potter needs a claybody that is strong and intact after firing. The accompanying glaze must fit the body; it must be neither too tight (craze or crackled), nor too loose (splinters off and/or cracks the claybody in its effort to remain on the clayform). This optimal result depends on the interaction of four kinds of minerals: clay minerals, feldspars, micas, and quartz. First and foremost are the clay minerals, primarily kaolinite (found in kaolins, ball clays, and fireclays), but also livesite (primary mineral in ball-kaolins), montmorillonite (primary mineral in bentonites),

102

chlorite (Sheffield Slip clay), and illite (Albany Slip clay and Red Art clay). Ceramic clays contain some or all of these clay minerals. They also contain varying amounts of feldspar, quartz, and/or mica. Separate amounts of these latter three minerals are often included in the claybody. How much of each is added depends on the amount already present in the clays of the claybody. Each mineral makes a unique contribution toward the creation of a claybody that fits the glaze surface and does not crack or break apart before, during, or after the kiln firing. Before firing, a ceramic clay is a mixture of clay minerals, quartz, feldspar, and/or mica. After firing, the ceramic clay is a combination of mullite crystals (two molecules of silica bonded with three molecules of alumina; named after the Island of Mull, Scotland[5]), which gives the fired ware its strength, and three kinds of silica-free quartz, cristobalite (finely divided, highly reactive silica), and silica glass (fast cooled, nonshrinking, noncrystalline, melted silica).

The strength of a claybody lies in the formation of as many mullite crystals as possible. Its weakness lies in an excess of any of the three kinds of silica. An excess of either cristobalite or free quartz will crack the claybody and/or cause pieces of the glaze to shiver off the pot. An excess of nonshrinking silica glass may result in slumped ware and/or crazing of the glaze, which weakens the ware. Thus, the essence of a good claybody lies in the right proportion of mullite crystals and the three kinds of silica. The achievement of such a claybody requires knowledge about the kinds and grain sizes of the minerals contained in the clays of the claybody. This information appears in the mineralogical analysis of a clay and is more important than its chemical analysis (which gives only the bare bones of oxide percentages), because irrespective of amount, the same oxide will behave differently in a claybody depending on its mineral source and consequent particle size. For example, the silica in kaolinite, montmorillonite, quartz, and amorphous silica each have a different particle size. In addition, the free silica contained in certain kaolinite clays may have a finer particle size than in other kaolinite clays. (Kentucky Stone is an example of such a clay; see pp. 122–123.) Montmorillonite silica and amorphous silica are finely divided and more reactive than kaolinite or most other forms of silica. They both convert more easily to the cristobalite form

of silica, which has such serious consequences for the glaze fit and fired strength of the claybody. On the other hand, there may be enough mica minerals in the montmorillonite clay to inhibit large amounts of cristobalite formation. Only mineralogical analyses give this kind of information.

Despite the importance of the mineralogical analysis, it is rare for a supplier to provide it. The customary data that accompanies a clay shipment from a supplier on the request of the customer usually contains only a chemical and particle size analysis and rarely provides information about the minerals that make up the particular clay. Of the more than 20 analysis sheets I obtained, only one contained a mineralogical analysis, and even this one did not provide percentages. (It is possible to compute the approximate mineralogical structure of a material from its chemical analysis by following prescribed formulas; see Robinson 1981, 9:79; Tichane 1990, 55–57.) In addition, a comparison of the actual chemical analysis of a material with its theoretical chemical analysis would indicate the presence of certain minerals. Thus, 1%–2% potash in a chemical analysis suggests the presence of 15%–30% mica (Tichane 1990, 14–15).

THE FIRING PROCESS

The importance of accurate mineralogical information becomes obvious when we look at the reactions of the claybody in the fire.

The metamorphosis of a fragile claybody and powdery glaze surface into a durable, hard, permanent form provides a fascinating account of the unseen action within a firing kiln. Just as the toys in the Nutcracker Suite ballet leap and dance after the toy maker closes the door, so too, do the clay, feldspar, mica, and quartz molecules spring to life after the kiln door is shut and the heat rises. Though the exuberant reactions of these materials are hidden from view, the final fired result is proof of the extraordinary transformations catalyzed by the fire.

When the temperature of either a bisque or glaze firing reaches 1063°F, the first important silica reaction takes place. Free silica, or quartz, expands about 1% as it converts from alpha quartz to its beta form. This change will reverse itself during the cooling process; as the kiln cools down to 1063°F, free silica will contract 1% as it converts back to its alpha form. At this point in the glaze firing, the glaze is already rigid and set. Hence, the contraction of free silica in the clay minerals at 1063°F will affect the glaze fit of the fired ware. If the glaze is too tight, as evidenced by craze marks on the glaze surface, this would mean that there is insufficient quartz in the clay minerals of the claybody. Additional free silica in the form of Potters' Flint can be added to the claybody so as to increase the amount of body contraction at the 1063°F cooling temperature. The particle size of the added free silica or of the free quartz contained in the clay minerals is of crucial importance, because a fine-grained quartz will more easily change into noncontracting silica glass or sudden-contracting cristobolite[6] and thus will not provide the necessary contraction for the proper glaze fit. As mentioned previously, some kaolinite clays (e.g. Kentucky Stone) possess sizable quantities of fine-grained, highly reactive free silica, which easily converts to cristobalite and thus plays havoc with the glaze fit. Ordinary oven heat does not reach the temperatures of silica conversions; thus, free silica expansion and contraction will occur again only if the ware is refired.

Mullite continuously forms from about 2000°F and upwards. Once this initial temperature is reached, the clay minerals in the claybody release their excess silica in the process of forming mullite. The kaolinite clay mineral, which is the primary mineral in kaolins (EPK kaolin), ball clays (Kentucky Stone), fireclays (A. P. Green), contains six molecules of silica joined with three molecules of alumina. It releases four reactive silica molecules in the process of forming mullite. The remaining two molecules of silica, joined with the three molecules of alumina, make up the strong, needlelike mullite crystal, which provides the strength of the fired ware.

Other clay minerals, such as Illites (e.g. Albany Slip clay and Red Art clay) or Montmorillonites (bentonites), would release even greater amounts of highly reactive silica molecules during the firing process.

Both the released silica molecules and the fine-grained particles of free silica are highly reactive. In the absence of sufficient feldspar or mica minerals, they will convert to

alpha and then beta cristobalite silica. The conversion to alpha cristobalite results in a 3% expansion. This expansion does not stress the claybody for the following reasons:

1. It is offset by the shrinkage of the claybody due to the ongoing contraction of the clay pores.

2. The body is still pyroplastic.

3. The conversion to alpha cristobalite is a continuous, gradual process that continues up to the end of the firing and does not occur all at one time.

The reverse of this process, during the cooling cycle, is a different story altogether. As the kiln cools down to about 450°F–500°F, the cristobalite formed in the claybody undergoes a 3% reversible contraction. As the glaze and other claybody materials have long since become rigid, the sudden movement produced by substantial amounts of cristobalite can cause cracking and glaze-shivering problems. In addition, oven temperatures often reach the temperatures of cristobalite contraction and expansion, at which point the cristobalite in the fired ware will, once again, expand and contract. This sudden movement within a fired, rigid clay form can shatter the ware and, even worse, destroy your dinner!

Hence, it becomes desirable to avoid formation of excess amounts of cristobalite. This can be accomplished by converting the reactive, fine-grained silica into inert, nonshrinking and nonexpanding silica glass. The melting power of the feldspar and mica minerals in the claybody activates this conversion. (See Chapter 1, pp. 38–40.)

Feldspar begins its melt at about 2200°F (cone 4), and thus functions as a melter in the stoneware claybody. It draws some of the glassmaker silica into the melt to create highly desirable silica glass, which, in the right proportion, renders a claybody vitreous and strong. The more silica glass formed, the less silica there is available for the formation of cristobalite, which threatens the strength and glaze fit of the ware. Thus, feldspar, although no longer the star player, is an important part of a claybody. Mica functions as a melter in a claybody in the same manner as feldspar.

Here we see the importance of the ratio of the four minerals that make up the claybody—clay, feldspar, mica, and quartz. The amount of each that appears in the various clays of a claybody affects the strength and glaze fit of the fired claybody. Wherever possible (depending on the availability of mineralogical data) this ratio will be a focal point in the description of specific ceramic clays.

CLAYS, CLAYBODIES, AND THEIR FUNCTION AS GLAZE MATERIALS

Most clays result from the transformation of feldspathic materials, and therefore, it is not surprising that their chemical structure contains that familiar feldspathic trinity of glassmaker silica, adhesive-glue alumina, and melter oxides of sodium, potassium, magnesium, calcium, etc. However, the proportion of these oxides will differ depending on the kind of clay mineral contained in each clay. This proportion is crucial and ultimately determines the ceramic function of the clay. Some clays contain large amounts of the clay mineral kaolinite. Their oxide ratio would reveal a high proportion of alumina and a correspondingly low amount of melters and iron impurities. Kaolinite clays become the core of light-burning claybodies and are a cheap source of alumina for glazes. Other clays contain large amounts of illite and similar iron-bearing clay minerals. Their oxide structure contains high amounts of silica and melters and, except for iron and other impurities, bears a close relationship to the feldspars. Although these clays can, and often do, function as iron-bearing claybodies at the lower firing temperatures, they make wonderful, dark-colored glaze cores at stoneware temperatures. *Figures Intro.3–4, pp13–14.*

It is possible to transform any clay, even one with high amounts of kaolinite, into a glaze core by adding melters and quartz. It is important to remember that the terms *claybody* and *glaze core* are not absolutely fixed. They do not, once and for all, identify specific ceramic materials. They are functional terms, and the materials that they identify often change their roles, and thus their classification, depending on the firing temperature and the balance of the materials in the formula. For example, the fireclay A. P. Green is not always just a claybody material. With the

right addition of melters and quartz, it can function as a glaze material (see *Fireclay Tests p. 119*). A porcelain claybody made up of kaolin clay, quartz, and feldspar may become a fine white slip, and even a glaze, with the additions of water and/or melters. Similarly, most clays can become part of a glaze or claybody if you adjust the temperature of the fire or if you add melters. *Figure Intro.4, p14.* This is a magical area where all things are possible once you know the oxide ratio and mineralogical structure of your materials. There are no constants—no absolute rules in the ceramic process, other than KNOW YOUR MATERIALS AND TO YOUR OWN SELF BE TRUE. If you achieve this knowledge, then you can conjure glaze-cores out of claybodies, transform clay-glaze cores into claybodies and have a wonderful time in the process.

Though the discussion of clays that follows stresses their general and most typical function, never forget that these are NOT the only ways to use these materials. Try less-traveled paths and attempt some creative transformations of your own.

CONCLUSION

Clay minerals are transformations of igneous rocks. These transformations occur in response to their new, ever-changing environment at the surface of the earth. They are as complex and varied as the crust of the earth itself. We can never ponder this fact too long. The chemical reactions that follow, which trace the alteration of actual soda and potash feldspars into three different clay minerals, show the complexity of the clay-making process and the resulting clay minerals (Birkeland 1984, 69; Blatt 1992, 32).

Potters, do not throw up your hands and avert your eyes in horror at these equations. They are NOT given to instruct in technical chemical reactions; on the contrary, they are repeated here because merely to view them shows more vividly than words the remarkable length and intricacy of the clay-making process. Consider this awesome process as you handle that humble, unassuming, yet incredibly complex material known as clay.

$$2KAlSi_3O_8 + 2H^+ + 9H_2O \rightarrow Al_2Si_2O_3(OH)_4 + 4H_4SiO_4 + 2K$$
[Orthoclase]   [Kaolinite]

$$3KAlSi_3O_8 + 2H^+ + 12H_2O = KAl_3SiO_{11}(OH_2) + 6H_4SiO_4 + 2K^+$$
[Orthoclase]   [Illite]

$$2NaAlSi_3O_8 + 2H^+ + 9H_2O = H_4Al_2Si_2O_9 + 4H_4SiO_4 + 2Na^+$$
[Albite]   [Kaolinite]

$$8NaAlSi_3O_8 + 6H^+ + 28H_2O = 3Na_{0.66}Al_{2.66}Si_{3.33}O_{10}(OH_2) + 14H_4SiO_4 + 6Na$$
[Albite]   [Smectite]

Smectite: Bentonites
Kaolinite: EPK Kaolin, Grolleg, Ball clays, Fireclays
Orthoclase = Potash Feldspar
Albite = Soda Feldspar

KAOLINS
Figure 2.3–2.4, 2.6 pp146–147, 149.

The word *kaolin*, which is used interchangeably with *china clay*, comes from the Chinese word *kao-ling*, meaning high ridge, and refers to a mountainous region in China that contains large deposits of high-temperature, white-burning clay minerals (*Ceramic Industry* 1998, 48). These clay deposits were the secret ingredient of the incomparable Chinese porcelains. Although the Chinese had already perfected the porcelain-kaolinitic claybody by the end of the twelfth century, Europe did not discover the existence of this kind of clay until the seventeenth century!

The term *kaolin* describes high-temperature,[7] white-burning clays, which contain predominant amounts of the clay minerals kaolinite, or livesite. These clay minerals are composed of a single tetrahedral silica sheet and a single octahedral alumina sheet, with an interlayer of water molecules. Both kinds of clay minerals form as a result of the decomposition of feldspathic rocks and granites from high-temperature gases and water vapors. Some of these clay deposits, such as the Cornish kaolins of southwestern England (Grolleg) remained at the site of the parent rock and are therefore described as "residual." Certain of our domestic kaolin deposits (EPK kaolin, Florida; Pioneer kaolin, Georgia) were transported by water and other weath-

ering forces and deposited in lagoons and lakes far from their original sites. These are known as secondary, or sedimentary kaolins (Ibid., 76–77). Transportation by water sifts out and deposits the finer particles of the minerals. Thus, many secondary kaolins such as Georgia kaolins (Pioneer, Ajax P, 6 Tile) and Florida kaolin (EPK) have a finer particle size and are more plastic than the residual kaolins of North Carolina. On the other hand, Cornish kaolins such as Grolleg are as fine grained and plastic as the secondary, or sedimentary, kaolins. A further characteristic of sedimentary or secondary kaolins is that they are less pure than residual kaolins. Water-transported minerals tend to pick up more impurities than the residual kaolins. Consequently, secondary kaolins are said to burn less white than the residual English kaolins such as Grolleg. (This difference in color is not apparent from a comparison of their chemical structures, which appear to be fairly similar with respect to iron and titanium impurities.) The fine-grained particle size, high plasticity, and green strength of the secondary kaolin EPK is suggestive of ball-clay minerals, and some texts refer to this secondary kaolin as a white-burning ball-kaolin. (There is, however, a considerable difference in particle size between ball clays and plastic kaolins—81% of ball clay particles are less than 2 microns; only 55%–65% of plastic kaolin particles are less than 2 microns.)

KAOLINS—CHINA CLAY

TRADE NAME	Grolleg	EPK	Pioneer, Six Tile (Georgia Kaolins)
GEOGRAPHICAL SOURCE*	England ECC International	Florida Feldspar Corporation	Georgia Dry Branch Kaolin Company

GEOLOGICAL SOURCE*
(Grolleg) Hydrothermal: steam, boron, fluorine and tin vapors attacking soda feldspar in granites and pegmatites more than 250 million years ago.

CHEMICAL STRUCTURE

	Kaolinite**	Grolleg***	EPK†	6 Tile‡	Pioneer‡	HWF‡
SiO_2	46.54%	48.00%	45.73%	45.5%	45.68%	45.5%
Al_2O_3	39.56	37.00	37.36	38.1	38.51	39.0
Fe_2O_3		0.68–0.82	0.79	0.3	0.44	0.3
TiO_2		0.02	0.37	1.4	1.43	0.5
MgO		0.30	0.98	0.5	0.14	0.1
CaO		0.06	0.18	0.4	0.24	0.1
K_2O		1.67–2.03	0.33		0.14	0.5
Na_2O		0.1	0.59	0.04	0.04	0.1
P_2O_5			0.236			
L.O.I. (H_2O)	13.96	12.2	13.91	13.8	13.52	13.5
P.C.E.			C/35	C/34–36	C/34–36	

MINERALOGICAL ANALYSIS: Quartz, feldspar, and trace minerals determined by X-ray diffraction. Alkali in feldspar subtracted from alkali in chemical analysis, and balance calculated as mica. Remaining mineral is kaolinite.**
Grolleg: (blend of different sources) Kaolinite 81%, mica 15%; feldspar 1%; other materials 3%.**
EPK, 6 Tile, Pioneer: Kaolinite 97%***

*ECC International 1987, "China clay production"; *Technical Data* 1998.
**Hosternan 1984, c-7.
***ECC International 1987.
†The Feldspar Corporation 1998.
‡Dry Branch Kaolin Company, 1998.

CLAYBODY FUNCTION

Plastic kaolins are the most important clay mineral of the porcelain claybody for the following reasons:

First, porcelain must be a white-burning ware, and the kaolins possess the lowest iron and titanium content of all the plastic clay minerals.

Second, plasticity is an essential requirement for most porcelain claybodies, and the sedimentary, water-transported kaolins and certain Cornish kaolins are highly plastic because of their fine-grained particle size. On the other hand, because of their fine particle size, these clays tend to absorb more water than the larger particle size residual kaolins. High water absorption can cause slumping and cracking problems, and thus, plasticity occurs at the cost of structural strength. Note that the domestic kaolins possess greater structural strength than the Cornish kaolins, due to the lower potash content of the former. The 1998 chemical data sheet lists 1.85%–2.05% potash for the Cornish kaolin, Grolleg. According to Robert Tichane, the presence of 1%–2% potassium in the Cornish kaolins indicates a mineralogical content of 15%–30% mica, and "this contamination is significant because it will have a large effect on the softening point of English kaolin bodies" (Tichane, 1990, 14).[8] On the other hand, the fired color of a porcelain claybody containing the Cornish kaolin Grolleg is said to be whiter than the fired color of a claybody with the domestic secondary EPK kaolin. Once again, it is a trade-off. (Dry Branch Kaolin Company in Dry Branch, Georgia, produces a coarse-particle kaolin, HWF, which is said to equal in whiteness the English kaolins [see p. 107]. Note its low titanium and iron content. J. David Sagurton, National Accounts Manager, Dry Branch Kaolin Company, August, 1998.) Grolleg (a combination of separate Cornish kaolins) is a highly stable product that has not changed substantially in chemical structure for 25 years. The only change appears to be a slight decrease in its refractoriness as indicated by its higher absorption and contraction rate. The manufacturer, ECC International, is one of the world's largest suppliers of ceramic materials, which would tend to guarantee stability.

Third, true porcelain is a high-fired ware that becomes translucent at stoneware temperatures of 2300°F or higher.

Kaolins are refractory clays, with a melting point in excess of 3100°F. However, with the assistance of additional materials, they will achieve the required state of hardness and translucency at high stoneware temperatures.

The mineral content of kaolin consists mainly of kaolinite, with low amounts of feldspar, quartz, and mica minerals. The high proportion of refractory silica and alumina contributed by kaolinite and the correspondingly low amount of feldspar and mica minerals enables the porcelain claybody to withstand high firing temperatures of 2300°F or higher. (Remember that feldspar and mica minerals function as melters in the high-fired claybody.) Unfortunately, as mentioned previously, it is not possible to use kaolin alone as the entire porcelain claybody. Although our kaolins are surely plastic enough to provide good working properties, their fine-grained particle size, which results in plasticity, may create slumping problems. In addition, the low feldspar and mica content of the kaolins would bring about serious glaze-fit and vitrification problems because of the formation of too much cristobalite and too little silica glass. The ancient Chinese potters were blessed with plastic kaolinitic deposits that naturally contained the necessary additional materials for a porcelain claybody. We, on the other hand, need to add feldspars, quartz and, in some cases, ball clays, to our kaolins in order to achieve a suitable porcelain claybody. (See examples of porcelain claybodies, p. 113.) The right proportion of the these minerals results in a workable, plastic porcelain claybody that combines whiteness, vitrification, translucency, and good glaze fit with structural and fired strength. Once again, proportion is everything.

Note that the term *porcelain* originally included only those white-burning, high-fired wares of 2300°F or higher with the property of translucency and a high-fired musical ring. Indeed, the sound of the ware when struck by a hard object was extremely important to the ancient Chinese potters, who, by this means, created musical instruments out of their pots. Today, the property of hardness, produced by the firing of high-silica, high-alumina wares to temperatures of 2300°F or higher, is no longer a prerequisite for the achievement of "porcelain." The term *porcelain* has expanded to include lower fired, white-burning wares, and even includes those that are not translucent. (It is now possible to achieve translucency even at earthenware temperatures; under this expanded definition, all such white-burning, translucent wares could be called porcelain!) Yet, no matter how white or translucent such wares may appear, the touch, feel, and ring of the higher-fired porcelain will always be different. For the sake of clarity, in this book the term porcelain is reserved for high-fired (stoneware temperatures), white-burning ware.

GLAZE FUNCTION

The refractory nature of the kaolins, together with their comparatively low iron and melter content, make them the best source of additional alumina available to potters. The exact function of additional alumina in glazes, together with the tests that illustrate this function, will be discussed in detail in Chapter 4. For now, it is enough to note that 25% kaolin in a glaze provides about 10% alumina (Hamer and Hamer 1986, 7). This amount of alumina helps to produce the stony-matt glaze surface. Ten percent kaolin contains about 4% additional alumina, which, alternatively, increases the transparency and gloss of the glaze surface and/or reduces the glaze fluidity, depending on the rest of the glaze materials (Ibid.). Although it is true that feldspars also con-

tain a fair amount of alumina (15%–23%) and are iron-free, the comparatively high amounts of silica and melters, which they also contain, make them a less-pure source of alumina than the kaolins. In addition, the plastic properties of kaolin clays contribute good suspending qualities to the liquid glaze mixture. Thus, kaolin, unlike feldspar and other nonplastic materials, aids in the physical application of the glaze. Note that different clays may produce different glaze consistencies because of the variation of alkaline melters (soda, potash, or calcium) contained in their chemical structure.

"Florida and North Carolina kaolin, for example cause different consistencies. Chemical changes taking place during storage may release alkaline agents capable of deflocculation; or the opposite is possible where flocculation occurs during storage."

(Behrens 1974, 40)

Because the function of kaolin in glazes is similar to that of the ball clays, the tests that show the role of kaolin in glaze surfaces are compared with the tests of the ball clays. In this way, the relative advantages and disadvantages of each clay can be compared and evaluated (*see pp. 114–115*).

BALL CLAYS[9]

Figures 2.1, p144; 2.6, p149.

Ball clays are fine-grained, highly plastic clays that formed over 40 million years ago from sedimentary deposits of clay materials and volcanic ash. Eighty-five percent of our domestic commercial deposits of ball clay are found in the Gulf of Mexico coastal plain region, which extends through Mississippi, western Tennessee, western Kentucky, Arkansas, Louisiana, and eastern Texas. It is believed that these ball clay deposits formed during the middle Eocene epoch (54–38 million years ago) when the climate of the coastal plain region was warm and humid, with heavy rainfalls, and the landscape was flat and interspersed with many flowing streams. Sediments of clay materials, transported long distances by the action of flowing water, and windblown volcanic ash were deposited in still lagoons protected by barrier beaches and eventually formed rich, high-grade, commercial deposits of ball clay. Major deposits are located in a small area of Kentucky and Tennessee. These deposits formed as a result of the erosion of the Appalachian mountains during the Upper Cretaceous period, more than 65 million years ago (*Ceramic Industry*, 1998, 88).

The term *ball clay* originated from the English practice of rolling the clay into 30–50 pound balls before shipping the clay to various points of destination—hence the name ball clay. In general, the term *ball clay* describes fine-grained, highly plastic clays that may contain varying amounts of relatively disordered kaolinite (33%–90%), quartz (14%–60%), mica (4%–40%), iron and titanium (1%–2%), organic matter (3%–16%), and small amounts of illite (trace–30%), montmorillonite clay minerals (trace–20%), and feldspar (trace–8%) (Tichane 1990, 19; Grimshaw 1980, 295; Hosterman 1984, C8). The mineralogical analysis of three ball clays produced by the Kentucky-Tennessee Clay Company from the Gulf coastal plain region described in *The Geological Survey Bulletin* 1558-C (Hosterman 1984) and used in many tests, stoneware glazes, and claybodies in this book, shows kaolinite to be the major clay mineral and quartz to be the major nonclay mineral; this corresponds with the Geological Survey's account of the best commercial grade of ball clay deposits (Ibid., C9). The highest grade of ball clay contains highly reactive clay-sized quartz rather than quartz sand. It does not contain silt or more than low amounts of iron oxide minerals; all ball clays contain a fair amount of carbonaceous, organic matter. However, even the best grade ball clay usually contains more silica, melters, iron, titanium, and organic matter than the kaolinite clays. Increased additions of these minerals (particularly the fine-grained quartz) will affect the glaze fit, fired strength, workability, and fired color of the claybody. Carbonaceous, or organic, matter in the form of humic acid can interfere with the workability of the claybody and break the bonds between the clay molecules, which create plasticity. As little as 1%–2% humic acid is sufficient to cause this deflocculated condition (Tichane 1990, 22). (A new shipment of a claybody, which contained sizable amounts of two different kinds of ball clay, felt unpleasantly sticky, had little throwing strength, and did not settle out of its slip so as to permit reconditioning. The supplier attributed the changed condition of this claybody to the presence of increased organic matter in the ball clay.) For these reasons, additions of ball clay to a claybody should be carefully considered.

Organic matter and other impurities in the ball clay cause it to appear dark gray, black, pink, or brown in its raw state. This organic material will burn out during the firing process and will not affect the fired color of the ball clay, which is usually almost white.

The primary identifying characteristics of a ball clay are fine particle size, plasticity, and high greenware strength. An important consequence of ball clay's fine particle size is its high water of plasticity rate.[10] *The Geological Survey Bulletin* 1558-C reports that 40%–65% water of plasticity is needed to make most ball clays plastic and workable (Hosterman 1984). Our tests on O.M.-4 Ball clay recorded a 38%–41% water of plasticity rate. Compare this to the 27% rate for EPK kaolin, 26% for Pine Lake fireclay, and 21% for Gold Art fireclay! Clays that require high amounts of water to become plastic have high greenware and fired shrinkage rates. Ball clays listed in the Kentucky-Tennessee Clay Company Data Sheet showed a firing shrinkage rate of 14%–17%. An acceptable shrinkage rate for a stoneware claybody is usually 12% or less. The release of considerable amounts of water during the drying and firing process can create grave warping and slumping problems for the ware.

BALL CLAY

TRADE NAME	Old Mine-4, Tenn-5, Tenn-9
GEOGRAPHICAL SOURCE	Western Kentucky, Western Tennessee; Kentucky-Tennessee Clay Company
GEOLOGICAL SOURCE	Sedimentary clay and volcanic ash material deposited by water; formed primarily during middle Eocene epoch approximately 46 million years ago, and Upper Cretaceous Period more than 65 million years ago.

CHEMICAL STRUCTURE*

	Old Mine-4	Tenn-5	Tenn-9
SiO_2	55.9	55.2	55.5
Al_2O_3	27.2	27.0	29.7
Fe_2O_3	1.1	0.9	1.0
TiO_2	1.2	1.5	2.1
MgO	0.4	0.4	0.3
CaO	0.4	0.4	0.2
K_2O	1.1	1.2	1.1
Na_2O	0.2	0.2	0.2
L.O.I.	12.5	13.2	10.4
			0.22 carbon
			0.02 sulfur

% Total Shrinkage

	Old Mine-4	Tenn-5	Tenn-9
Cone 5	13.0	12.5	11.5
Cone 10	14.5	15.5	14.4
P.C.E.	32	32	33

Particle Size

	Old Mine-4	Tenn-5	Tenn-9
% finer than .5 micron	35	31	33
% H_2O of Plasticity	37	37	31.2

MINERALOGICAL ANALYSIS**

	Old Mine-4	Tenn-5	Tenn-9
% K Spar	5.9	6.5	2.4
% Na Spar	2.5	1.7	2.5
% Kaolinite	66.7	62.8	73.8
% Free Quartz	19.0	21.1	16.6
% Organic	3.3	5.0	1.1

There are hundreds of different ball clays. Examples listed above are the ball clays used in claybodies and glazes in tests for this book.

Range of P.C.E. of various ball clays can be cone 18 to cone 33. According to Hosterman 1984, *Geological Survey Bulletin* 1558-C, average melting temperature is between 1670°–1765°C, or 3038°F–3209°F.

*Data sheet Kentucky-Tennessee Clay Company, 1998.
**Letter to the author of 4 October 1988, Kentucky-Tennessee Clay Company; No material change 1998.

Hence, ball clays rarely make up a large part of a claybody and are often used in no more than 10%–25% increments.

Porcelain bodies often utilize both the extreme plasticity of the ball clay and the pure whiteness of the kaolin and combine two parts kaolin to one part ball clay. The balance of the formula usually consists of equal amounts of Flint and feldspar, neither of which shrink or absorb water. Because these nonplastic, nonshrinking materials usually make up about 50% of the total, porcelain claybodies require a higher proportion of plastic ball-clay minerals than do stoneware claybodies. (Nonplastic materials usually do not exceed 10%–15% of the total stoneware claybody.) On the other hand, porcelain bodies seek to achieve maximum whiteness and translucency. Most ball clays contain a greater amount of iron and titanium than are found in secondary kaolins. Titanium and iron will gray or yellow the fired claybody color, and there is some evidence that they interfere with translucency (Tichane 1990, 24). For this reason, and despite the need for plasticity, a porcelain claybody in which whiteness and translucency are the goals should restrict the amount of ball clay in its formula.

There are many different kinds of ball clays on the market, all with varying chemical and mineralogical structures, particles sizes, and deformation rates (P.C.E.). Note that the deformation rates can range from cone 18 to cone 33! The Kentucky-Tennessee Clay Company, which is one of the largest suppliers of ball clay, lists 23 different kinds of ball clays. Of these, O.M.-4 and Tenn-5 and 9 appear in many of the glaze and claybody tests throughout this book. The claybody used by the New York 92nd Street Y for at least 20 years contained 23% O.M.-4. Fifty percent of this claybody contained low-shrinkage, larger particle size materials such as fireclay, grog, and feldspar, which permitted the incorporation of a relatively high proportion of ball clay (see p. 111). The balance of low- and high-shrinkage materials resulted in a fired shrinkage rate of about 12%. This rate would be an acceptable shrinkage for a high-fired stoneware claybody.

Even though the chemical structure of two ball clays is similar, if their particle size or mineralogical content is different, the resulting claybody may suffer great changes from an inadvertent substitution of one for the other (Robinson 1988, 76; Tichane 1990, 76).

"The particle size of ball clays is notoriously variable and strict control over the raw material has to be exercised, because slight changes can alter its properties considerably. For this reason it is inadvisable to use a different type of ball clay in the manufacture of a ceramic body, without first assessing its behavior."

(Grimshaw 1980, 414)

The data sheets supplied by the clay companies do not always reflect the current shipment. Hence, it is not possible to rely exclusively on their printed material. In the absence of current chemical and mineralogical data, it is important to make tests of each new supply of ball clay, both alone and in combination in the claybody. Although this can be a tiresome task, it can save the potter from a kiln load of ruined ware.

It is not clear as to whether or not there exist adequate future reserves of the highest commercial grade of ball clays.

"Because of the lenticular[11] nature of the clay bodies, close-spaced drilling, detailed sampling, mineralogic analyses, and ceramic testing are needed before adequate future reserves of ball clay can be proved."

(Hosterman 1984, C15, 20)

The demise of a particular ball clay, or material changes in its mineral composition (both of which are real possibilities for the future) require substitutions and hence corresponding changes in the claybody formula. This, in turn, requires knowledge about the minerals in all of the clays of the claybody. The aim of this book is to provide this knowledge for some of the most commonly used ceramic clays and to alert the potter to the dangers of blind, untested substitutions.

Note: At the time of printing of this second edition, the Legislative and Public Affairs Committee of The American Ceramic Society warns of the possibility of dioxin contamination (a human carcinogen) in ball clays. It is recommended that all clay manufacturers prove their clay products are free from dioxin contamination. See UpFront, *Ceramics Monthly*, October 2000, 12, or visit www.ceramics.org for the latest information.

WHITE STONEWARE

In addition to porcelain claybodies, the secondary kaolins and ball clays are the core of the popular "white stoneware" bodies. The difference between pure porcelain and these claybodies has to do with the properties of translucency and color. Although the "white stoneware" is in fact white, we all know from our experience with paint color charts that there are an infinite number of shades of white. White stoneware is in fact less white than porcelain and has a yellow or gray cast, depending on the kiln atmosphere—yellow if fired in an oxidation atmosphere and gray if fired in a reduction atmosphere. White stoneware is not translucent. The reasons for these differences lie in the lessened quartz-silica content and/or the addition of low amounts of fireclay, which coarsens the fired surface and contributes a low amount of iron to the color.

EXAMPLES OF WHITE-BURNING PORCELAIN AND WHITE STONEWARE CLAYBODIES USING SECONDARY KAOLINS AND BALLCLAYS

CONE 9-10 OXIDATION, REDUCTION

Kaolins	Temple	Tiso A	Tiso B
EPK Kaolin		19	25
Grolleg	55		
Ball Clays			
(Ky. O.M.-4)		45	25
Melters			
Cornwall Stone			25
Potash Spar	22		
Nepheline Syenite		10	
Talc		10	
Glass Maker			
Flint	23	15	25
Suspenders			
Bentonite	2	1	
Macaloid	1		
Color			
Oxidation	White	Off-white	Off-white
Reduction	Blue-white	Gray-green	Gray-green

CONE 5/6 OXIDATION

	M.O. III	Cushing Porcelain	Cushing White Stoneware
Kaolins			
Grolleg	50	20	
EPK Kaolin			30
6 Tile		20	
Kingsley Kaolin			10
Ball Clay			
Ky. Ball		5	
Tenn. Ball			30
Melters			
Nepheline Syenite	15	30	10
Potash Spar	7		
Dolomite	8		
Talc			8
Gerstley Borate		5	
Glass Maker			
Flint	20	20	12
Color			
Oxidation	White-yellow	White	White-yellow
Traits	Translucent. Self-glaze at Cone 9/10R.	Translucent. Nonplastic.	Not translucent.

GLAZE FUNCTION

KAOLIN AND BALL CLAYS
Figure 2.1, p144; 2.3, p146; 2.4, p147; 2.6, p149

The kaolins and ball clays are used in glazes for the following reasons:

1. They are a cheap source of alumina.

a. Substantial additions of alumina create the stony-matt surface in glazes. The secondary kaolins are the preferred source of material for the creation of the stony-matt surface. Ball clays are also a good source of alumina, but they contain less alumina (30% instead of 38%), more silica (50% instead of 45%), and more iron, titanium, and melters than do the kaolins. The tests performed with ball clay and kaolin at the cone 9/10 reduction temperatures indicate that additions of ball clay create satin-matt, rather than stony surfaces. Ball clay would be a good choice for the creation of a satin-matt surface provided the color change caused by increased iron and titanium in the ball clay would not be objectionable (*see test, p. 115*). The ball clays commonly used in our glazes and claybodies (Kentucky O.M.-4, and Tenn-5 and 9) contain more than 1% iron and 1.2%–2% titanium; the kaolin (EPK), contains only 3/4% iron and less than 0.3% titanium. Thus, ball clays may cause color changes that would not occur with the purer secondary kaolins. It is important to know the amount of titanium in a claybody, as titanium can have almost as much effect on the glaze surface as iron oxide. When we tested a porcelain claybody with O.M.-4 ball clay instead of the secondary kaolin Grolleg, the claybody color appeared gray instead of white, and the celadon glaze was green instead of blue. The reason for these color changes was not just the increased iron content of the substituted ball clay, but also its increased titanium content.

b. Low amounts of alumina may actually increase gloss and transparency due to alumina's tendency to inhibit large crystal formation. (See Chapter 4 for further discussion.)

The quantity needed to create a transparent or, alternatively, a stony surface, depends on the rest of the glaze ingredients and the firing temperature. In some glazes the addition of 10% kaolin increased the shine and transparency of the glaze. In other glazes, this same amount increased the opacity of the surface.

2. Low additions of both kaolin and ball clays will check the flow of a high-gloss glaze and prevent it from running off the pot onto the kiln shelf. Here again, the more refractory kaolins have a greater effect than the ball clays in achieving this result. (*See tests p.115*, in which 10% additions of ball clay actually increased the crazing and flow of the glaze, compared to additions of 10% kaolin, which reduced both the crazing and the flow.)

3. Kaolins and ball clays each function as suspending agents in glaze mixtures. Glazes that include at least 10% additions of kaolin and/or ball clays are more likely to remain in suspension over a long period of time than are glazes without such additions. It is for this reason that most glazes include at least 10% kaolin and/or ball clays in their formula. The finer particle size of the ball clays makes them more successful as suspending agents than the kaolins.

4. The fine particle size of the secondary kaolins and ball clays make them a good base for a white or colored slip. Ball clays alone or with the addition of a few materials such as cobalt, iron, or manganese oxides etc. create a decorative slip when mixed with water. I found that when thinly applied, the ball clay slip bonded closely with the stoneware claybody in the leather hard state, and created interesting white, slip-trailed patterns. The fired color of the ball clay slip at cone 9/10 was a pleasing soft white with a slightly creamy cast. *Figure 2.1, p144.* Here again, the finer particle size of the ball clay increases its bonding power and makes it a preferred choice for a slip.

COMPARATIVE COLOR TESTS OF KAOLIN AND BALL CLAY SLIPS

Kaolin Slip (EPK)	White color; eggshell, cracked surface. Color of celadon glaze over slip is blue-gray. Surface is transparent gloss.
Ball Clay Slip (Tenn. #5)	Yellow-white color; smooth surface. Color of celadon glaze is gray. Surface is opaque, satin-matt.

These tests show the different surface color that results from using a ball clay slip compared to that of a kaolin slip. The higher titanium and iron content of the ball clay account for the yellower cast produced with the ball clay slip and the gray, muted color of the celadon glaze over the ball clay slip. The resulting surfaces—opaque, satin-matt produced with the ball clay slip, and the gloss, transparent surface produced with the kaolin slip—are puzzling. Due to the higher alumina and lower melter content of kaolin, one would expect a less glossy and more matt surface from the kaolin slip. Note the opposite results obtained when these clays are added to a feldspar glaze (p. 115). These tests bear repeating to determine if this is a conclusive result or due to a firing temperature variable.

COMPARATIVE GLAZE TESTS of CLAYS

Cone 9–10 Reduction

Kona F-4 Feldspar	80%		Blue-gray-green.
Whiting	20		Transparent.
			Gloss. Craze.
Add		EPK	BALL (O.M.4)
10%		Color same.	Greener color.
		Flow cut.	Flow increased.
		No craze.	More craze.
		Less gloss.	Higher gloss.
20%		Lighter color.	Increased green.
		Flow cut.	Similar flow.
		No craze.	No craze.
		Satin-matt.	Lower gloss.
		Opaque.	Opaque.
30%		White color.	Yellow-green, iron
		Pinholes.	spots. Smooth,
		Satin-matt.	surface. Satin-matt.
40%		White color.	Similar to above test.
		Pinhole, crawl.	Lighter color.
		Dry surface.	Slight lowering of gloss.

Conclusions

1. Kaolin is more effective than ball clay in the production of a stony-matt surface. Ball clay produces beautiful satin-matt surfaces. Note that the addition of 10% kaolin began to cut the gloss, whereas the equivalent ball clay addition increased both the gloss and flow of the surface. Pinholing began with 30% additional kaolin and the stony-matt surface appeared at 40% kaolin. In dramatic contrast, ball clay produced satin-matt surfaces at both the 30% and 40% additions.

2. The ball clay and kaolin tests revealed marked color differences. If a white, light, or unchanged color surface is desired, then kaolin would be the best source of additional alumina. On the other hand, if deep-green, yellow-green, olive, or any other dark shade is the goal, then ball clay would be the preferred choice. Note that 10% ball clay changed a blue-gray surface to a deep green. Increased amounts of ball clay in 10% increments continued this green hue, although the cast became yellower as 20%–40% additions were made. Kaolin, on the contrary, merely lightened the color; a white surface appeared at the 30%–40% additions. Note that at the 10% kaolin addition, no real color change was apparent, in contrast to the deep green color change that occurred with 10% ball clay addition.

COMPARATIVE GLAZE TESTS of CLAYS

Cone 5–6 Oxidation
Jacky's Clear II[12]

Figures Intro.6, p16; 4.7, p271.

	Nepheline Syenite	50%	High gloss.
	Colemanite	10	Transparent where thin.
	Wollastonite	10	Semi-white where thick.
	Zinc oxide	5	No craze.
	Tenn-5 Ball	5	
	Flint	20	
	Bentonite	2	
Add	EPK KAOLIN		TENN-5 BALL
10%	White color.		White color.
	Crawl begins.		No visible crawl.
	Semigloss.		Semigloss.
	Satin surface.		Less opaque than EPK.
20%	Whiter color.		Whiter color.
	Lowered gloss.		Same as above.
	Satin surface.		Less satin than EPK.
	Opaque.		No visible crawl.
	Increased crawl.		Even coverage.
30%			Increased whiteness and opacity. Similar to EPK 10% test but no satin surface.

Conclusions

1. Unlike the 9/10 reduction tests, these ball clay tests did not produce the satin-matt surface. On the contrary, even where the surface increased in gloss, there was no satin

surface. Note that the EPK test at 10% produced a satin-matt surface.

2. The higher melting temperature of EPK kaolin was evident in the crawling tendency of the 20% and 10% tests. None of the ball clay test showed any signs of crawling.

3. There were no marked color differences. Both ball and kaolin produced a white opacity, but the kaolin test achieved this earlier at the 10% addition. Hence, EPK kaolin is the preferred source of additional alumina for the achievement of soft, white opacity at this firing temperature. In view of the different results obtained at the cone 9–10 firing temperature, I found the harsh, superficial, white surface produced by the ball clay to be surprising. Although the 30% addition did not crawl and finally produced a whiter opacity, the surface appeared paintlike and without depth, in contrast to the satin-matt opacity of all of the kaolin tests.

4. The crawling tendency of the kaolin, which marked the surface of these tests and which began at the 10% addition, could be overcome by applying a less-thick coat or by using calcined kaolin. Calcined kaolin has been prefired to about 1066°F, at which point the chemical water contained in the kaolin has burned out. The removal of this water during the glaze-firing often causes a high alumina glaze to separate and crawl over the surface of the clay form. The sluggish flow of alumina prevents the melted glaze from moving into the spaces left by the eruptions of the water vapor. Removing the chemical water from the high alumina material prior to glaze-firing eases this problem.

FIRECLAYS
Figure 2.1, p144; 2.5, p148; 2.8, p151.

Fireclays are the most important clays for the stoneware claybody and are frequently used in greater quantity than ball clays. In many stoneware claybodies, the fireclay is truly the essential core of the body. Three important characteristics of a fireclay are as follows:[13]

1. Coarse particle size.

2. High resistance to heat; minimum firing range of 2700°F–2900°F. Note that the name fireclay comes from this refractory characteristic. Fireclays are an important ingredient in the manufacture of refractory bricks.

3. A mineral content of three primary materials:

a. coarse-grained crystalline quartz or sandstone (up to 50%). A high silica content in the chemical analysis of a fireclay indicates the presence of sizable amounts of this kind of free silica.

b. fine-grained kaolinite or livesite clay mineral, which contributes to plasticity (15% or more).

c. hydrous mica of variable composition and grain size. This is the main source of alkali in the fireclay. Because hydrous mica has an average alkali content of 6.7%, a mere 2% alkali in a fireclay could indicate the presence of at least 30% hydrous mica!

In addition, there may be auxiliary minerals of limestone, nodular ironstone (pyrite), or manganese. Iron or manganese impurities in a fireclay can cause brown or black spots or even bloats to appear in both claybody and glaze surfaces at the higher firing temperatures. Thus, brown-black speckles appeared on the surface of tests in which PBX fireclay replaced 50% of A. P. Green fireclay. These dark speckles indicated the presence of iron and/or manganese nodules in the PBX fireclay *(see chemical analysis, p. 117)*. A claybody that replaced A. P. Green fireclay with equivalent amounts of Pine Lake fireclay displayed fine, black speckles on the surface of the fired ware. Note that the actual iron content listed in the chemical analysis of Pine Lake was not higher than the iron content in the chemical analysis for A. P. Green. The larger particle size of the iron mineral in the Pine Lake would not be apparent from its chemical analysis.

Fireclays absorb less water, shrink less and more evenly, and impart strength to a claybody, before, during, and after firing. Hence, it is not surprising that a plastic fireclay makes up the heart and soul of a stoneware claybody. Some stoneware formulas rely heavily on the fireclay to the exclusion of all other clays. For example, a former stoneware claybody at the New York 92nd Street Y and Greenwich House Pottery studios (New York) contained more than 60% fireclay. For many years, the standard hand-building and sculpture claybody at Greenwich House Pottery contained 80% A. P. Green fireclay. *Figure 2.1, p144; 2.5, p148.*

The harsh touch of the fireclay's grainier, coarser, particles

FIRECLAYS

TRADE NAME   **GEOGRAPHICAL SOURCE**
A.P. Green   Missouri; Harbison-Walker
Gold Art   Ohio; Cedar Heights Clay, Division of RESCO Products, Inc.
Pine Lake   Ohio
Hawthorne Bond   Missouri; Laguna Clay Co., CA (supplier)
PBX   Missouri

GEOLOGICAL   Sedimentary deposits in deep fresh water.
SOURCE   Ancient swamps metamorphosed into layers of coal and clay.

CHEMICAL STRUCTURE*

	A.P. Green	Gold Art	Pine Lake	Hawthorne	PBX (Old)	PBX(1986)
SiO_2	50.0–54.0%	56.72%	58.6%	55.23	45.7%	66.0%
Al_2O_3	29.0–33.0	27.30	27.3	38.11	31.10	24.0
Fe_2O_3	1.5–2.5	1.42	1.5	1.55	6.10	2.0
TiO_2	1.5–2.3	1.80	1.9	2.05	1.70	1.50
MgO	0.1–0.6	0.42	0.1	0.85	0.23	0.10
CaO	0.1–0.6	0.22	0.3	0.15	0.02	0.10
K_2O	0.5–1.5	1.90	0.8	1.48	0.11	0.50
Na_2O		0.20			0.03	0.03
SO_3		0.12 (other)				
L.O.I.	10.1–12.0	10.03	9.5	13.0	15.80	15.30
P.C.E.	30–31.5	29–31	29–30.0	32–33		

MINERALOGICAL STRUCTURE

MINERALOGICAL STRUCTURE	General**	Gold Art***
Crystalline quartz or sandstone (Coarse-Grained)	0.500%	20–24%
Kaolinite-Livesite (Fine-Grained)	Rarely less than 15%	Principal mineral occurs in ratio of 4:1 with Illite.
Hydrous Mica (Illite) (Fine-Grained)	Variable, 30% or more	Kaolinite + Illite make up 65%–70% of clay.
Additional Materials Carbonaceous Ironstone Titanium Chlorites Amorphous Silica and Alumina (cements rest of materials)	Variable Up to 10%	Other common minerals: muscovite (3%), sericite (less than 1%), siderite, pyrite (1/4%), and rutile (trace). Less-common minerals include tourmaline, zircon, apatite, and limonite (trace).

*Manufacturers' Data Sheets; Tichane 1990, 325; Val Cushing Glaze Class, Alfred University, 9/24/86.
**Grimshaw 1980, 300–301.
***Steven D. Blankenbeker, Technical Services, Materials Engineer, Cedar Heights Clay, letter to author, 18 October 1988, 12 August 1998.

may be softened by adding some ball clay. Fireclays such as A. P. Green contain sizable amounts of fine-grained clay minerals (such as livesite), which contribute to their general workability. (Even without the addition of ball clay, I found it possible to throw with A. P. Green alone on the wheel.)

Coarse-grained particles of sandstone and iron in the fireclay enhance the fired surface of the stoneware claybody. A. P. Green fired alone at 9–10 oxidation creates a soft, pleasing yellow-white grainy surface. The surface changes to a speckled, orange-red-brown in the reduction atmosphere. Thus, the high-fired reduction claybody owes its characteristic toasty, warm, orange-brown appearance to the presence of the fireclay. *Figure Intro.8, p18; 2.1, p144.*

The high iron content of certain fireclays such as PBX[14] (no longer available) make them a desirable addition for color and surface effects. This is particularly important in the oxidation atmosphere, where, admittedly, it is more difficult to achieve surface effects of warmth and depth. The dangers of bloats from iron and manganese nodules in these fireclays are lessened in an oxidation atmosphere.

Although fireclays are refractory materials and have slumping temperatures of cone 30 and higher, they are successfully combined with lower melting clays to produce handsome cone 5–6 oxidation claybodies. The combination of 50% A. P. Green fireclay with 25% Cedar Heights Red Art clay and 25% Jordan clay produced a warm, toasty, red-brown claybody at cone 5–6 oxidation. (See claybody formulas, p. 140; *Figure 2.3, p146; 2.4, p147*.)

As is true for all ceramic materials, the kind and particle size of the minerals in the fireclay are more important than the chemical analysis. Calcium minerals of varying particle size frequently appear in fireclays. If the particle size of the calcium deposits is not fine, they can cause bloats and blisters to appear on the surface of the ware. Ten years ago, Pine Lake fireclay substituted for A. P. Green because of bloating problems caused by calcium nodules. A few years later, Pine Lake fireclay caused similar problems in the fired ware. After extensive testing by both potters and the manufacturer, the problem was resolved by using a finer particle size for the fireclay (Kaplan 1981, 19–22; Powers 1981, 23–25). As a general rule, when ordering a fireclay, it is advisable to order the finer grind.

Once again, it is difficult to keep abreast of the mineralogical changes that frequently occur in new supplies of the same fireclay. For example, seven years ago it was reported that A. P. Green had changed over the years from a granular, relatively unplastic fireclay to a fine-grained plastic clay (Robinson 1988, 81). Tests of A. P. Green made more than five years ago confirmed this change; the fireclay proved to be surprisingly plastic and fine-grained. If increased plasticity results from additions of fine-grained kaolinite or livesite, there can be serious changes in the fit of the glaze on the fired ware. Finer-grained clay minerals release increased amounts of highly reactive silica during the firing process, which in turn converts freely into cristobalite. Excessive amounts of cristobalite can destroy the glaze fit and fired strength of the claybody (Ibid.). Although finer-grained, more plastic fireclay would surely minimize the risk of calcium bloats that coarser-grained fireclays contain, the very fineness of the particle size could result in the equally dangerous consequence of increased cristobalite formation.

It can never be mentioned too often that all clay minerals are highly vulnerable to change. The manufacturer will closely monitor all mineralogical changes in the clay if its predominant industrial use demands stability and a standardized product. This is not usually the case with fireclays, as they are used primarily in the manufacture of refractory bricks, steel, and foundry plugs. Minor mineralogical changes (such as calcium, iron, and/or magnesium nodules), which could be disastrous for the potter, would not necessarily affect their industrial usage. Studio potters do not command a sufficient percentage of the total market to ensure this kind of standardized product. The manufacturer of A. P. Green fireclay continues to warn nonindustrial ceramic customers (potential sales in this market are very low) that this fireclay is not recommended for ceramic use. "We sell this material...to use as a cheap heat-set brick mortar. In most cases it is too variable in trace element chemistry and too coarse and variable in particle size to be of any use to potters and artists. We do not recommend it for these applications..." (Dennis Schubert, 9/15/98, Harbison-Walker, A. P. Green Research). For these reasons, fireclays such as A. P. Green, which often make up more than 60% of the stoneware claybody, pose the greatest risk for the potter. In order to minimize this risk, one needs accurate information about the mineralogical and chemical structure of the fireclay. Because such data may be difficult to come by, it is advisable to test each new supply of a claybody that contains a sizable proportion of fireclay. In the final analysis, close contact with a knowledgeable and reliable manufacturer and/or supplier is the potter's best protection against the variability of fireclays (See Zamek 1989, 22).

GLAZE FUNCTION

The coarse particle size of most fireclays, together with their high firing temperature, limits their use as a glaze material. For the most part, fireclays function as claybody materials and rarely appear in everyday functional glazes. On the other hand, a combination of fireclay and melter (such as Wollastonite, Whiting, or wood ash), produces warm yellow-brown, stony, coarse-grained surfaces at the cone 9–10 firing temperatures. This surface can be used very effectively on sculptural forms. In addition, our comparative tests of different clays in feldspathic glazes show that some fireclays (such as the now unavailable PBX) provide an untapped source of unusual surface effects.

COMPARATIVE TEST OF CLAYS: FIRECLAYS

Cone 9–10 Reduction

Kona F-4 feldspar	80%	Blue-gray-green.
Whiting	20	Transparent. Gloss. Craze.

Add:	A. P. Green	PBX
10%	Color same. Grainer.	Greener-orange flecked with iron spots; grainer texture. Satin-matt surface. Less fluidity. No craze.
20%	Lighter color. White flecks. Opaque; semigloss. Grainy texture. Less fluidity.	Pale-blue gray, flecked with brown iron spots. Opaque, satin matt.
30%	Lighter color. Otherwise similar to 20% test.	Slightly greener-browner. Thin coat dark-brown, Otherwise similar to 20% test but no grainy texture.
40%	Same as 40 Ball Clay.* Yellow-green. Iron spots. Satin-matt.	Dark brown; silver spots Satin surface; Beautiful Test!

These tests show the larger particle size of the fireclays together with the iron and manganese impurities that they contain. Note that only in the 40% A. P. Green test did a satin-matt texture finally appear. However, in the PBX tests, a satin-matt surface appeared from the onset of 10% additions. Note that the all-pervasive grainy texture disappeared with a thin application in the 30% addition.

* See EPK/Ball Clay test, p.115. See also note 15.

Our tests indicate that it is possible to produce interesting textures, colors, and opacities in glaze surfaces with the addition of fireclays. This is particularly true of the high-iron fireclays. *Experiment with all of your ceramic materials and never be afraid to use them in unorthodox ways.*

GOLD ART FIRECLAY

Figure 2.1, p144.

Gold Art fireclay lies buried in the middle two feet of a six-foot seam of lower Kittaning fireclay. The layer closest to the seam abounds in iron sulfide or pyrite (FeS_2). This clay is mined for the maximum plasticity possible with the least sulfur content. However, "plasticity is sacrificed when sulfur is reduced" (Steven D. Blankenbeker, Technical Services, Ceramic Engineer, Cedar Heights Clay, July 9, 1998). Prior to the sale of the Cedar Heights Company to Resco Products Inc. in 1984, a single shipment of Gold Art fireclay could contain 0.6% sulfur. The toxic nature of sulfur fumes caused potters who fired their kiln indoors to avoid this clay. After the sale to Resco Products Inc., Gold Art clay was obtained exclusively from the middle layers of the clay deposits. By this means, the sulfur content has been reduced to 0.1%. Conceivably, the iron content of the post-sale clay should be less than the older shipments. However, a comparison of the Cedar Heights Company's chemical data sheets does not reflect this difference. Instead, the iron content of pre-1984 gold Art is listed as 1.23%, compared to the post-sale Gold Art iron content of 1.4%! Post-sale Gold Art clay appears to be less refractory than the older deposits; the more current data sheets reveal a P.C.E. range that begins at cone 29 instead of cone 31.

Gold Art fireclay accounts for only 5% of the total market of the Cedar Heights Company, and only eight days out of the entire year are devoted to the mining of Gold Art clay. (Although the clay itself is mined in six days, note that it takes three months to move the 60 feet of clay above the two feet of Gold Art.) Despite this small market, the Cedar Heights Clay Company seeks to maintain a stable mineral content in its clay product by means of careful excavation procedures (Conversation with Materials Engineer, October 1988, July–August 1998).

GOLD ART FIRECLAY

TRADE NAME	Gold Art Clay
GEOGRAPHICAL SOURCE	Cedar Heights Clay, Division of Resco Products Inc. Oak Hill, Ohio
GEOLOGICAL SOURCE	Sedimentary deposits deep in fresh water. Ancient swamps metamorphosed into layers of coal, below which are layers of clay.

CHEMICAL STRUCTURE*

SiO_2	56.7
Al_2O_3	27.3
Fe_2O_3	1.4
TiO_2	1.8
MgO	0.4
CaO	0.2
K_2O	1.9
Na_2O	0.2
other	0.1
L.O.I.	10.0%
P.C.E.	Cone 29–31

Mineralogical Structure

Kaolinite	4 to 1	} 65%–70%	Other Common Minerals
Illite	Ratio		Muscovite 3%
			Pyrite 0.5%
			Sericite 1.5%
			Siderite 0.5%
			Rutile 2.0%
Iron Oxide	7%		Less Common Minerals**
Primary Quartz	Up to 25%		Tourmaline
			Zircon
			Apatite
			Limonite

*Technical Data of Cedar Heights Clay, 8/98
Steven D. Blankenbeker, Technical Services, Materials Engineer, Cedar Heights Clay, letters to and conversations with author, October 1988, July–August 1998.
**Less than 5%.

GOLD ART CLAY AS A SUBSTITUTE FOR JORDAN CLAY

Gold Art fireclay contains a larger particle size and a lower melter and iron content than Jordan clay (see below). Comparative tests with Jordan clay and Gold Art clay showed that Gold Art clay was far less plastic. On the other hand, because of its fireclay characteristics, it possessed considerably more tooth and strength than Jordan clay. Gold Art fireclay is also more refractory; its P.C.E. (slumping indicator) is more than three cones higher than Jordan's (cone 29–31 compared to cone 23–26). It is important to take note of these properties in the event of substitution. Small additions of a more-plastic, lower-firing clay could compensate for the less-plastic and more-refractory Gold Art clay.

COMPARATIVE CHEMICAL STRUCTURE OF GOLD ART AND JORDAN CLAYS		
Chemical Structure		
	Gold Art	**Jordan***
SiO_2	56.7	68.07
Al_2O_3	27.3	19.68
Fe_2O_3	1.4	1.90
TiO_2	1.8	1.11
MgO	0.4	0.61
CaO	0.2	0.05
K_2O	1.9	2.21
Na_2O	0.2	0.04
S	0.1	
L.O.I.	10.0%	6.19%
P.C.E.	Cone 29–31	Cone 23–26
* New England Ceramic and Kiln Supply Technical Data		

Note that Gold Art clay can contain up to 25% quartz minerals. This high free-silica content combined with its low melter content could create glaze-fit problems in the event of claybody substitutions. It has been reported that these problems do, in fact, occur when Gold Art clay replaces Jordan in a claybody (Robinson 1988, 81). In addition, subtle color differences could result due to Gold Art's higher titanium and lower iron content.

It is interesting to note that despite the above differences, Gold Art clay was often substituted for Jordan clay in claybodies without further adjustments. In many of these cases, glaze-fit problems were absent. The cone 5–6 Portchester claybody is a good example.

PORTCHESTER 5–6 CLAYBODY	
Figure 2.3, p146; 2.4, p147; 2.6, p149.	
A. P. Green Fireclay	50
Red Art clay	25
Jordan clay	25
Kona F-4 Feldspar	10
Grog (medium)	10

Long before the demise of Jordan clay, many potters used Gold Art clay instead of Jordan clay in this claybody. Although the substitution of Gold Art clay for Jordan clay produced a less-vibrant color (the beautiful, warm, red-orange smooth surface of the Jordan claybody changed to a paler, browner, coarser surface), it did not create any serious glaze-fit problems. One reason for this could be the lower stoneware firing temperature; at cone 5–6 temperatures, serious cristobolite production has not yet begun. Hence, at these temperatures, it may still be possible to make substitutions without suffering the glaze-fit consequences that may occur at the higher stoneware temperatures.

Most important in any consideration of substitutions is the statement of the Cedar Heights Clay Company Materials Engineer that the Company values ceramic customers (despite the fact that they are such a small percentage of its business) and strives to maintain the quality of its ceramic clays. This emphasis on a standardized, reliable product, together with its lower melting temperature and reduced sulfur content, make Gold Art clay one of the brighter stars on the ceramic horizon.

STONEWARE CLAYS

China possesses large deposits of a white-burning, high-fire, plastic material known as "Petuntze," which naturally contains the right proportion of clay, quartz, and feldspar minerals for a high-fire claybody. The potters of the Song Dynasty created their incomparable pots with this material, and for hundreds of years, a mystified Europe sought vainly to learn their secret. Although Petuntze is uniquely Chinese, the United States possesses various deposits of high-fire clays that can be used alone as a claybody. These clays are known as "stoneware" clays. The term is potter-nomenclature and does not describe specific clay minerals. The name stoneware clays includes a combination of clays that belong to some of the clay mineral categories previously discussed, such as secondary kaolins, ball clays, fireclays, and illites. It describes clays that possess the properties of good fired color, plasticity, low cristobalite formation, and a vitrification range of cone 7–10. Clays that singly fulfill all of these requirements are rare indeed. Jordan clay, a ball clay from Poplar, Maryland, met most of these specifications. Jordan was a plastic, fine-grained ball clay with a high silica content (68%), a melter content of almost 3%, and a relatively high iron content (almost 2%). The iron mineral created a wide color spectrum at the various firing temperatures. At bisque temperatures, (cone 08–06, 1738F°–1830°F) Jordan became a rosy, salmon color that, in addition to its fine-grained particle size, made it a useful terra sigillata clay slip.[19] At cone 5–6 oxidation (2150°F–2232°F), Jordan turned a pale orange. At cone 9/10 oxidation (2300°F–2340°F), Jordan became a dull beige. In the reduction atmosphere, at cone 9–10 temperatures, Jordan turned a deep chocolate brown. This wide spectrum of color, together with its other favorable attributes, made Jordan clay an essential ingredient of numerous cone 5/6 and 9/10 claybody formulas (*see pp. 139–140*).

The Jordan clay mines in Poplar, Maryland, have closed down, and the land has been sold to real-estate developers. Hence, Jordan clay is no longer available. Potters seek acceptable substitutes for this stoneware clay, which often constituted 25% or more of many stoneware claybodies. Mention has already been made of Gold Art fireclay as a possible substitute for Jordan clay. Some other possible substitutions for Jordan clay will now be considered.

FOUNDRY HILL CREME

Foundry Hill Creme is a ball clay that, like Jordan, functions as a stoneware clay in a claybody. It is the product of H. C. Spinks Clay Company in Paris, Tennessee, which operates three processing plants—two dry plants and one slurry plant. Their clays are used in the manufacture of a wide range of industrial products. Standardized control of their clay products is maintained by a technical staff of their R&D center and by a new production quality control lab in Gleason Tennessee.

In comparing New Foundry Hill Creme to old Foundry Hill Creme, which is no longer available, note the new clay's lower silica content (59.4 compared to old Foundry Hill Creme's 66.21 and Jordan's 68.07), its higher alumina content (25.6 compared to the old clay's 20.48 and Jordan's 19.68), and the resulting higher P.C.E. of cone 31 (old Foundry Hill Creme: cone 30; Jordan: cone 23–26). Additions of feldspar or melter could be necessary if New Foundry Hill Creme replaces the old clay or any other similar stoneware clay in a claybody.

KENTUCKY STONE

Kentucky Stone is a highly plastic, fine-grained ball clay, which has replaced Jordan clay in some stoneware claybody formulas. Comparative tests of Kentucky Stone with Jordan clay showed it to be more plastic than Jordan, but with less strength. Pots made with Kentucky Stone held their shape for but a brief period of time before slumping.

Kentucky stone has been described as a "quartz-rich" clay (Robinson 1988, 81), although it contains slightly less silica than Jordan clay (67.2% compared to Jordan's 68.07%). However, Kentucky Stone now contains slightly more melter than Jordan clay (3.1% compared to 2.91% for Jordan clay). Despite this fact, note again the higher P.C.E. for Kentucky Stone. and this difference could necessitate small additions of feldspar or melter to a claybody in which Kentucky Stone replaces Jordan clay.

Kentucky Stone clay is mined by the Kentucky-Tennessee Company in Mayfield, Tennessee. This company is known for its standardized, reliable ball clays, which it

EXAMPLES OF STONEWARE CLAYS

TRADE NAME
Jordan
Kentucky Stone
Foundry Hill Creme

GEOGRAPHICAL SOURCE
Poplar, Maryland
Mayfield, Kentucky; Kentucky-Tennessee Clay Company
Paris, Tennessee; H.C. Spinks Clay Company Inc.

GEOLOGICAL SOURCE Sedimentary Ball Clays (see p.111)

CHEMICAL STRUCTURE*

	Jordan	Kentucky Stone*	Old Foundry Hill Creme* (no longer available)	New Foundry Hill Creme*
SiO_2	68.07	67.2	66.21	59.4
Al_2O_3	19.68	20.8	20.48	25.6
Fe_2O_3	1.9	1.3	1.6	1.0
MnO			0.3	
TiO_2	1.11	1.4	0.53	1.4
CaO	0.05	0.3	0.50	0.2
MgO	0.61	0.5	0.51	0.3
K_2O	2.21	1.3	0.67	0.7
Na_2O	0.04	0.1	0.55	0.2
SO_3			0.09	856 (ppm)
P_2O_5			0.09	
L.O.I.	6.19	7.1	8.21	9.8
P.C.E.	C-23—26	C-27	C-30	C-31
$\%H_2O$ of Plasticity		26.5	28	

MINERALOGICAL ANALYSIS**	**Kentucky Stone****	**Old Foundry Hill Creme*****
Kaolinite	48.0%	31.1%
Free Quartz	38.7%	36.7%
K Feldspar (potash)	7.7%	
Na Feldspar (soda)	1.7%	
Montmorillonite		14.2%
Mica		12.5%
Organic Matter	0.8%	2.6%

*Kentucky-Tennessee Clay Company 1998; H.C. Spinks Clay Company Inc., 1988, 1998.
**Bradley K. Lynne, Product Manager, Kentucky-Tennessee Clay Company, letter to author, 4 October 1988; No change 1998.
***Mark Alexander, Assistant Technical Director, H. C. Spinks Clay Company Inc., letter to author 3 November 1988.

ships throughout the United States. The excellent reputation of this company is an important consideration in the selection of Kentucky Stone for a claybody.

GLAZE FUNCTION

It should be apparent by now that most ceramic materials can function as glaze materials, provided appropriate adjustments are made to the glaze formula. Although Jordan clay is no longer available, the following tests of Jordan clay are of interest here because they show how a stoneware clay combines with a feldspar and calcium melter to produce satiny glaze surfaces. Note the similarity to the ball clay tests. Jordan clay is, in fact, a ball clay, and its kinship to the ball clays is revealed in the following test series.

Kona F-4	80%	Gray-blue.
Whiting	20	Transparent. Gloss. Crazed.
Add:	**Jordan**	**Ball O.M.-4**
10%	Pale gray green. Semigloss. Matt-shine. Crystalline formation.	Greener color. Slightly higher gloss. Larger crazing. Increased flow.
20%	Lighter color. No craze. Flow cut. Semigloss.	Increased green color. No craze. Similar flow. Lower gloss. Opaque surface.
30%	Gray green. Satin matt.	Yellow green. Smooth, satin surface.
40%	Lighter gray green. Stonier but otherwise similar to test II of Cornwall, Whiting and 20% Ball clay (p. 263).	Similar to 30% test but slightly lower gloss. Lighter color, but subtle difference.

Two things come to mind in comparing these two tests:

1. Both Jordan and ball clay produce a greenish color.
2. Both Jordan and ball clay produce a satin-matt surface. Surprisingly, Jordan, which has a higher melter content than ball clay, already lowers the gloss at the 10% addition, whereas ball clay does not produce this result until the 20%

addition. Particle size might account for this result. As a general rule, the larger the particle size, the more refractory the material. Thus, even though ball clay has less melter than Jordan, its smaller particle size could cause it to activate the melt more vigorously than Jordan.

According to the Manufacturers Data:

| Particle size finer than .5mm | Jordan 30.5% | O.M.-4 55% |
| P.C.E. | c/23–26 | c/32 |

These tests should be repeated in order to rule out firing temperature variables.

3. Although both Jordan and ball clay produce a green color, Jordan produces a gray-green shade and ball clay produces a yellow-green at the 30% additions. It is possible that the combined effect of lower iron (0.08) and higher titanium (1.6) in the ball clay has caused the yellower cast of the 30% clay tests; correspondingly, the higher iron (1.9) and slightly lower titanium (1.1) content of the Jordan clay accounts for the gray-green shade. According to the manufacturer's Data Sheets, O.M.-4 has changed its iron and titanium content over the years. At the time these tests were made:

Old Data Sheet of O.M.-4
TiO_2 1.6–1.8
Fe_2O_3 0.8–1.0

Current Data Sheet of O.M.-4
TiO_2 1.2
Fe_2O_3 1.1

In any event, irrespective of the reasons, the important point is that different color results were obtained with these two clays.

CONCLUSION

The passing of Jordan clay has destroyed many of our long-working claybodies and underscores the danger of heavy dependence on a single clay material. Jordan's abrupt disappearance bears witness to the ephemeral nature of our clay materials and the inevitability of change and loss. For an illuminating account of problems inherent in mining procedures in the excavation of clays, see nine articles in *Studio*

Potter 1981, 4–33. Note the excellent suggestion made by the late Richard P. Issacs of Hammill & Gillespie, Inc. that potters request test samples before purchasing large supplies of particular claybodies, clays, or other ceramic materials (Ibid., 33). This is a practical method for avoiding the disasters inherent in the variability of clay minerals. As a further safeguard, potters should investigate the major industrial usage of the ceramic clays in their claybodies so that they can evaluate their reliability. This kind of knowledge, together with pre-purchase test samples, is the most practical method for avoiding future disasters.

IRON-RICH, MELTER-RICH CLAYS

ILLITES AND CHLORITES
Intro.4, p14; 2.1, p144; 2.6, p149.

Thus far, the clays discussed have contained large amounts of the clay mineral kaolinite. The chemical structure of these clays contained low amounts of iron and melters and relatively high amounts of alumina. Hence, they would function primarily as high-temperature claybody materials and as an economical source of alumina in glazes.

We come now to a group of clays that are higher in iron, silica, and melter materials. The predominate clay mineral in these clays is illite and/or chlorite.[16]

Illite clay minerals are a fascinating example of the transformation of mica into each of the three major groups of clay minerals, kaolinite, montmorillonite, and illite. The structure of illite is similar to that of muscovite mica, except for the fact that illite contains less potassium and more silica and water. It is believed that illite clay minerals are an intermediate stage in the alteration of muscovite mica to kaolinite or montmorillonite. As a result of physical and/or chemical weathering, muscovite mica can change into illite. The illite clays in the Hudson River Valley Region (Albany Slip clay) are a good example of the result of this process. Further weathering action transforms muscovite-illite into either kaolinite or montmorillonite. The relative instability of the illite clay mineral—the fact that it is usually in the process of becoming something else—is reflected in its disorderly crystalline structure and complex chemistry.

Clays such as Albany Slip in which illite is the major clay mineral are often secondary deposits—transported by the weathering forces of glaciers, rain, snow, etc. to streams, riverbeds, and glacial lakes. Consequently, these clays have picked up many additional minerals in the course of their travels and possess a complex and variable mineralogical composition, which in addition to illite, includes fine-grained alpha quartz, limonite (iron oxide), chlorite, calcite, and dolomite. Accordingly, the chemical composition of illite clays is free-ranging and, in addition to silica, alumina, titanium, iron, calcia, magnesia, soda, and potash, may include barium, strontium, sulfur, lithium, manganese, and lead! The high melter, iron, and silica content of illite clays makes them the essential core of the earthenware claybody. They function as a melter and/or colorant in the stoneware claybody. Because the illites are rich in iron and melters, they provide the potter with a wealth of glaze core materials. Their silica-alumina-melter oxide ratio is often similar to that of a feldspar, and consequently, illites behave comparably to feldspar when fired. For example, Albany Slip clay melts to a shiny glass at cone 9–10 temperatures and to a soft, satin-matt at cone 5–6 temperatures.

On the other hand, illite clays differ sharply from feldspars in their iron content, which can constitute as much as 8% of their total chemical structure. In contrast, feldspars contain as little as 0.2% iron. Manganese minerals frequently appear in illite clays and are rare occurrences in feldspars, if at all. Thus, illite clays are a natural choice for the achievement of rich, dark surfaces. Red-black to purple-red surfaces, flecked with silvery, iron spots are produced at cone 9–11 reduction temperatures with the illite clay Albany Slip. In the oxidation atmosphere, the surface changes in color to a glossy black-red-brown; at 5–6 oxidation temperatures it becomes a yelow-brown-green satin-matt. *Figure Intro.4, p14.*

Surfaces produced by illite clays are often unpredictable, due to their variable mineral content. A safer approach would be to doctor a colorless feldspar with additions of dark-burning (and often highly toxic) metallic oxides. Note that it is these very same mineral impurities in the illite clay minerals that produce ceramic surfaces of incomparable depth and complexity. Thus, once again, the price of a gemstone surface may be an occasional failure.

ILLITES AND OTHER IRON-BEARING CLAYS

TRADE NAME	**GEOGRAPHICAL SOURCE**
Cedar Heights, Red Art Clay	Ohio, Cedar Heights Clay, Division RESCO Products, Inc.
Albany Slip Clay	Hudson River Valley. No longer available. Albany, Kingston
Sheffield Clay	Sheffield. Mass., Sheffield Pottery Inc.
Blackbird-Barnard Clay	Pennsylvania, Hammill & Gillespie, Inc.
Ocmulgee Red Clay	Macon, Georgia *No longer available.*

GEOLOGICAL SOURCE

Cedar Heights, Red Art Clay	Shale, prehistoric lake bed.
Albany Slip Clay	Wisconsin Ice Age (10–15,000 years ago); shale and slates, deposited into glacial Albany lake.
Sheffield Clay	Wisconsin Ice Age; deposits in glacial lake of altered biotite mica, hornblende, and other high-iron magnesium rocks.
Blackbird-Barnard Clay	Glacial detritus
Ocmulgee Red Clay	Sedimentary deposition by currents of water (alluvial)

IDEAL CHEMICAL STRUCTURE*

$(OH)_4K_2(Si_6Al_2)(MgFe)_6O_{20}$ ILLITE

$(Mg \cdot Fe^{+2} : Fe^{+3})_6 \, AlSi_3O_{10}(OH)_8$ CHLORITE

MINERALOGICAL STRUCTURE

Cedar Heights Red**	Illite 40%, Kaolinite 10%, Free Quartz 30%, Mixed Layered Clays 15%, Red Iron 7%.
Albany Slip***	Illite, Free Quartz, Limonite, Chlorite, Calcite, Dolomite, Kaolinite.
Sheffield***	Chlorite (primary), Biotite Mica, Alpha Quartz.

*Hosterman 1984, C8; *Glossary of Geology* 1974, 123.
**Cedar Heights Clay Company 10/18/88, 7/98.
***Murtagh.

CHEMICAL STRUCTURE†

	Red Art	**Albany**	**Sheffield**	**Barnard‡**	**Ocmulgee**	**Lizella Red**
SiO_2	64.27	57.40	51.02	59.70	53.860	69.60
Al_2O_3	16.41	14.66	19.58	10.87	21.820	21.40
Fe_2O_3	7.04	5.20	8.46	14.65	6.400	3.90
TiO_2	1.06	0.8–1	0.97	0.67	1.196	1.10
CaO	0.23	5.78	3.20	0.27	0.370	0.77
MgO	1.55	2.68	3.53	0.75	0.793	0.85
K_2O	4.07	3.25	4.05	2.04	1.285	1.40
Na_2O	0.40	0.8–1	1.20	0.12	0.358	0.85
MnO	—	—	—	3.40	0.230	—
Ba,Sr,Pb,Li,Mn	—	0.13	—	—	—	—
L.O.I.	4.92	9.46		7.48	8.9	8.50

†Technical Data: Cedar Heights Clay 1998; Sheffield Pottery Inc. 1998; Hammill & Gillespie 1989, No material change 1998; Burns Brick Co. 1989; Lizella Clay Co. 1998.
‡Also present BaO, PbO, ZrO_2, SrO, NiO.

ILLITES AND OTHER IRON-BEARING CLAYS (CONT.)			
COLOR-TEXTURE			
Cone	**Red Art**	**Albany**	**Barnard**	**Ocmulgee**
04 o	Deep salmon. Matt.	Pale pink- brown. Matt.	Chocolate. brown. Matt.	Salmon. Matt.
5–6 o	Red-brown. Semi-gloss.	Brown. Split. Bloated. Semigloss.	Deep brown-black. Matt.	Red-brown. Matt.
9–10 o	Mahogany, Brown, Glassy.	Black. Gloss.	Black. Melted matt.	Purple-red- brown. Matt.
9–10 r	Mahogany, Bloated. Glassy.	Black with brown spots. Glassy.	Blue-black. Melted matt.	Black. Gloss. Pitted.

o = oxidation; r = reduction

ALBANY CLAY—AN OBITUARY
Figures Intro.3–Intro.4, pp13–14.

The Wisconsin Glacier deposited the crumbled remains of 500-million-year-old shales and slates from the Taconic mountain range into the Albany glacial lake 15,000 years ago. Illite was the major clay mineral in these deposits. The glacier also transported to this lake the weathered remains of 425-million-year-old limestones and dolomites from the weathered Helderberg mountain range. Albany Slip clay derives from these deposits (Murtagh).

Albany Slip clay has been a favorite glaze material for hundreds of years. We are all familiar with the glossy, brown-black surface of Albany Slip clay on the bean pots, crocks, and storage jugs of the early nineteenth and twentieth centuries. The low fusion point (cone 3) and high silica content (57.64%) of Albany Slip clay made it a natural glaze core. Alone, without the assistance of additional melters, it melted to a brilliant brown-black glass at temperatures of 2300°F in an oxidation atmosphere and produced a mahogany, glossy surface in reduction atmospheres. At cone 5–6 oxidation temperatures, it created a greenish, brown, semigloss surface. Albany Slip clay functioned as a glaze core at earthenware temperatures with the aid of additional melters. Thus, it was an extremely useful material for the production of richly colored, dark surfaces at various firing temperatures.

Albany Slip clay was not generally used as a claybody material because of its variable impurities content. These impurities often included a fair amount of sulfur together with trace amounts of barium, strontium, lithium, manganese, and lead. Impurities such as sulfur can give depth and luster to a glaze surface; when they appear in a claybody, they can cause blisters and bloats. Tests of a cone 5–6 claybody with 20% Albany Slip clay produced a smooth, red-brown claybody. None of these tests contained bloated or blistered surfaces. However, these were only preliminary tests. Further tests were not performed due to the closure of the Albany Slip mines.

According to Industrial Mineral Products Inc., the producers of "true Albany Slip clay," the mine had reserves for another 50 years. (The data sheet obtained in 1986 from Industrial Mineral Products Inc. stated that these "unique glacial clays are widespread in the Hudson Valley but clay of uniform properties is limited to a small area in the city of Albany...reserves for another fifty years are in sight in just one section of our holdings.") However, the real estate boom of the 1980s caused the land to become more valuable than the clay supplies that it contained. The result was the sale of the land to real estate developers, and, once again, potters have lost a historic and invaluable glaze material.

The manufacturing process of Albany Slip clay was highly complicated, and this may have hastened its economic demise. Albany Slip clay was mined in Albany, trucked to Bowmanston, Pennsylvania, for refining and grinding, and finally shipped to Hammill & Gillespie in New Jersey for distribution to ceramic suppliers. At each step of the way, a fee was tacked onto the price of the clay, with the result that Albany Slip clay ended up three times the price of Red Art clay and more expensive than imported Grolleg clay. The 1987 supply sheet produced by Hammill & Gillespie quoted Grolleg porcelain clay at $429 per unit, Cedar Heights Red Art at $187 per unit, and Albany Slip clay at $530 per unit!

The demise of Albany Slip clay confronts the potter once again with the ephemeral nature of clay and the necessity of finding acceptable substitutes. Slip clays that are similar, although not identical, to Albany Slip clay are mined in other parts of the country. A slip clay dug by a student in the Hudson Valley, New York, area produced a red-brown, smooth, stony-matt surface at cone 9–10 reduction temperatures. The surface lacked the luster and sheen produced by Albany Slip clay, possibly because it had not gone through the refining process. The efforts of three potters to replace Albany Clay are described in *Ceramics Monthly* 1988 (October) 49–50. Two of these potters found Red Art clay to be a possible substitute when combined with additional ceramic materials. The third potter has found a satisfactory replacement in deposits of sedimentary Redearth, known as Alberta Slip and supplied by IMC in Medicine Hat, Alberta, Canada.[17]

An artificial mixture of ceramic materials that duplicated the chemical structure of Albany Slip clay did not produce the brilliance nor the glossy depth of color provided by the natural clay material (Ibid., 49).

Sheffield Pottery Inc. of Sheffield, Massachusetts, offers its iron-bearing slip clay as a possible substitute (see pp.134–135).

Potters who have not been able to find an acceptable substitute for Albany Slip should take heart. For as long as it lasts, a limited supply of Albany Slip is now available. Great American Wheelworks in New Baltimore, New York, has acquired a 20-ton supply and offers a run of the mine 50-pound bag for $4.50 per pound (see classified advertising section, *Ceramics Monthly*, February 2000, 103).

Glaze Function
Figure Intro.3, p13.

Cone 9–10 Reduction

Albany Slip	50%
Nepheline Syenite	50%

Black, satin-matt surface flecked with lustrous silvery spots if fired in semi-oxidation atmosphere to cone 9 temperature on stoneware claybody. Same result obtained on porcelain claybody, but yellow gloss where glaze is thin, instead of brown-red. A dense, mahogany colored, satin-matt surface results when fired in reduction atmosphere to cone 10–11 temperature on both stoneware and porcelain claybodies.

The iron and other impurities (such as sulfur) in Albany Slip clay create the silvery black surface (subject, of course, to the conditions of atmosphere and firing temperatures). Note that this surface is highly variable. We fired the same glaze in three kiln firings; only one kiln produced the black, silvery surface on all the tests. The other two kilns resulted in a dense mahogany colored surface.

Despite the fact that the amount and kind of impurities can vary in each new supply of slip clay, the silvery black gemstone surface achieved in one kiln firing makes this simple combination of slip clay and feldspathic material worth trying over and over again.

Cone 5–6 Oxidation

Albany Yellow		Yellow satin-matt.
Albany Slip	42.25%	
Nepheline Syenite	22.75	
Gerstley Borate	17.50	
Dolomite	17.50	
Tin Oxide	3	

Albany Slip clay shares a glaze-core role with Nepheline Syenite, as it does in the cone 9–10 reduction test. The Albany Slip provides the yellow color[18] and the satin-matt texture of this glaze, and thus modifies the white, gloss, contributed by Nepheline Syenite. If you omit Albany Slip, the surface becomes a shiny, white opacity. If you omit Nepheline Syenite, the surface becomes a mustard-colored, stony-matt.

CEDAR HEIGHTS RED ART CLAY
Figure 2.1, p144; 2.6, p149.

Red Art clay consists of clay-shale deposits located in Ohio. A 1988 mineralogical analysis provided by the Materials Engineer of the Cedar Heights Clay Division of Resco Products Inc. states that the mineral content of Red Art clay is about 40% illite, 10% kaolinite, 15% mixed layered clays, and 7% red iron oxide. This clay is the exclusive property of the Cedar Heights Clay Company. It is a relatively inexpensive clay. The 1987 Hammill & Gillespie supply sheet listed its price as $187 per unit, compared to $530 per unit of Albany Slip clay. Red Art clay appears to be extremely stable in composition. The chemical structure that appears in a current chemical data sheet is basically unchanged from the chemical analysis of 10 years ago. It is larger in particle size than Albany Slip clay, contains more iron and potash and has far less of the alkaline earth melters such as calcium and magnesium. Albany Slip clay contained a sizable amount of calcium (5.78%) and magnesium (2.68%). Red Art contains a mere 0.23% calcium and 1.55% magnesium. Red Art has a higher firing temperature than Albany. At cone 5–6 oxidation (three cones past the slumping point of Albany) it retains its shape and displays a deep brick-red color. At comparable temperatures, Albany has lost its shape and has turned a glassy, dull brown.

Red Art clay has a relatively high shrinkage rate. At cone 01, its shrinkage is almost 13% with a 9.93% absorption rate. At cone 1 its shrinkage is almost 14%. At cone 3, the shrinkage is well over 14%, with a 1.12% absorption rate.

Our workability tests showed that it was not as agreeable to work with as Jordan, Gold Art, or A. P. Green fireclay. It felt unpleasantly sticky to the touch, dried very rapidly, and cracked apart at rims and seams.

Red Art clay makes up the core of many earthenware claybodies and often appears as an important auxiliary clay in stoneware claybodies. For example, Red Art clay constitutes 25% of the handsome cone 5–6 oxidation Portchester claybody. The high shrinkage and absorption rates of this clay require the addition of coarser-grained, more open clays, such as A. P. Green fireclay, which makes up 50% of this claybody, and Gold Art fireclay (25%). In addition,

more plastic clays such as Jordan (25%) are needed to counteract the poorer workability of Red Art clay. The warm orange-red-brown surface, which was the significant characteristic of this claybody when Jordan was present in the formula, derived from the high iron content of Red Art clay (7%) and Jordan clay (1.9%). Gold Art has less iron than Jordan (1.4%), and this could account for the duller, browner color of the claybody with the Gold Art substitution. A small increase in the Red Art content might help to restore the warm, red-brown shade of the claybody.

According to chemical analysis sheets, this iron content of Red Art clay has not varied over the last 20 years and still remains at 7%. The color of Red Art clay fired alone is described by the manufacturer as "medium brick red" at cone 04, "rich red" at cone 02, and "dark red" at cone 1. My tests of Red Art clay, which were made in the oxidation atmosphere, produced a pale brick color at cone 06; at cone 5–6 the color became a deep brown-red, and at cone 9–10 the color changed to a deep mahogany.

According to the manufacturer's data, Red Art clay has a long firing range (P.C.E. cone 13), and our tests corroborated this fact. It was only at the cone 9–10 temperature that the surface of Red Art clay became glassy and bloated. Although Red Art clay reached a maximum vitrification point at cone 3 (absorption of 1.12%), our tests showed that at cone 5–6 it held its shape without apparent bloating or bubbling, and even the cone 9–10 test did not show any slumping. Amazingly enough, two small pinch pots of Red Art clay retained their shape and form even at cone 9–10 reduction firing temperatures. These small pinch pots presented a blistered, metallic surface and stuck to the kiln shelf, but their round, full shape remained intact. This is a tribute to the extraordinary firing range of Red Art clay.

Glaze Function

Red Art clay contains 64% silica (of which 25%–30% is primary quartz), low melter content (6% total, mostly potassium, which reflects the original muscovite origins of the illite clay mineral), and a high iron content of 7%. Until the disappearance of Albany Slip clay, Red Art clay was not often included in a glaze formula. However, since the demise of Albany Slip clay, Red Art clay may well appear in some glaze formulas that previously contained Albany Slip clay. If Red Art clay is properly deflocculated, it produces a brick-colored terra sigillata, and as such, it has successfully substituted for Albany Slip clay in a glaze (Rowan 1988, 48). The deflocculating process of terra sigillata results in the selection of the finest particles of Red Art clay, which will then fuse at a lower temperature than the original Red Art clay. In this manner, and despite the fact that Red Art clay is more refractory than Albany clay, a successful substitution of Red Art clay for Albany clay has been accomplished.

Long before the demise of Albany clay, tests made by my students clearly revealed Red Art's potential for creating interesting color and surface effects in glazes.

CONE 9/10 REDUCTION

Kona F-4	80%	Blue-green-gray. Transparent. Gloss.
Whiting	20	Crazed.
Red Iron	1/2%	

Add Red Art	
10%	Deeper green color.
20%	Same as above; slightly increased fluidity and gloss.
30%	Satin-matt surface. Noncraze. Gray-green-yellow color. No fluidity.
40%	Denser, stonier surface. Gray cast predominates.

Amounts of 10%–20% Red Art clay added to this simple feldspathic glaze deepened the green cast and slightly increased the fluidity and gloss of the glaze. 30%–40% additions matted down the glassy surface and dulled the color. The result was a soft, dense, gray, satin-matt surface, which contrasted strongly with the glassy blue-gray of the original glaze.

Additions of 5%–20% Red Art clay to Jacky's Clear glaze resulted in a greener cast and a fatter, softer surface.

CONE 5-6 OXIDATION

Jacky's Clear II
Figure Intro.6, p16.

Nepheline Syenite	50%	Transparent.
Colemanite or Gerstley Borate	10	Craze.
Wollastonite	10	High gloss.
Flint	20	Yellow cast.
Tenn. Ball Clay	5	
Zinc oxide	5	
Bentonite	2	

Add Red Art	
5%	Slight deepening of yellow cast; larger craze.
10%	Increased yellow cast; increased craze. Trapped bubbles where glaze thick. Increased flow.
15%–20%	Deeper yellow color, trapped air bubbles throughout entire surface. Pitted surface; slight craze; increased flow.
20%–30%	Yellow-brown. Satin-matt surface, No craze visible.

The addition of 30% Red Art clay to this gloss, transparent glaze creates a soft, yellow-brown, satin-matt surface. Lesser additions deepened the yellow cast of the original glaze. Thus, even at the lower temperatures in the oxidation atmosphere, Red Art clay can function effectively as an iron-bearing glaze material.

I have found Cedar Heights Red Art clay to be useful in glazes in less-traditional ways. One can apply thin slabs of Red Art to the stoneware claybody, and then apply a gloss glaze over all or some of the attached Red Art clay slab. The result, when fired to cone 9/10 reduction, is a bubbled texture and a dark-brown to black lustrous surface. A smooth, highly colored surface appears at cone 5/6 oxidation, which contrasts strongly with the original stoneware body, both glazed and unglazed. Although all of the various clays described thus far provide interesting textural contrasts to the regular stoneware claybody, Red Art clay proved to be the most dramatic in the series.

Thin slabs of Red Art clay sandwiched between different colored clays, such as fireclay, kaolin, and the stoneware claybody, created a fascinating surface when the total clay sandwich was cut and rolled, at both 5–6 and 9–10 oxidation and reduction temperatures. The deep reddish-brown, or mahogany Red Art color provided a dramatic contrast to the white, sandy, and buff-colored clays. The resulting multi-striped patterns were reminiscent of southwestern landscapes. *Figures 2.3–2.4, pp146–147; Figure 2.6, p149.*

Thus, although Red Art clay is primarily used as a claybody material, it has interesting possibilities as a glaze material. With the aid of chemical and mineralogical data, you can explore this potential. Once again, knowledge of your materials frees you from the rigid bonds of prior use and propels you toward the achievement of unique ceramic surfaces.

OCMULGEE RED CLAY—AN OBITUARY

Ocmulgee Red Clay was a sandy, coarse, high-iron clay that was mined by the Burns Brick Company for the manufacture of bricks and pipes.[21] As of 1989, ceramic suppliers made up less than 1% of the total market (Burns Brick Company, 1989). Unfortunately, this clay is no longer available.

With the exception of its high-iron and manganese content, its chemical structure appeared closer to fireclay than to earthenware. In particular, note that the total melter content of Ocmulgee Red clay (2.81%) was more than 50% lower than Red Art clay (6.25%) and in this respect was closer to Gold Art fireclay's low melter content of 2.67%. The same resemblance to a fireclay held true for Ocmulgee's lower silica content (53.56% compared to 64.27% silica content of Red Art clay and 56.72% for Gold Art fireclay). Yet, despite these similarities and because of its high iron content, Ocmulgee Red clay was often classified with Red Art clay. This illustrates the difficulty of fitting ceramic materials into neat, tidy pigeonholes. Classification of clays as earthenware, stoneware, or fireclay, although useful as an organizational tool, can, if taken too literally, inhibit further experimental efforts at different firing temperatures. The point, after all, is not whether a clay is labeled "earthenware," "stoneware," or "fireclay," but rather *how* it functions in a particular claybody at a specific firing temperature.

Claybody Function

At low stoneware temperatures, Ocmulgee Red clay displayed a rich, iron-red color and a coarse, sandy texture. At cone 5–6 oxidation, Ocmulgee Red clay became a deep red-brown. The broad firing range of this clay extended to the higher stoneware reduction firing temperatures. At cone 9–10 reduction, although the color of the surface blackened, all of the small pinch pots held their shape without any visible bloats or deformation. According to the manufacturer, the range of the clay was from cone 5 to cone 10. At cone 8–9 its color was dark blue-gray in reduction and dark red in oxidation. At that temperature there was almost no absorption (Neil Struby, Burns Brick Company, letter to author, November 1989).

The following tests illustrate in specific detail the unique properties of Ocmulgee Red clay as a claybody material.

TESTS

I

Cone 5–6 Oxidation

Ocmulgee Red clay	85%	Brick-red color.
Ball clay O.M. #4	15	Grainy texture.
Bentonite	3	

II

Cone 5–6 Oxidation

Ocmulgee Red clay	60%	Orange-salmon color.
A.P.Green	20	Black speckles.
Ball clay O.M. #4	20	Smoother texture than I.

Cone 9–10 Oxidation

Purple-red-brown color. Specks of black, some glassy.

Claybodies I and II plastic and good for hand building, but too coarse for wheel.

III

Cone 9–10 Oxidation

Ocmulgee Red clay	1 lb.	Warm orange-brown.
Ball clay O.M. #4	1/2 lb.	White speckle.
Jordan clay	1/2 lb.	Grainy texture.
Spodumene	1/8 lb.	

Claybody plastic and good for wheel.

Claybody Colorant

Ocmulgee functioned as a colorant in certain high-fire stoneware bodies. The stoneware, hand-building, sculpture claybody used by Greenwich House Pottery replaced Red Art clay with 4% Ocmulgee Red clay. A similar percentage of Ocmulgee Red clay was part of a stoneware claybody used at the Portchester Clay Art Center.

Glaze Function

A brown-green, black, mottled, glossy surface, streaked with black and brown crystals resulted from applying Jacky's Clear (gloss transparent glaze) over an Ocmulgee Red clay pinch pot at the cone 9–10 reduction temperatures. Similar results appeared in an oxidation firing at the same temperatures. In this test, the combination of the transparent Jacky's Clear glaze and underlying Ocmulgee claybody (p. 131, test III) created a deep, rich, black, glossy surface flecked with red-brown iron spots. This suggested that Ocmulgee Red clay would function successfully as an iron-bearing glaze core in combination with feldspars and additional melters.

Conclusion

The preceding tests performed with Ocmulgee Red clay at cone 5–6, cone 9–10 oxidation, and cone 9–10 reduction temperatures revealed a far-ranging potential as both a claybody and glaze material. Alas, this potential will never be realized because of its disappearance from the ceramic scene. The substitute recommended by Sheffield Pottery Inc. is Lizella Red clay, produced by Lizella Clay Company in Lizella, Georgia. Note the lower iron content (3.90 compared to 6.40), the absence of manganese, and its higher silica content (69.6 compared to 53.9). It also contains a somewhat higher melter content. Hence, a substitution of Lizella for Ocmulgee in a claybody or glaze would require some adjustments.

BARNARD CLAY

Barnard clay is processed exclusively by Hammill & Gillespie "from crude clay stockpiled in the woods of Western Pennsylvania" (Dorna L. Isaacs, Vice-President, Hammill & Gillespie Inc., letter to the author, 28 March 1989). The chemical analysis sheet describes Barnard Clay as a "siliceous kaolin type clay containing iron and manganese." Trace amounts of barium, lead, zirconium, nickel, and strontium are also said to be present. It is said to burn "dark brown at cone 02 to black at cone 7."

Over the years, the chemical structure of Barnard clay has remained unchanged, and this indication of stability, together with its high-iron and manganese content, makes Barnard clay a valuable ceramic material for the potter.

Barnard clay is classified as a glacial and boulder clay. The powerful grinding action of giant glaciers on the earth's surface produced a superfine, sticky clay mixed with sand, rocks, and gravel. After these impurities are removed from Barnard clay, it looks and feels like fine, dark-brown soil.

Despite the fine particle size of Barnard clay, it is remarkably unplastic. Small pinch pots of Barnard clay cracked and fell apart during the forming process. Because this clay lacks the all-important property of plasticity, its primary ceramic function in both claybodies and glazes is that of a colorant. In this capacity, as a carrier of iron and manganese, Barnard clay creates handsome black glazes and claybodies at cone 5–6, and 9–10 temperatures, in both oxidation and reduction atmospheres. *Figures 2.2; p145; 2.6, p149.*

Claybody Function

Barnard clay has functioned as a colorant in the handbuilding, sculptural claybody of Greenwich House Pottery, known as T-1. The following formula represents one of the prior variations of the T-1 claybody (see pp. 139–140; *Figures 2.5, p148; 2.8, p151*).

Fireclay (A. P. Green)	55%
Kentucky Ball clay	8
Barnard clay	4
Bentonite	3
Grog	30

This claybody, when fired in a properly reducing atmosphere, produced a toasty, warm orange color. In recent years, Red Art clay, and subsequently, Ocmulgee Red clay, have replaced Barnard clay as the colorant material.

The following formula, with 10% Barnard Clay and 2% Cobalt Oxide, resulted in a smooth, black, plastic claybody. This claybody was best fired at cone 4–5. When fired to cone 6, its surface sometimes contained tiny bloats and blisters. *Figure 2.2, p145.*

Jordan clay	40
Ball clay	15
Barnard clay	10
Custer Feldspar	25
Rotten Stone	20
Cobalt Oxide	2

The mysterious, meteor-like surface of Barnard clay fired to stoneware temperatures reveals the potential of this clay for the creation of dramatic, sculptural surfaces. Thin slabs of Barnard clay can be sandwiched between different colored clays such as porcelain, Red Art, and a basic stoneware claybody to produce a multilayered surface of white, buff, red, and black layers (see p. 130). The blackness of the Barnard clay contrasts dramatically with the white porcelain and the brown, red-brown, or buff of the other clays. *One note of caution:* Barnard clay fuses by the time cone 4 is reached, and for this reason, in order to protect the kiln shelf, the bottom clay layer should be a more refractory clay, such as porcelain or the stoneware claybody. Even when surrounded by more refractory clays, Barnard clay can migrate upwards and outwards toward the surface of the multilayered clay form. At cone 9–10 temperatures, Barnard clay bubbled and solidified into an irregular, beaded pattern, breaking out of the orderly black lines that it formed at the cone 5–6 temperatures. Thus, at the higher stoneware temperatures of cone 9–10, even greater care must be taken, and it is advisable to surround thin Barnard clay layers with thicker slabs of more refractory clays. Though extra care must be exercised at both 5–6 and 9–10 temperatures when using Barnard clay in this manner, the resulting decorative surfaces are extraordinary and well worth the trouble. *Figure 2.6, p149.*

WARNING: *Barnard clay contains 3.4% manganese dioxide, which is highly toxic.[22] Manganese is particularly dangerous in a claybody where there is close contact between the hands of the potter and the clays of the claybody. The clay dust of a claybody that contains manganese can be harmful to breathe. It goes without saying that adequate safety precautions should be taken, including an industrial mask, and most important of all, a dust-free, clean, and properly ventilated work area.*

Barnard Clay as a Slip

Barnard clay functions successfully as an intermediate layer of dark color between glaze and claybody—a clay slip. Although the results are hardly as dramatic, this method of using Barnard clay is more controllable than the clay sandwich described above. A soft brush or dip of Barnard clay mixed with water upon a cone 5–6 oxidation claybody results in a stony black surface. At cone 9–10 reduction, the surface becomes a metallic, silvery black. This result is especially handsome on carved surfaces. A transparent glaze applied over a coat of Barnard slip will take on the dark colors of Barnard clay at all firing temperatures. When the finest particles of Barnard clay are utilized by means of the terra sigillata process, a fine, black semigloss surface results at earthenware temperatures.

Glaze Function

The manganese and iron minerals contained in Barnard clay make it a useful material for the creation of dusky, black glaze surfaces. Barnard clay fired alone produces a coal, blue-black, cratered, and bubbled surface at cone 9–10 reduction temperatures. The interiors of the craters and bubbles reveal a bluish-black, metallic, silvery surface. At cone 5–6 oxidation temperatures, Barnard clay displays a black, dry, stony, cracked surface. However, when applied to a white claybody and fired to a high cone 6 temperature, Barnard clay produced a handsome, smooth, black, metallic surface.

Glaze Colorant

Sanders Celadon—the most stable of all the celadon glazes tested—contains 3%–5% Barnard clay for the achievement of the gray-blue-green color. This muted celadon color is similar to the color of ancient Japanese celadon glazes. *See Figure 1.5, p82.*

"Barnard clay is similar to many of the rocks and clays containing iron and manganese used by Japanese potters. It may be used effectively for underglaze iron decoration and should be tried as a possible substitute for #2 Yusha, Okawachi stone, Kamo River stone and #2 Shaju."

(Sanders 1982, 242)

CONE 9-10 REDUCTION

Figure 1.5, p82.

Sanders Celadon

Gray-green, transparent, gloss.

Kona F-4	44%
Whiting	18
Flint	20
EPK	10
Barnard clay	3–5

Barnard clay functions in this glaze as a colorant—a carrier of iron and manganese. It provides the gray-green shade of the celadon.

Omit Barnard clay	Surface is bluer, more
Substitute 1/2%	brilliant in color and less
Red Iron Oxide	gray-green.

Comparative Tests of Red Iron Oxide and Barnard Clay

Nepheline Syenite	90%		
Whiting	10		
ADD			
Red Iron Oxide	1%	2%	3%–4%
	Gray blue color. Brown where thin. Matt with puddles of blue-gray glass.	Green color. Brown where thin. Matt with puddles of green glass.	
Barnard Clay	Similar to Red Iron, but less matt. Lighter in color.	Similar to 1% test. More matt and lighter in color.	Similar to 1% test, but progressively lighter and more matt.

Conclusion: Barnard clay created gray-blue, progressively lighter and stonier surfaces. Red iron oxide created glassier, greener surfaces in 2%–4% amounts.

Glaze Core

The number of glazes that use Barnard clay as a colorant source of iron and manganese at all firing temperatures are many and require no further documentation. More unusual is the fact that Barnard clay functions effectively as an iron-bearing glaze core at both 5/6 and 9/10 reduction and oxidation temperatures.

CONE 9/10 OXIDATION AND REDUCTION

Barnard clay	85%
Wollastonite	15

Oxidation	Opaque, black-brown. Semigloss.
Reduction	Opaque. Silvery, metallic, black. Semigloss.

CONE 5/6 OXIDATION

Barnard clay	85%	High gloss. Dense black. Smooth,
Colemanite	15	uncrazed surface.
Lithium Carb.	2	

CONE 9/10 OXIDATION

Kona F-4	50%	Rich, brown. High gloss. Very fluid.
Dolomite	25	
Barnard Clay	25	
Cone 6/7 Oxidation		Yellow-brown. Satin-matt.

SHEFFIELD SLIP CLAY

Sheffield Slip clay contains the chlorite clay mineral as opposed to illite, which is the chief clay mineral in Albany Slip clay. Chlorite is a primary mineral in high-iron, high-magnesium igneous rocks known as pyroxenes and amphiboles. It also forms in response to environmental pressures acting on pre-existing igneous rocks. It is this latter form of chlorite that is of interest here.

The chlorite clay mineral in Sheffield slip clay, like illite in Albany Slip clay, comes from ancient parents. High-iron, high-magnesium rocks of the Berkshire and Taconic mountain ranges (500–600 million years old), together with underlying 600 million year old limestones and dolomites, altered to chlorite 15,000 years ago in response to the pounding, scouring forces of the Wisconsin Glacier (Murtagh, 1–2). The complex, ancient parentage of chlorite is reflected in its mafic (magnesium-iron) chemical structure.

As is the case with Barnard and Albany clays, Sheffield

Slip clay is most useful as a glaze or colorant material. It is higher in iron and alumina and lower in silica than Albany Slip clay. Additions of materials that combine melting power with glassmaking, such as boron or boron and soda frits, are said to compensate for some of these differences and will enable direct substitution of the Sheffield substitute clay for Albany Slip clay.[23] Sheffield Slip is also used in some of the prepared claybodies provided by Sheffield Pottery, Inc. It is an ingredient of cones 01 to 06 earthenware claybodies and cones 6 to 10 stoneware claybodies (see *Ceramic Supply Catalog* 1998, Sheffield Pottery, Inc.).

NONCOMMERCIAL CLAY DEPOSITS

Most potters obtain iron-bearing clay from commercial suppliers. But when we dig clay from the bottom of a pond or from the edge of a stream and use this clay in a glaze, we connect directly with the source of our ceramic material. One of the most beautiful glazes I have ever seen came about in just this way; a student dug up some clay from a Long Island shore; sieved it when dry; combined it with 50% Nepheline Syenite and water; and applied it to a pot, which was then fired to cone 9–10 in a reduction atmosphere. The result was a silky, golden-orange surface reminiscent of Korean ware of the early Yi Dynasty. This glaze remains a unique, one-of-a-kind surface, which belongs solely to the digger of the Long Island clay; it can never be reproduced on a large-scale production basis. Those who fear the instability of unrefined clays and depend on the security of ultra-refined commercial clays should bear in mind that these commercial supplies may also vary from bag to bag. Then, too, it may well be these unpredictable, volatile impurities in the unprocessed clay that create the depth and excitement of the glaze surface. For all the pots lost, it is surely worth everything to achieve that one, unique, gemstone surface. This can only come about by boldly exploring unrefined materials and risking some losses. Native, iron-bearing illite and chlorite clays are a splendid source of original and exciting glaze materials and can restore the lost connection between potters and the source of their materials. It is this loss that haunts our modern potters and weakens the vigor of their work. *Figure 2.7, p150.*

MONTMORILLONITE CLAY MINERALS

Montmorillonite clay minerals are born in a watery environment from the ash and debris of fluid volcanic eruptions, which are richer in magnesium and iron and lower in silica than the thicker, deeper, granitic magmas. The periodic flooding of the seas during the last 60–70 million years of the earth's history transforms much of this volcanic ash into montmorillonite clay minerals.

The chemical structure of this class of clay minerals is charged with energy and is ever changing. It is understandable that these clay minerals have inspired the "Clay as Life" theory. Montmorillonite is made up of two tetrahedral silica sheets and one octahedral alumina sheet. There are no outside hydroxyl (OH) bonds, such as exist in kaolinite, to hold these layers together. Widespread substitutions and replacements take place within the interlayer space of the montmorillonite clay molecule, including the incorporation of organic molecules such as amines and proteins. Here is where all the action lies. Because of the weak bonds holding the layers together, water molecules are able to push between these layers, creating a greatly expanded lattice network. In addition to the incorporation of water molecules, some silicon is exchanged for aluminum, and these atoms are further exchanged for higher charged magnesium and iron ions, which create a negative charge deficiency in the molecule. In order to achieve stability, the negatively charged ions attract wandering positively charged sodium (Na^+) and/or calcium (Ca^+) ions to restore the balance. Positively charged organic molecules of amines and proteins may also become part of the interlayer montmorillonite structure at this point (Grimshaw 1980, 147). The theoretical result of this exchange activity is two kinds of montmorillonite minerals—sodium bentonites, which incorporate large quantities of water molecules into their lattice structure, and expand to 15–20 times their volume when immersed in water, and calcium bentonites, which swell only 0–5 times their volume when mixed with water (Hosterman 1984, C12). Calcium bentonites can be made to swell to 700%–1000% of their volume when combined with a 0.25% solution of Na_2CO_3. Although the demarcation between sodium, high-swelling bentonite, and calcium, lower-swelling bentonite is clear in theory, in point of fact the line tends to blur; some bentonites contain sizable amounts of soda, potash, and calcia.

BENTONITE CLAYS

Bentonite is the general name given to clays that contain predominant amounts of montmorillonite clay minerals. The name chosen once again refers to the original site of discovery. These clays were first discovered in 1896 in Fort Benton, Wyoming, and Wyoming is still a primary source of commercial bentonite clay deposits (Grimshaw 1980, 309; Bates 1969, 130; *Ceramic Industry* 1998, 70, 71).

Claybody Function

The most important characteristic of the montmorillonite clay mineral is minute particle size. Montmorillonite clay minerals contain the finest particle size of all the clay minerals. The unique traits of bentonite clays, such as high water absorption, excessive shrinkage, and rubbery plasticity, all result from this single characteristic.

The particle size of bentonite clay is one tenth that of kaolinite clay. In addition, because of their loosely bonded silica and alumina layers, bentonite molecules readily make room for intrusive water molecules. The practical consequences of this are that bentonite takes on excessive amounts of water (10 lbs. of bentonite can absorb 25 lbs. of water without becoming a liquid).[24] This, in turn, causes high shrinkage and frequent cracking during the drying and firing process. Every clay pot we made of bentonite clay cracked apart during the drying and/or firing stages due to the pressure of large amounts of escaping water absorbed by its minute clay particles. The extreme plasticity of this clay caused it to feel unpleasantly rubbery and sticky. When water was added to the bentonite clay, complex, fine-grained particles of bentonite bonded with the water molecules and formed an expanded lattice structure that was three times its original volume. We combined 900 grams of bentonite with 100 grams of colemanite, and enough water was added to make this combination the consistency of light cream. This liquid mixture was then poured into a quart-size container. Within a short period of time the volume of the mixture had doubled, and then tripled, and we ended up with three quarts of liquid! Here was dramatic proof of one of the most important characteristics of the sodium group of montmorillonite clay minerals—the tendency to bond with water molecules, which visibly results in swelling and expansion of volume. As we watched the volume of the bentonite clay mixture increase before our eyes, we saw visible proof of the "Clay as Life" theory. No one could doubt the life and energy in the expanding bentonite clay mixture.

For all of the above reasons, bentonite clay is not used in large percentages in claybody mixtures. In low percentages (2%–3%), it contributes plasticity to a claybody. Bentonite clay is a common ingredient of porcelain and stoneware claybodies, and the plasticity it provides is equal to that of 10% ball clay. (Note that an ill-advised addition of 4% bentonite to a porcelain claybody caused cracking and splitting of the clay form during the drying and firing stages.) Bentonite functions most effectively in porcelain claybodies, where often 50% of the materials in the claybody are nonplastic. Not only does 2% bentonite add plasticity, but it also increases the greenware strength of the clayform.

Glaze Function

The fine particle size of bentonite clay minerals, together with their attraction to water molecules, gives rise to important suspension properties when low amounts are immersed in a liquid solution. For this reason, 2%–3% bentonite is often routinely added to most glaze solutions for the purpose of keeping in suspension all of the various components of the glaze batch. Bentonite is commonly added in dry form to the dry mix of glaze ingredients before water is added to the batch. (It is difficult to incorporate the dry particles of bentonite into a wet glaze slop, because they float on the surface in much the same manner as oil floats on water.) Val Cushing recommends that you "mix bentonite into a slip before adding it to a body or a glaze—otherwise you will lose much of its benefits and advantages" (Some notes on bentonite, Cushing 1986). Dry, powdered bentonite added to the dry glaze ingredients does not consistently provide adequate suspension, as indicated by the frequent necessity of adding magnesium-sulfate (Epsom Salts), or alternatively (depending upon the glaze batch), a deflocculant such as soda ash.

Another, and less usual, approach is to consider bentonite as a glaze core. In such a case, the more fusible Black Hills bentonite provided unusual textural surfaces. The combination of 10% colemanite and 90% bentonite created a glossy, brown, crawled texture that had interesting possibilities for sculptural animal forms. Note that the high water content of the bentonite glaze liquid may cause small blobs of glaze to spit and break off during the firing. Thus, it is advisable to distance pots with this glaze from the rest of the ware.

Before attempting the above, it is helpful to know exactly what kind of bentonite material is being used. Only the more fusible bentonites will work as a glaze core in this way. Once again, a knowledge of your clay minerals is essential.

Kinds of Bentonites

As is the case with most clay minerals, there exist various kinds of bentonite clays, all of which contain different chemical structures and working properties.

As mentioned previously, there are two basic types of bentonites—those that are high in soda and swell in water, and those that are high in calcium and do not swell as much in water. Each has different commercial and ceramic uses as lubricant, sealer, viscosity, and suspender materials. *Ceramic Industry* lists a Black Hills bentonite with low calcia (0.46%) and high potash/soda content (2.90%) for use in electrical porcelain bodies (*Ceramic Industry* 1998, 71). But note that the Wyoming bentonite offered to potters by Hammill & Gillespie in 1989 contained considerably more calcia. Spinks Gel is a sodium-based Western Bentonite produced by H. C. Spinks Clay Company with far lower calcia content (see chemical data, p.138).

In addition to bentonite, there are other kinds of montmorillonite clays that are said to have superior plasticizing and suspending functions. (For a definitive article on these materials see Jeff Zamek "Additives for Glazes and Clay Bodies," *Ceramics Monthly*, December 1998, 61–62, 86–91.) Macaloid is a montmorillonite clay that is high in magnesium and calcium and low in aluminum and iron. It comes from a magnesium-silicate known as Hectorite ore. R. T. Vanderbilt Company, Inc. produces it under the name of Veegum T and Veegum Pro. (A compound known as Veegum Cer is also produced for the specific purpose of improving the hardness and durability of the raw glaze coat.) Macaloid has twice the plasticity of bentonite clay; a mere 1% addition will provide the same plasticity as 2% bentonite. It is commonly added to porcelain claybodies because it contains less than 1% iron and will not dim the whiteness of the porcelain claybody. For best results Macaloid should be mixed with hot water before it is added to either a claybody or a glaze.

Another kind of montmorillonite clay is sold commercially under the name of Bentolite. There are two kinds: Bentolite L, described as "a low viscosity, low swelling form which is used where maximum suspension is required with minimum viscosity build," and Bentolite H, "a high viscosity sodium modified form which in small quantities can increase gel rate and give maximum viscosity build." Both materials are white-burning and contain less than 0.4% iron oxide (Technical Data Southern Clay Products, ECCA Company, Gonzales, Texas).

The melting temperatures and the fired color of different kinds and different supplies of bentonite can show considerable variation. For example, pinch pots made from the bentonite of 10 years ago, which were fired to 2300°F in a reduction atmosphere, melted to a glassy, brown-black, speckled glass. Prior and subsequent pinch pots obtained from different supplies of bentonite retained their shape despite some cracking, and displayed an orange-brown, stony, coarse-grained texture. As stated previously, the chemical structure of bentonite reported by various ceramic texts and chemical data sheets reflects bewildering variations. (*Ceramic Industry Materials Handbook* contains the statement that "Bentonite also lowers the P.C.E. of white-ware mixtures. . .Black Hills bentonite softens at 1900°F with complete fusion at about 2440°F."; *Ceramic Industry* 1998, 71.) As shown on the bentonite chart on page 138, the chemical analysis of this Black Hills, Wyoming bentonite differs from the chemical analysis of Wyoming bentonite contained in Grimshaw 1980, 310. (The formula of bentonite, which appears in several standard ceramic texts [e.g., Nelson 1984, 322] is actually the chemical structure of the Black Hills bentonite). And finally, the 1989 Hammill & Gillespie data sheet for K-4 Western bentonite shows, once again, a different chemical structure. Thus, as in the case of most of our ceramic materials, the name "bentonite" does not guarantee freedom from variation. Bear in mind also that bentonite comes in different mesh sizes (325 and 200) and the coarser grades will not be as reactive in the melt as the finer grade. Check the mesh size that is printed on the bag of all of your ceramic materials and make comparative tests of all new supplies. These precautions will help to pinpoint the cause of subsequent surface changes that may occur in your glazes and claybodies.

BENTONITE

TRADE NAME	Bentonite, Bentolite L, Bentolite H, Bentone, Macaloid, Veegum®.
GEOGRAPHICAL SOURCE	Wyoming (most important commercial source), South Dakota, Texas. R.T. Vanderbilt Company Inc; H. C. Spinks Clay Company, Inc.
GEOLOGICAL SOURCE	Alteration of volcanic ash to montmorillonite clay material in marine environment during and after Cretaceous period (100–50 million years ago). Volcanic ash derives from ancient eruptions high in magnesium and/or iron. Also results from hydrothermal alteration and surface weathering of igneous rock high in magnesium and iron minerals.

CHEMICAL STRUCTURE

	Western K-4* (Wyoming K-4)	Black Hills, Wyoming** (refined, moisture free)	Spinks Gel***	Veegum®T†
SiO_2	59.40%	64.320%	64.7	53.74
Al_2O_3	17.15	20.70	17.6	0.97
Na_2O_3	2.99	2.90 (KNaO)	2.5	1.74
K_2O	0.26		0.5	0.10
CaO	4.09	0.46	1.3	4.54
MgO	2.36	2.26	1.8	23.82
TiO	0.10	0.11	0.2	
Fe_2O_3	3.86	3.49	4.4	0.15
S		0.35		Li_2O 0.98

P.C.E. Black Hills, WY   1900°F (softens)
Swelling Type   2440°F (fusion)

Particle Size: Colloidal—less than 1 micron.

MINERALOGICAL STRUCTURE

Bentonite: Montmorillonite Clay Mineral ($Al_2Si_4O_{10}(OH)_2 \cdot nH_2O$)
Mg may replace part of Al; Na and Ca ions present in interlayer space.‡
Macaloid: Hectorite (Montmorillonite Clay group.) 90%

*Hammill & Gillespie data sheet 1989.
**Ceramic Industry 1998, 71.
***H. C. Spinks Clay Company, Inc. 1998.
†R. T. Vanderbilt Company, Inc. 1998.
‡Blatt 1992, 37.

CLAYBODY FORMULAS
Examples of Claybodies Using Fireclay, Ball Clay,
Iron-Bearing Clays, and Bentonite Clay*

CONE 9-10 REDUCTION

	Greenwich House Pottery (NYC) T-1	92nd St. Y (NYC) Stoneware	Parsons School of Design (NYC) Stoneware
A.P. Green	55	36	40
Gold Art Fireclay			20
Ball Clay (OM-4, Tenn-9 or 5)	8	23	20
Jordan		23	
Barnard	4		
Red Art			5
Feldspar (Custer, or K-F4)		9	8
Flint		9	
Bentonite	3	2	
Grog	30(medium)	3–6 (fine)	3–1/2 (fine)
Macaloid		1	
Red Iron Oxide		1/8–1/4 %	

Color-Texture	Toasty orange red-brown. Coarse-grained.	Red-brown. Speckled.	Gray-brown. Speckled.

*Not current.

CONE 9-10 OXIDATION

	OIV	R.S.1
Pine Lake Fire	1.25 lb.	5 lb.
Ocumulgee Red	1 lb.	
Jordan		2.5 lb.
Ball	.5 lb.	1.5 lb.
Spodumene	.25 lb.	
Rotten Stone		1 lb.
Color-Texture	Coarse, orange brown with white speckles.	Pale brown. Smooth texture.

OIV claybody was very effective under a black-brown, glossy, high-iron glaze. It is doubtful that it would prove to be equally successful with a low-iron glaze. Spodumene, which functions as a melter and is partially responsible for the burnt-orange color, tends to influence the glaze surface as well and can muddy the clarity of lighter colors.

EARTHENWARE CONES 04–02

	Parsons School of Design NYC	
Red Art	67	40
Ball	20	20–25
A. P. Green	15	20–25
Flint	15	
Talc	3	
Barium Carbonate	0.5%	0.5%
Fine Grog		10

Color-Texture	Bright, brick red color. Smooth texture.

Claybody Formulas (cont.)

CONE 4–5–6 OXIDATION

	Portchester Clay Art Center, NY	Crafts Students League NYC			
	Red-Brown	Toast	Fine Red-Brown	Fine Brown	Fine Black
Jordan	25	30		40	40
A.P. Green* ⎫ Fireclay North American ⎭	50	30			
Pine Lake					
Ball Clay		15	16.7	15	15
Ocumulgee			66.7		
Albany			16.7		
Red Art	25	5			
Barnard		5			10
Feldspar	10 (Kona F-4) (Custer)	5 (Kona A-3)		25	25
Rotten Stone				20	20
Flint		10			
Grog 325 Mesh	10				
Cobalt Oxide					2

Red-Brown: Toasty, red-brown claybody. Has some plasticity problems. High shrinkage rate. Dense throwing properties. Medium grain.

Toast: Pale, brown, smooth-textured claybody. Highly plastic.

Fine, Red-Brown: Smooth, fine grained, pale red-brown. Needs further testing to determine if bloating occurs due to presence of Albany Clay.

Fine Brown: Smooth, fine-grained, chocolate-brown color. Needs further testing.

Fine Black: Silky, deep black with greenish, metallic cast. Bubbles at cone 6. Highly plastic with good throwing properties. Needs further testing.

Note: Fine Red-Brown, Brown and Black claybodies contrasted with a white claybody create handsome, decorative effects for tiles or other decorative wall pieces.

*or Pine Lake

Glaze Function

COMPARATIVE TESTS of ILLITES and OTHER IRON BEARING CLAYS

Cone 9–10 Reduction

Kona F-4	80.0%	Green-blue gloss.
Whiting	20.0%	Transparent.
Red Iron		Crazed
Oxide	0.5	

ADD

	Red Art	**Albany**	**Barnard**
10%	Deeper blue-green.	Deeper blue-green than control but lighter than Red Art.	Greenish, mustard, yellow. Opaque, uncrazed matt. Increased flow; puddles and roll of moss-green glass.
20%	Deep green-yellow glass. Increased flow.	Pitted surface. Greener color. No visible crazing.	Yellow-brown matt. Puddles of brown glass. No visible crazing.
30%	Satin matt surface. Gray-green color. No visible crazing.	Gray-green gloss. Deeper color than control. Less runny than control.	Dark brown matt with puddles of black glass. Yellow crystal flecks in black glass.
40%	Similar to above but deeper green color	Deep yellow-green color. Increased flow.	Deeper brown, opaque, matt. Rim flecked with black, glassy specks. Increased fluidity. Fat roll on outside of pot. Yellow crystals in black puddles of glass.

The test results reflect the different chemical structure of these clays. Albany clay contains approximately 5% red iron; Red Art clay contains 7% red iron oxide, and Barnard contains 14% red iron oxide, plus 3.4% manganese. The tests show Albany to have the lowest coloring power. Cedar Heights Red Art has a slightly higher coloring effect, and Barnard clay has the strongest colorant power. With respect to fusion, note that according to the chemical analysis, Albany clay has the highest melter content and Barnard clay has the lowest. Once again, test surfaces reflect these difference. Barnard clay tests show a matt-shine surface with less fluidity than Albany. However, note that the fluidity of Red Art clay is greater than Albany's in the 10%–20% tests, despite the lower melter content of Red Art clay. It is possible that uneven heat conditions in the kiln are responsible for these results. The tests should be repeated on larger test forms so as to rule out these variables.

| **Cornwall Stone** | 90% | Milky, gray-white gloss. |
| **Whiting** | 10 | Large overall crackle. |

ADD	1/2%	1%	2%	3%	4%
Red Iron Oxide	Milky, blue green. Semitransparent. Gloss. Crackle.	Deeper blue green. More transparent. Higher gloss. Crackle.	Moss green color. Transparent. Gloss. Crackle.		
Barnard Clay		Pale blue gray. Fat, semitransparent. Less crackle.	Grayer blue. Same as 1% test. Same as 1% test.	Brighter blue-gray. Same as 1% test. Same as 1% test.	Greener cast. Same as 1% test. Same as 1% test.
Albany Clay		Similar color to test. Barnard test, but more transparent.	Same as 1% test.	Same as 1% test.	Almost same as 1%. Slightly bluer shade of gray.

CONCLUSION: Barnard and Albany clays create cooler, grayer, and less-brilliant colors in the feldspathic glaze surface than does red iron oxide. Albany produces the palest, grayest color of all three iron-bearing clays. Barnard clay at the 4% test produces the greenest color. Both Barnard and Albany tests show less transparency and less brilliance of color than do the red iron oxide tests. Titania, alumina, and other minerals contained in the clays tend to bleach and lighten the brilliant blue-green shades produced with pure red iron oxide.

CONCLUSION

Clays are the most ephemeral of all our ceramic materials. They appear, disappear, and reappear in different forms with new mineral and chemical structures. Like human beings, each clay is unique; once gone, the same clay rarely appears again.

The consequences for the potter of these everchanging conditions can be harsh indeed. Most potters do not extract their own clay from the earth, nor are they lucky enough to live near the site of their clay mines. Neither do most potters have the equipment to refine, process, or mix clays on a large scale. The result is that we have become dependent on the supplier for our clays and our claybodies. Unfortunately, suppliers are often not able to give their customers timely notice of ongoing changes in their clays and claybodies. A kiln of cracked and shivered ware may become the first and only notice to the potter of such changes. My own research for this book amply corroborates the clogged lines of communication between the potter and the supplier. In general, I found that suppliers are not as knowledgeable about their materials as the manufacturers or, if they are, do not always communicate their knowledge in time to avoid problems. On the other hand, I found many of the manufacturers eager to help and most knowledgeable about their materials. When a major supplier refused to give me any information about a particular ball clay because I was not a customer (one ton minimum!), the manufacturer of the ball clay, Kentucky Tennessee Company, sent me all the available data and also shipped five pounds of the clay free of charge. The same degree of cooperation and knowledge was extended by the Cedar Heights Clay Company of Resco Products, Inc., ECC International, Ceramic Division, as well as many others. Hence, it would be advantageous to order ceramic clays directly from the manufacturer. The price would be lower, the analysis sheets more current, and, best of all, the potter could learn much from direct contact with the manufacturer's materials engineers. Unfortunately, this practice is not a practical solution for many potters because it necessitates clay-mixing equipment. Potters without such equipment, who work with a custom-mix claybody, are of necessity dependent on the supplier for mixing their individual clay-bodies. These potters are the most vulnerable and in the highest risk category because they usually do not represent an important percentage of the supplier's market. Unless a supplier is well informed about changes in the specific clays of custom-mix claybodies, such changes in clay may lead to disastrous consequences for the potter. Depending on how fast the potter uses the clay supplies, these consequences may occur as late as one year from the time of the original shipment. For this reason, the safest practice for a potter without clay-mixing equipment would be as follows:

1. Expend time and energy in finding a knowledgeable supplier, and no matter how inconveniently located, use this supplier. The increased transportation costs will be more than balanced by the decrease in ruined ware.

2. Wherever possible, use a claybody prepared for a number of potters rather than for one or two. This could mean using the supplier's claybody as opposed to a custom-mix claybody. The advantage is simply a question of numbers. A supplier who sells a claybody to a large number of customers will expend time, money, and resources in testing both the final claybody and the individual clays that it contains. In addition, the supplier will pay close attention to possible changes in the clay materials. The individual potter does not usually have comparable resources at hand. From a practical standpoint, using a standard claybody may well be the only effective protection against a kiln of ruined ware. Unlike glaze surfaces, which first and foremost should be unique and personally expressive, a claybody must be safe, durable, and free of cracking and shivering problems. Twenty years ago, in my happy ignorance of the treacherous variability of clays, I urged potters to create their own claybodies, and wherever possible, to dig their own clays. Today, sadder, older and wiser, I think that the most sensible approach would be to use a claybody that has been prepared, tested, and guaranteed by a reputable supplier. Save your originality and uniqueness for the structure, form, and glaze surface of the piece.

3. If you use a custom-mix claybody, obtain a test sample before purchasing a new supply, and test carefully to be sure it conforms to a prior shipment. Find out about the mining

practice and industrial usage of each clay in your claybody. If the primary industrial usage does not require a high standard of uniformity (such as, for example, a foundry clay used as a plug) consider substituting a different clay.

If precautions are not taken, no matter how extraordinary or unique your claybody may be, there will come a time when all will be lost due to the everchanging, unpredictable nature of clays.

Notes

[1] The following account of the origin of clay minerals is based on: Birkeland 1984; Grimshaw 1980; Frye 1981; Cardew 1973; Cairns-Smith and Hartman 1986; Gross 1987; Kuenen 1950.

[2] μ = micron or one millionth of a meter. μm = one thousandth of a micron.

[3] Frye 1981, 71–78.Grimshaw 1980, 285–288.

[4] The following account of the firing process is based on: Cardew 1973, 61–80; Lawrence 1972, 111–127; Robinson 1981 9: 76–80; 1988 16:73–82.

[5] Frye 1981, 675.

[6] See p. 103.

[7] Melting point over 3100°F.

[8] Molochite, a kaolin grog calcined up to 2780°F, provides structural strength for the Grolleg claybody at the expense of translucency. It is the exclusive product of ECC International. (http://www.eccl.co.uk/ceramics/cermolo.htm)

[9] Geological material based on Hosterman 1984; *Ceramic Industry* 1998, 88.

[10] The water of plasticity rate is the percentage of water that must be added to the dry clay to render it plastic and workable.

[11] Double convex lens-shape.

[12] Claybody: Portchester 5–6. See p. 140 for formula.

[13] Grimshaw 1980, 296, 301.

[14] Unfortunately no longer available.

[15] See EPK/Ball clay test, p. 115.

[16] The following geological information is based on: Murtagh, "A comparison review of the geological origins of the Sheffield glacial clay deposit (Massachusetts) with the Albany slip glacial clay deposit (New York)." Courtesy of Sheffield Pottery Inc., Sheffield, Mass.)

[17] A sample may be obtained from the Archie Bray Foundation Clay Business, 2195 Country Club Avenue, Helena, MT 59601.

[18] The yellow color is due to the 8 1/2% calcia-magnesia content, which bleaches and yellows the 5% iron content of Albany slip. Note that the substitution of Albany Clay for the potash glaze core in a high-iron, cone 9–10 reduction glaze yellowed the original black color of the surface. See Appendix B1.

[19] Terra sigillata, an ancient Mediterranean technique, (1st century B.C.; see Hamer and Hamer 1986, 314) refers to a slip-coated earthenware with a high-gloss finish. The gloss results from the decanting, burnishing, or pregrinding of the clay slip so as to utilize only the finest particles.

[20] Or Gerstley Borate.

[21] For an interesting account of the geology and mining procedures of Ocmulgee Red clay, see *Studio Potter* 1981, (December) 10-1:16–18.

[22] See Material safety Data Sheet Chemetals Inc.; Alexander, William, Ceramic Toxology.

[23] Comparison tests of Albany Slip Clay and the Sheffield substitute in a recent cone 9–10 reduction firing proved remarkably similar.

[24] Hamer and Hamer 1986, 23.

[25] Be sure to glaze one-half of the test pot with the regular glaze—this is your control.

Figure 2.1 Tests: Clay Pillows.

Four vertical rows, left to right in descending order:
Row 1: Y claybody: Oxidation cone 06, 6, 9–10, reduction cone 9–10.
Row 2: Red Art clay: Oxidation cone 06, 6, 9–10, reduction cone 9–10.
Row 3: Gold Art clay: Oxidation cone 06, 6, 9–10, reduction cone 9–10.
Row 4: A. P. Green Fire Clay: Oxidation cone 6, 9–10, reduction cone 9–10. O.M.-4 Ball Clay, reduction cone 9–10.

Figure 2.2 Black Bowl and Slab with Pillow Chopstick Rests by author.

Bowl: Fine black claybody, cone 5–6 oxidation; Interior: Ron's white matt glaze. Blue color result of cobalt in claybody.

Slab with Pillow Chopstick Rests: Fine black claybody, cone 5–6 oxidation.

Figure 2.3 Slab-rolled, multilayered clay platter by Sue Browdy.

Cone 5–6 oxidation. Portchester red-brown claybody, Grolleg porcelain 9–10 claybody, stoneware 9–10 claybody.

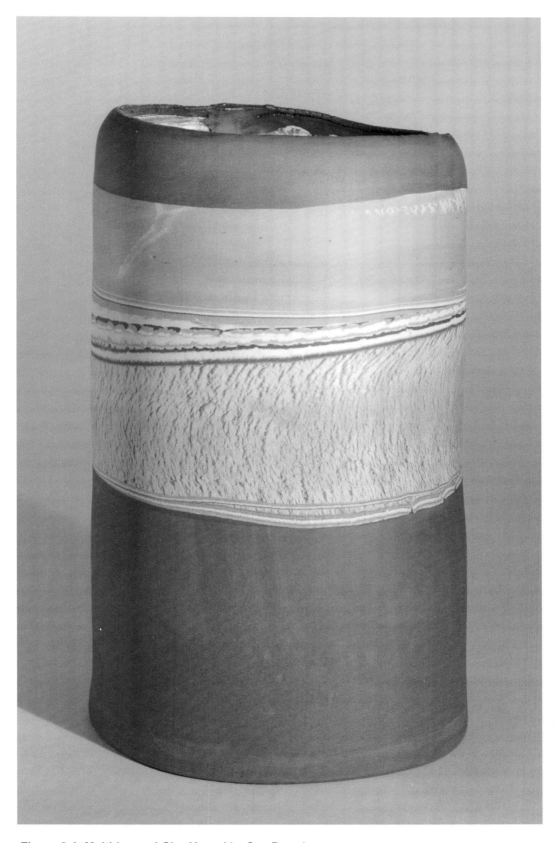

Figure 2.4 Multi-layered Clay Vessel by Sue Browdy.

Cone 5–6 oxidation. Portchester red-brown claybody, Grolleg porcelain
9–10 claybody, stoneware 9–10 claybody.

Figure 2.5 Slab-constructed Vessel by Steve Burke.

T-1 claybody. Cone 9–10 reduction. Neck: saturated iron glaze. Body: unglazed.

Figure 2.6 (top) Layered Clay Tests.

Clockwise from front center: Layered clay tests of:

porcelain 9–10, stoneware 9–10, Barnard, and O.M.-4 ball clays;

test 1–3: cone 9–10 oxidation, 9–10 oxidation, 9–10 reduction.

test 4: Portchester 5–6 claybody, porcelain 9–10, and Red Art clay: cone 5–6 oxidation.

Figure 2.6 (bottom) Chopstick Holder by author. (Photo by Chris Dube)

Cone 5–6 oxidation. Layered clays of porcelain 9–10, Portchester red-brown 5–6, and stoneware mix 6–10.

Figure 2.7 Terra sigillata Jar by Sue Browdy. **Bowl** by Jacqueline Wilder.

Terra sigillata of found Maine river clays applied to leather-hard stoneware 9–10 claybody; bisqued to cone 09, and sawdust fired.

Figure 2.8 T-1 Planter by author.

Cone 9–10 reduction. Unglazed. (A.P.Green Fire Clay 55%).

151

LAB III CLAYS

1. Find out from your supplier what clays are in your claybody. Get the chemical and mineralogical data in these clays from your supplier. Note especially the shrinkage and absorption rates, and the temperature at which they slump (P.C.E.).

2. Measure out 500 dry grams of each of these clays, and add enough water to each 500-gram batch to produce plasticity. Make four pinch pots out of each of these clays. Record the exact amount of water by using a measure to add the water to the dry 500-gram batch. Record the workability properties of each clay. Measure the height, width, and depth of each pot in its plastic state and in its bone-dry state.

3. Fire the four pinch pots of each clay to all of the available firing temperatures (bisque, 5/6 oxidation, 9/10 reduction, and oxidation, etc.) Measure the height, width, and depth of each pot in its fired state and compare these measurements to the plastic and bone-dry measurements. Weigh the fired pots, immerse them in boiling water for five minutes, and then weigh them again. The difference between the saturated weight and the dry weight divided by the dry weight and multiplied by 100 will give the percentage of water absorption for each fired state. Repeat steps two and three with your claybody.

4. Roll out thin slabs of as many different clays as are available to you in your studio and fire these slabs to your available firing temperatures. (Be sure to mark the name of the clay and the firing temperature on the back of each slab.) Look long and hard at the resulting color and texture of these fired slabs at the various firing temperatures, and consider how you could use this range of color and texture for decorative surface effects.

For example, attach thin slabs of one or more of the different clays to a clayform of your claybody and bisque to usual temperatures. Glaze one-half of the attached slab or slabs. Fire to your customary firing temperature. Note the color and texture contrast between the glazed and unglazed surfaces of the clay slabs and the claybody.

Roll out thin slabs of some or all of the clays. Layer the individual slabs and place these layers on top of a slab of your claybody. Cut the resulting clay sandwich into vertical pieces. Roll the layered face of each vertical piece and note the marbleized pattern that appears. Fold the marbleized section in half, continue to roll, and observe the resulting changes in pattern. Continue to roll until a marbleized pattern appears that you wish to keep. (This is the hardest part—once you continue to roll, the prior pattern is lost forever—the skill is in knowing when to stop!) Line each marbleized section with a slab of your own claybody and fire each section to all available firing temperatures. **Note:** porcelain clay may lift up from the clay sandwich on drying or during the firing unless pressed into the other clays with the even, powerful pressure of a slab roller.

USING CLAYS AS GLAZE MATERIALS

1. Select a glaze with which you are familiar. Substitute each one of the clays listed below for the clay in your glaze formula.

EPK	Ocmulgee
Ball Clay	Cedar Heights Red Art
Grolleg	Barnard-Blackbird
Fireclay	Any other clay that you may have on hand.

Fire a test of the resulting glaze with the substituted clay to the appropriate firing temperature for the glaze. Glaze one-half of test pot with original glaze—this is your control. *Do this for all tests.*

2. Increase the amount of clay in your glaze formula by increments of 10% up to 40%. (There will be four tests, ranging from 10%–40%). Note the different textures and colors that result from these clay additions.[25]

3. Add water to one or more of the clays listed in step 1, so as to make a thin slip (light cream consistency). Dip one-half of a greenware test pot into this slip and then bisque and glaze the entire pot with your glaze. Repeat the same procedure with a bisque test pot. Note the differences, if any, that result from applying the slip to the greenware and the bisque pot. Note the various colors and textures that result from using the different clay slips.

4. Explore the potential of the following clays as glaze cores:

a. Substitute Ocmulgee, Cedar Heights Red Art, or Barnard clay for the feldspar in your glaze.[25]

b. Take one of the clays listed in step 1 and add one or more of the melters discussed in Chapter 3 in 10%–40% increments. (There will be four tests of each clay-melter combination, ranging from 10%–40% additions of melter.) Apply these mixtures to four bisque pots; fire to customary firing temperature for the claybody, and compare the color and textural differences that result as the melter increases from 10%–40%.

Chapter 3 | AUXILIARY MELTERS

INTRODUCTION

Thus far we have described feldspathic and clay materials in which the glassmaker oxide (silica) and the adhesive oxide (alumina) are bonded with various melter oxides (soda, potash, calcia, magnesia, and/or lithia). Because of this oxide trinity, each feldspar and clay material can form the very core of the glaze surface hence, the term *glaze-core*. Sometimes this compound material will be enough in and of itself to achieve the final, desired surface—but in most cases we are not that lucky. In order to perfect the glaze, it is usually necessary to add separate amounts of glassmaker and/or melters and/or adhesive oxides. The glaze-core transformations of a ceramic surface described in Chapters 1 and 2 are but a prelude to the wide range of surfaces that result from separate additions of these oxides. This chapter and the following one deal with these accessory materials and their special contributions to the glaze surface.

We begin with the auxiliary melters—materials added to the glaze-core to help it achieve a greater melt and fusion. Again, there are many different kinds of melters, but at stoneware temperatures, it is the limestone or calcium-based melters that are most important for the stoneware glaze. The internal oxide structure of limestone is mostly calcium carbonate, which turns into calcium oxide during the firing process. For this reason Whiting (calcium carbonate), Wollastonite (calcium silicate), Dolomite (calcium-magnesium carbonate), Gerstley Borate (calcium-sodium-boron), and Colemanite (calcium-boron) receive primary attention in this chapter.

At the start, we need to clarify the meaning of certain words used to describe glaze surfaces, such as *transparent, opaque, gloss,* and *matt*. All solid earth materials possess a characteristic, crystalline structure. The atoms of each material have arranged themselves into a specific crystal shape that is typical of that material and is a primary means of its identification. Heat transforms a solid into a liquid by breaking up these atomic arrangements. Whether or not a particular material resumes its crystalline structure after passing from a liquid, melted phase into a cooled solid depends on the temperature of the fire and the length of the cooling period. Most earth materials reform into a characteristic crystalline structure after being fired in a potter's kiln. Quartz (silica), germania, boron, and phosphorus are among the few materials that do not. In their newly cooled, solid state, these materials resemble "frozen liquid," or glass. Glass is by definition "transparent" because it does not possess a characteristic crystalline structure. Its crowded atoms solidify in a more or less random disorder. Thus, light is able to travel freely through it in all directions. This fact has great significance for the potter. Crystals absorb, trap, refract, and reflect light rays. If light is trapped and absorbed by minute, dense crystalline formations, the resulting surface will appear to be both opaque and nonshiny and is called stony-matt. On the other hand, if light freely penetrates the glaze surface without obstructions, the surface will be transparent. Transparent surfaces are usually, (but not always) shiny— that is, light reflects off the surface, as well as moving through the glaze layers to the underlying claybody. Between the two extremes of a glassy transparency and an opaque-matt surface lie a host of transitions. If some of the crystalline formations are larger than others, light may be refracted off their varying shapes and sizes in such a way as to create an opaque but shiny surface, or a semi-opaque, satiny, semigloss surface. We call this latter surface satin-matt. The final visual appearance of the glaze surface—its shine,

155

opacity, and color—is determined by the amount, size, and shape of the crystalline formations. Thus, the possible variations are limitless!

Besides crystalline formations, there are other phenomena that interrupt the free flow of light and create an opaque effect. Unmelted particles of a material, trapped bubbles of gas suspended in the melt, and, finally, glasses of differing refraction rates suspended inside each other can all accomplish this result. Opacity created by crystalline formations may differ markedly from opacity created by unmelted particles of a glaze material or, again, from opacity created by trapped gas bubbles. Often the opacity of the surface has multiple causes; it may be caused by a combination of any two, three, or even all of the above factors.[1]

Thus, a glaze surface may appear transparent or opaque, glossy or nonglossy, depending on the way in which light travels through the glaze surface. The transparency and gloss of this surface can be changed to one of opacity, semigloss, or nongloss by increasing the amount of silica, alumina, and/or melter oxides contained in the glaze materials. There will, of course, be vastly different visual effects depending on which particular oxide creates the changed surface. In addition, because each oxide can be obtained from a score of different ceramic materials, there will be further differences in result, depending on which material provides the particular oxide in the glaze.

Our goal in this book is to try to recognize the unique opacifying and glassmaking properties of each member of the trinity and of a few materials from which they derive. If we accomplish this—to the point of an educated guess—then we will be able to unravel the complex, interwoven strands of a particular glaze surface and understand how it came to be. We can then use this knowledge to create different kinds of glaze surfaces. This is a lofty goal, to be sure, and one that is never completely attainable, but its pursuit will bring rich rewards in greater knowledge and control of materials.

It makes no difference where you begin. That is, you could start with the materials that provide additional glass, additional adhesive, or additional melters. At the end, you will arrive at the same point in your understanding. The journey you are undertaking is a closed circle—you may start at any point in the circle, but until you make a complete revolution and return to the original starting point, some confusion and lack of comprehension will be inevitable. Bear with it; once the complete revolution has been made, the mists will begin to clear.

We are going to enter the circle with the melter materials because once you understand how a few melters work on the feldspathic glaze cores, it is within your power to achieve remarkable stoneware glazes. There is nothing like instant success to lure you deeper and deeper into the fascinating and never-ending maze of ceramic exploration. This kind of seduction is the very aim and purpose of the book.

CALCIUM OXIDE MINERALS

Calcia or calcium oxide is the most important auxiliary melter of the stoneware glaze. The many tests we have performed over the years show clearly that it is calcium that creates both the satin-matt, jade-like surface, and the blue-gray-green color of the stoneware celadon glaze. An analysis of Song Dynasty pot shards reveals the presence of substantial amounts of calcium. Without calcium, the incomparable celadons of the Song Dynasty could never have existed. The Song potter perfected the celadon glaze in an attempt to reproduce the magical and highly revered gemstone of jade. It is no coincidence that nephrite, which is one of the two forms of jade, is a silicate of calcium, iron, and magnesium. Thus, the celadon glaze, which reached its zenith over 800 years ago, already existed millions of years earlier in the form of the natural mineral nephrite. *Figures 3.1, p223; 3.3, p225.* The survival of Song Dynasty celadon vessels for eight centuries without surface decomposition or color loss is due in part to the hardness and durability of the calcium melter. Compare the surface of these pots to the lead-based ware of the Han and Tang Dynasties. The bright greens of the lead surface have in many cases changed to an opalescent silver, due to the decomposition of the soft, lead-based glaze. The celadon surfaces of the Song Dynasty ware contained high amounts of silica and calcium. Unlike lead oxide, which melts at about 932°F–1616°F, the melting points of both silica and calcia reach well beyond 3000°F. Together they form a strong eutectic bond at the high stoneware firing temperature of 2340°F to produce a surface that is hard and acid-resistant—a surface that has proudly resisted the deterioration of time. Vessels that contain this kind of surface are truly named STONEWARE.

Calcium makes up 3.6% of the earth's crust by weight and is the fifth most abundant element. It combines readily with various other elements (primarily carbon) and is never found alone in nature in a pure and elemental state. The rocks of limestone and marble; the minerals of calcite, aragonite, vaterite; and the soft, earthy deposits known as chalk are all made up of calcium carbonate ($CaCO_3$), which, in turn, ultimately derives from the skeletal remains of animals and plants, precipitates from seawater, or igneous rock solutions. It is these calcium carbonate materials that are the primary source of calcium oxide. The heat of the kiln fire breaks apart the calcium-carbon bonds at about 1517°F. The carbon atoms regroup with oxygen to form carbon monoxide and dioxide gases, which will leave the kiln. The remaining calcium atoms bond with oxygen to form refractory calcium oxide, which alone, without further eutectic bonds, will not melt until well past 4000°F. This disassociation process occurs whenever any carbon alloy is subjected to heat. Thus, all carbonate compounds will have changed into oxides by the end of the kiln firing.

CHARACTERISTICS OF THE CALCIUM OXIDE MELTER

Figure 3.2, p224.

1. *Calcium oxide is an active melter at cone 9/10 stoneware temperatures provided the amount of calcium oxide does not exceed approximately 25% of the total glaze materials.* When fired alone, it melts at about 4658°F, but when combined in the right proportion with silica and alumina, a eutectic bond is formed, and a vigorous melt occurs at stoneware firing temperatures. The following tests show the melting power of calcium oxide.

CONE 9–10. OXIDATION AND REDUCTION

FELDSPAR-WHITING TEST		
Feldspar (potash)	100%	Opaque white gloss. Crawled surface.
Add 10%–40% increments of Whiting (calcium carbonate)		
Feldspar	90%	Light blue-gray color (r)*
Whiting	10	Buff color (o). Smoother, less crawled surface.
Feldspar	80%	Darker blue-green-gray. (r)*
Whiting	20	Deeper buff. (o) Increased fluidity.
Feldspar	70%	Green yellow. (r)*
Whiting	30	Yellow-buff. (o) Semigloss opacity.
Feldspar	60%	Yellow-white. (r and o)
Whiting	40	Dry, opaque surface.

*Blue or green color appeared only in small (2-in.) tests due to fierce interaction of iron in claybody with glaze materials. For equivalent results in larger test pieces add 1/2% red iron oxide. See #3, p. 159.

Conclusion

These tests illustrate the tremendous importance of the proportion of the melter in the glaze batch. The surface color change of the originally white-colored feldspar to blue-gray (r) and buff (o), green-yellow, and ultimately yellow-white, is characteristic of increasing the amount of calcia from low, to medium, and finally to high. Whiteness is usually an indication of opacity caused by unmelted particles of material, and this is the condition of the feldspar without additional melters. As we add low amounts of Whiting to the feldspar, the unmelted particles begin to melt, and the surface begins to flow—hence the blue-gray color in reduction, the buff color in oxidation, and the increased transparency and fluidity of the glaze. (The blue-gray reduction color and the buff oxidation color is an indication of lessened opacity, increased transparency, and fusion because the underlying claybody color in reduction is gray and in oxidation is buff.) As we increase the amount of calcium (30%) the excess calcium ceases to behave as a melter. It bonds with silica to form calcium-silicate crystals, which create a waxy, semigloss, opaque surface. Finally, in even larger amounts (40%) calcium dries and matts the glaze surface. The excess calcium oxide particles remain alone and unbonded in the glaze melt, and thus create a dry, stony, and yellow-white surface. Even if the firing temperature rises, transparency does not necessarily return. As the temperature rises, the flow of the glaze increases, but the surface remains opaque, creating the condition known as a "running matt."

The addition of calcium carbonate to five opaque-matt, or semi-opaque, semimatt reduction glazes changed the orange-yellow-white color to blue-gray, thus indicating increased transparency.

SHINO CARBON TRAP

Figure 1.10, p87.		Orange-red-(white if thick). Opaque, semigloss surface.
Nepheline Syenite	45.0%	
Kona F-4	10.8	
Spodumene	15.2	
Soda Ash	4.0	
Ball clay	15.0	
EPK kaolin	10.0	

Add 10, 20, 30% Whiting:
    Orange-red color changes to blue-gray.

CASCADE

		Orange-white matt surface.
Custer spar	30 grams	
Spodumene	20	
Dolomite	22	
Whiting	2	
EPK kaolin	22	
Tin oxide	6	

Replace Dolomite with Whiting:
    Orange-white color changes to blue-gray.

CHARLIE D

		Gray white satin-matt.
Custer spar	20%	
Kona F-4	20	
Flint	20	
Dolomite	15	
Talc	13	
Whiting	2	
Ball clay	10	

Replace Dolomite and Talc with Whiting:
    Gray-white changes to gray blue.

RHODES 32

Figure 4.2, p266.		Orange-yellow-brown-white matt.
Custer spar	48.9%	
Dolomite	22.4	
Whiting	3.5	
EPK kaolin	25.1	

Replace Dolomite with Whiting:
    Surface changes from orange-yellow white to blue-gray.

MAMO MATT

Figure 4.4, p268.		Black-green-orange matt.
G-200 spar	49 grams	
Dolomite	19	
Whiting	4	
Kaolin	21	
Tin Oxide	8	
Copper Oxide	1.5	

Add 20%–30% Whiting:
    Color change to turquoise blue.

Replace Dolomite with Wollastonite (calcia-silica) and G-200 Feldspar with Cornwall Stone:
    Strong turquoise blue surface color.

2. *Despite the fact that calcium oxide is not as active a melter as it is in the higher stoneware temperatures, low amounts of calcium oxide will increase the melt of a feldspar at lower stoneware temperatures.*

CONE 5–6 OXIDATION

FELDSPAR-WHITING TEST

Kona F-4 Feldspar	100%	White. Opaque. Semigloss. Crawled.
Feldspar	90	Flow, gloss and even coverage begins. White (thick); Pink-purple (thin).
Whiting	10	
Feldspar	80	White. Silky. Satin-matt.
Whiting	20	
Feldspar	70	Increased opacity. Yellow-white. Stonier surface.
Whiting	30	
Feldspar	60	Yellow-white. Dry stony surface.
Whiting	40	

Conclusion

The white color of the surface with the addition of 10% calcium indicates that transparency is not as complete as in the cone 9–10 stoneware tests. Note that a thin layer of glaze does indicate transparency, because the pink-purple color reflects the underlying red-brown claybody. In general, these tests indicate that although 10%–20% of calcium carbonate increases the flow and coverage of the feldspar and produces waxy, satin-matt opacities, calcium carbonate does not create the same degree of transparency as occurs at the higher temperatures. It functions more effectively as an opacifier than as melter at these lower stoneware, oxidation temperatures.

3. *At the higher stoneware temperatures calcium oxide helps to bond the glaze surface with the underlying clay body by forming a buffer layer between the two that is both glaze and claybody.* This intermediate layer reduces the tension between the claybody and the glaze caused by their different expansion and contraction rates and thus reduces crazing and shivering problems. A test of pure calcium carbonate applied to the inside of a pot, and fired to cone 9–10 temperatures in a reduction atmosphere produced evidence of this buffer layer. *Figure 3.2, p224.* Although most of the calcium carbonate did not melt at the firing temperatures of 2380°F, small sections of the test pot revealed a green-yellow glassy surface. The presence of such a surface indicates that the bottom layer of calcium oxide had bonded with some silica, alumina, and iron contained in the underlying claybody to form a thin glaze layer. Unlike the violent action of soda, potash, and lithia melters on the claybody, which caused high shine, fiery purple-red colors, and cracked pots, calcium oxide is gentle and strengthening by comparison and produces the unique, skinlike character of the high-calcium, stoneware glazes (see Rhodes 32, p. 158 and Pavelle, p. 176). Note that in the oxidation atmosphere, the Pavelle glaze creates a skin-tight, stony, pink-yellow-ivory colored surface. Although less colorful than the yellow-orange color produced in a reduction firing, the surface appears wedded to the underlying claybody. Thus, calcium glazes in high stoneware oxidation and reduction firings bond with the underlying clay structure to become a natural extension of the claybody, rather than a surface glaze coat.

It is not clear from the tests performed thus far that calcium oxide fuses with the glaze and the underlying claybody at the cone 5–6 oxidation temperatures. The lower firing temperature would make this fusion unlikely in view of the refractory nature of calcium oxide. The only test that indicated a possibility of this kind of interaction is as follows:

CONE 5–6

FELDSPAR-WHITING TEST

| Feldspar (potash) | 70% | Yellow, opaque surface over red-brown (Claybody: Portchester c/5–6.) |
| Whiting | 30 | |

The yellow color of this test indicates the possibility of a weak interaction of the glaze with the iron in the underlying claybody. (In the oxidation atmosphere, low amounts of iron will generally create a yellow color in a feldspathic glaze base.) Thirty percent additions of calcium carbonate and more (up to 40%) bleached and yellowed the glaze surface just as it did in the cone 9–10 oxidation and reduction tests.

4. *Calcium oxide transforms a high-gloss surface into an opaque satin-matt at cone 5–6 and 9–10 oxidation and reduction temperatures.* At or below cone 5–6 temperatures, calcium materials combine readily with silica to form calcium silicate crystals, which have been identified as Anorthite—a calcium feldspar, and at high stoneware temperatures as Wollastonite, a calcium silicate mineral (pp. 166–168; Hamer and Hamer 1986, 196). Here is a fascinating example of the formation of new minerals from a glaze magma consisting originally of soda and/or potash feldspar, quartz, Whiting, and clay.

Cone 9–10 Reduction
Figure 1.5, p82.

SANDERS CELADON

		Gray-green. Transparent. Glossy.
Kona F-4 feldspar	44%	
Whiting	18	
Flint (silica)	28	
EPK kaolin	10	
Barnard clay	3%	
Add Whiting	20	Yellow-green. Satin-matt opacity. Semigloss.

SANDERS WOLLASTONITE

Figure 3.3, p225.

		Gray-yellow-green color. Satin-matt opacity. Jade-like surface.
Kona F-4 feldspar	36.5%	
Wollastonite	38.5	
Flint	21.0	
EPK kaolin	4.0	
Barnard Clay	3%	

Note: Increase of temperature to cone 10–11, (2380°F) caused the glaze to flow off the pot in thick, dense, opaque rolls. Where the glaze was thin, it was transparent and glossy. This fluid, opaque surface is an example of a "running matt."

At cone 5–6 temperatures in an oxidation atmosphere, calcium oxide functions more effectively as an opacifier and matting agent than as a melter. It produces satin-matt surfaces in much the same manner and for the same reasons as occur at the higher stoneware temperatures. The primary difference at cone 5–6 temperatures is that opacity is achieved with lower amounts of calcia.

Cone 5–6 Oxidation

NEPHELINE SYENITE GLAZE

Figure 4.7, p271.

1. Nepheline Syenite	90%	Opaque, white, high-gloss surface where thick. Transparent glass where thin.
Whiting	10	
2. Nepheline Syenite	80%	Opaque, white, satin-matt. Craze lines visible.
Whiting	20	
3. Nepheline Syenite	80%	Smooth satin opacity. More gloss than 2. Craze lines visible.
Wollastonite	20	

PORTCHESTER COPPER II (BASE)

Nepheline Syenite	56.7%	Surface displays smooth satin opacity but less gloss than 3. Craze lines not visible but strong crackle pattern revealed with rubbing of tea or artist ink.
Whiting	22.7	
Flint	17.5	
Tin oxide	3.1	
Omit Whiting		Stiff, paintlike gloss.

The Portchester Copper glaze displays an extraordinary surface. It is skinlike and smooth. Pots glazed with this formula are amazingly easy to clean—almost Teflon-like in the ease with which greasy and clinging food particles can be removed. An addition of 1% copper carbonate added to this formula produces a soft turquoise blue that is very dramatic on a white claybody. A 2% copper carbonate results in a deep green-black, which is handsome on both white and iron-colored claybodies. *Figure 1.6, p83.* When applied over or under a celadon glaze at the cone 9–10 firing temperature (reduction), it produces a brilliant copper-red gloss surface. Alone, without the celadon glaze, it produces a mottled red-green surface with areas of both matt and gloss. However, the smooth, satin surface that it produces at the cone 5–6 temperature (oxidation) outshines the reduction, cone 9–10 mottled red-green surface. Here is one instance where a glaze works more effectively at cone 5–6 oxidation than at the cone 9–10 reduction firing temperatures!

KATHERINE CHOY GLAZE

Figure 4.6, p270.

Nepheline Syenite	53.9%	Yellow, satin-matt
Whiting	11.7	flecked with areas
Lithium carbonate	4.7	of shine.
Zinc oxide	10.3	
EPK kaolin	17.9	
Flint	1.5	

Omit zinc oxide	Satin surface. Semi-opaque.
Omit Whiting	Opaque gloss. No satin surface.
Add 10% Whiting	Stony, dry surface. Yellower color.

These tests indicate that:

(a) Less than 12% calcium oxide was sufficient to create a satin-matt surface. Calcium oxide and not zinc oxide was responsible for the satin-matt surface, as the test without the zinc oxide still produced a satin-matt surface.

(b) The surface changed from satin-matt to opaque- gloss only when the calcium melter was entirely eliminated.

(c) An addition of 10% calcium melter destroyed the satin-matt surface and changed it into a dry and stony surface.

Note: A recent test on Grolleg porcelain cone 5–6 claybody produced a gloss surface!

Additional factors that affect the satin-matt glaze surface at cone 9–10 oxidation and reduction and cone 5–6 oxidation are as follows:

Changes in both firing temperature and cooling cycle may disturb the jade-like surface and render it transparent, glossy, and more fluid. The rise in temperature will cause more of the calcia and silica to enter the glaze melt, and less will be available for crystalline formations. The satin-matt surface of Charlie D (3% manganese, p. 158) changed into a high gloss, transparent surface flecked with visible, golden crystalline formations when subjected to a higher firing temperature of cone 10–11. *Figure 3.14, p236.* A slow cooling period at temperatures around 2200°F (temperature will vary depending on glaze composition and firing temperature) is necessary for the formation of calcium-silicate crystals (see p. 246).

Glaze application is crucial. A thick glaze coat (consistency of heavy cream) is necessary if a satin-matt surface is to be achieved. Thinner coats will produce a transparent, glossy surface.

The glaze formula should not contain high amounts of alumina. Alumina retards the formation of calcium-silicate crystals. Note the low kaolin content of Sanders Wollastonite glaze (p. 160). Kaolin is the primary source of additional alumina.

5. *Excess amounts of calcium oxide will opacify a high-gloss, transparent surface and turn it into an opaque, stony-matt.* This holds true for lower and higher stoneware temperatures in both reduction and oxidation atmospheres. (For examples of opaque, stony-matt surfaces produced by excess of calcium materials, see Alexandra's Glaze, p. 49; Tests, pp. 157–158) Calcium oxide is a refractory melter compared to soda and potash, which have melting points below 1700°F. Calcia requires temperatures of well over 4000°F in order to melt. This means that if there is more calcia in a glaze solution than needed to form calcium-silicate crystals, the excess will remain in the glaze melt in the form of unmelted calcium particles. Unmelted particles will obstruct the flow of light through the glaze layers, thus creating a dense, unreflective opacity. This kind of matt surface will be hard to distinguish from a glaze surface with unmelted silica or alumina particles—unmelted particles of most ceramic materials create an undistinguished, opaque, dry surface. If the firing temperature increases, so that the unmelted particles become an active part of the glaze melt, this opaque, dry surface can change to satin-matt. Once again, a hard and fast rule is hard to come by. What constitutes an excess of calcia in one glaze formula may not be the same for another. It all depends on the rest of the materials in the glaze batch and, of course, the firing temperature of the kiln.

6. *Calcium oxide reacts with iron oxide and feldspathic glaze materials to create unique glaze color effects in both oxidation and reduction atmospheres at cone 5–6, and cone 9–10 firing temperatures. An understanding of the important role that melters, glassmakers, and adhesives play in producing certain glaze colors is one of the aims of this book.*[2] The color and tone of a glaze surface depend to a large extent on the presence or absence of particular melters. In many glazes, it is not possible to achieve the desired color without including a certain amount of a particular melter. For example, it would

not be possible to obtain celadon blue, green, or gray colors at the cone 9–10 reduction firing temperatures without the presence of calcium oxide. Calcium oxide is the primary melter in the celadon glazes and helps iron oxide produce these celadon colors. *Figure 3.2, p224.* A celadon glaze usually contains from 1/2% to 3% iron and 10% to 20% calcium minerals. A series of cone 9/10 reduction tests with 14 different melters showed that the purest and most brilliant blue-green celadon color occurred with the calcium minerals Whiting (calcium carbonate) and Wollastonite (calcium silicate). Whenever the calcium mineral was combined with, or entirely replaced by, a different melter, such as lithium carbonate, soda ash, Gerstley borate, talc, magnesium carbonate, dolomite, or zinc oxide, the blue-gray-green color lessened and in some cases entirely disappeared.

Calcia has an equally strong effect on the deep red-brown-black color of high-iron glazes at cone 9–10 reduction temperatures. These test results show calcium oxide yellowing the deep red-brown-black color of a saturated iron glaze.

PAM'S SATURATED IRON GLAZE

Brown-black color.
High gloss.

Custer feldspar (potash)	52.4%
Whiting	10.1
Gerstley Borate	6.2
Flint	17.0
Kaolin	5.6
Red iron oxide	8.6

Replace Gerstley Borate with Whiting: Yellow color. Double the quantity of Whiting: Green-yellow-brown color. Replace Custer feldspar with Albany Clay:* Yellow-brown color.

*contains 5.78% calcia.

In a similar test, 10% Whiting was combined with 90% potash feldspar and 8.5% red iron oxide to produce a strong black-brown colored surface. As the Whiting was increased to 30%, the color became yellow-brown.

A similar yellowing and bleaching effect was observed in tests of glazes with little or no iron content. In these tests, the increased calcium melter changed the gray-white, or gray-blue surface of a cone 9–10 reduction glaze to a yellow-flecked surface.

TEMPLE WHITE

Figure 4.8, p272

Gray-white.
Satin-matt satin gloss.

Custer feldspar (potash)	34.7%
Dolomite	19.6
Whiting	3.1
Flint	18.9
Kaolin	23.6

Add 20% Whiting: Yellow flecks appear on gray-white surface.

This bleaching and yellowing effect of calcium also occurred in the oxidation atmosphere at low and high stoneware temperatures.

ELENA'S SATIN-MATT

Cone 8 (2257°F–2300°F)

Gray-white-yellow.
Satin-matt.

Custer feldspar	56 grams
Whiting	20
Flint	6
Kaolin	18
Zinc oxide	9
Rutile	7

Add 10%–30% Whiting: Yellow increased with each addition.

Low amounts of calcium material added to the following cone 5–6 oxidation glaze bleached and lightened the strong blue surface.

JACKY'S CLEAR (BLUE)

Bright blue gloss .

Nepheline Syenite	50%
Gerstley Borate/Colemanite	5
Wollastonite	10
Zinc oxide	10
Flint	20
Ball clay	5
Cobalt oxide	2%

Add 10% Whiting: Pale blue surface.

SUMMARY

1. *Importance of Quantity of Calcium Melter*

It must be evident from the tests that the key problem is one of proportion. In other words, only a particular amount of calcium melter will produce the desired surface result in a particular glaze and firing temperature. This principle of proportion will follow us throughout our exploration of glassmaker, adhesive, and melter materials and will hold true for all of them. In a certain proportion the material will obediently combine with the rest of the glaze materials to produce the expected result. If we exceed that amount and increase the amount of the material, the opposite result can occur. The precise amount needed to achieve the desired surface will, of course, vary, depending on the particular glaze formula, as well as the firing temperature and atmosphere of the kiln.

2. *Importance of Glaze Application and Firing Conditions*

As stated on p. 161 with respect to tests of satin-matt surfaces, it is necessary for the glaze coat to be thick to achieve the satin-matt surface. In addition, because the shape and form of the pot play a role in obtaining the desired thickness, the method of glazing also becomes important. The inside of a pot is usually glazed before the outside. Consequently, the inner walls and base pick up a thicker coat of glaze than the later-glazed outer walls, which have become damp from the interior glaze coat and will thus resist a thick application of glaze. Furthermore, because heat remains longer inside a vessel form, both the heating and the cooling period will be different for the inside and the outside walls. Then, too, there is that ever-intriguing question as to the role that fractional crystallization plays in ceramic glaze formation. According to this principle, certain portions of the glaze magma may well crystallize before the rest of the glaze materials in the melt. Crystalline materials are heavier and will sink to the bottom of the melt. In a vessel form, they would slide downward toward the center of the interior. Thus, even the glaze composition may be different on various portions of the same clay form. All of these factors can produce vast differences between the inside and the outside of a pot and could result in a satin-matt interior and a glossy, transparent exterior.

3. *Presence of Alumina in the Glaze*

The particular kinds of glaze materials in the batch recipe will obviously affect the final result. As stated previously, if the glaze batch contains a high amount of alumina it will be difficult to achieve a satin-matt, calcium-silicate crystalline surface no matter how much calcia is present. Alumina retards the formation of large crystals. The inhibiting tendency of alumina was dramatically revealed when we eliminated EPK kaolin, (primary source for additional alumina) from two standard celadon formulas. The shiny, transparent celadon surfaces changed to a satin-matt, semigloss surface in the absence of kaolin. Note that there is almost no additional alumina present in most of the standard formulas for macrocrystalline glazes, in which one or two enormous crystals seed themselves within a high-gloss, transparent matrix (see Sanders 1974, 14–39). We deal with this characteristic of alumina in more detail in Chapter 4 when we consider the oxide of alumina and its special contribution to glaze surfaces. For now, it is enough to know that a satin-matt, jade-like opacity can only be achieved through the formation of large, calcium-silicate crystals—and to achieve this kind of surface, the less alumina the better.

CONCLUSION

Despite the tremendous sensitivity of the high-calcium surface to the various conditions described above, its special, jade-like beauty more than compensates for these difficulties. Gemstone surfaces by their very nature are elusive and defy standardized repetition. This kind of rarity and uniqueness simply serves to enhance their value. Though one does not always achieve the desired glaze surface, the search itself can provide its own rewards in the form of increased knowledge and better understanding of ceramic materials. As you immerse yourself in the exploration of ceramic materials, you will find that joy comes not only from the successful end result, but also from its hot pursuit.

SOURCES OF CALCIUM OXIDE
WHITING (CALCIUM CARBONATE)
Figure 3.2, p224.

Calcium carbonate is the best source of calcium oxide for the potter.

The softest and the most easily pulverized form of calcium carbonate is the earth material chalk. Chalk is a marine deposit of the decomposed bones and shells of ancient sea animals. The finest grind of chalk, as well as all other soft forms of calcium carbonate, is packaged under the trade name of Whiting, and it is under this name that calcium carbonate finds its way into many of our stoneware glazes. It will emerge as calcium oxide after the conclusion of the firing. Some of the purest deposits of Whiting make up the white cliffs of England; Hammill & Gillespie sells Yorkshire English Whiting under the name of SnowCal 40. In this supplier's catalogue, domestic Whiting appears as Dolomite Dolocron 4013, a form of dolomitic limestone, which is less pure than the English Whiting, due to the presence of considerable amounts of magnesium carbonate. It is also much less expensive (see pp. 198, 201 for magnesia and dolomite). Hamada, the great Japanese potter, obtained his calcium carbonate from pulverized seashells. Whatever the source, Whiting is a reliable source of calcium oxide for the potter.

Cone 9–10 Reduction

The effect of 10% Whiting on a potash feldspar (K200) and Cornwall Stone at the cone 9–10 reduction temperatures with 1/2% iron oxide was compared with 14 other melters. With the exception of the Wollastonite test (see pp. 166–168), both the Cornwall Stone and K200-Whiting tests produced the smoothest surface and purest blue-gray color.

A more recent series of tests, which compared the effect of Whiting with other calcium melters, produced some interesting differences.

G-200 GLAZE		
		Gray-blue color. Black carbon spots on some tests. Gloss.
G-200 feldspar (potash)	90%	
Whiting	10	
Red iron oxide	1/2	
Bone ash	2	
Replace Whiting with Wollastonite		Light blue-gray: No carbon spotting.
Replace Whiting with Dolomite		Lighter blue-gray. Gold luster. Gold-gray-brown spots. No carbon spotting.

Conclusion

Whiting tends to attract carbon particles from the kiln atmosphere, unlike Wollastonite or Dolomite. In addition, because the underlying claybody color in all of these tests is dark gray, the grayer cast of these Whiting tests suggest that Whiting interacts more strongly with the claybody and creates more transparency than Wollastonite or Dolomite.

NEPHELINE SYENITE GLAZE		
Nepheline Syenite	90%	Dark gray.
Whiting	10	Matt. Glassy spots.
Replace Whiting with Wollastonite.		Gray-green gloss.
Replace Whiting with Dolomite.		Yellow-green gloss.

Conclusion

Whiting again creates the darker surface because it attracts carbon particles. Once again, it interacts with the dark-gray claybody to create a transparent surface. Note the matt surface in the Whiting tests interspersed with areas of glass. Whiting in combination with low-silica Nepheline Syenite helps to achieve a matt, but still transparent, surface. Wollastonite (which brings in added silica) provides an overall gloss surface. Dolomite increases the gloss of the surface and, once again, encourages a gold-yellow color.

WHITING (CALCIUM CARBONATE)

TRADE NAME Whiting

GEOGRAPHICAL SOURCE* England, Hammill & Gillespie, Inc.
U.S.A. Purest domestic source: St. Louis, Mo.; Kentucky.

GEOLOGICAL SOURCE* *European source:* Finest grind of aragonite (chalk-marine shell deposits).
Domestic source: Marble and sugar calcite ores; "Whiting" now includes any pulverized, fine-grained limestone material containing calcium carbonate, with lesser amounts of magnesium carbonate.

CHEMICAL STRUCTURE

	$CaCO_3$	$CaCO_3 \cdot MgCO_3$
	SnowCal 40 Yorkshire English Whiting**	Dolocron®***
Calcium Carbonate	97.05%	55.0%
Silica	1.85	
Alumina	0.35	
Ferric Oxide	0.10	0.3
Magnesia	0.25	43.0 ($MgCO_3$)
Sulphuric Anhydride	0.05	
Potash	0.01	
Soda	0.16	
Copper	5 ppm	
Manganese	350 ppm	
Phosphorus Pentoxide	425 ppm	
Moisture (H_2O)	0.1	0.1

*Ceramic Industry 1998, 177.
**Hammill & Gillespie, Inc. Technical Data 1993, 1998. Note that Dolocron® (dolomitic limestone with approximately 55% calcium carbonate and 43% magnesium carbonate) is listed by the supplier under the heading of Whiting and is less than half the cost of the English material.
***Specialty Minerals Inc., courtesy of Hammill & Gillespie, Inc. 1998.

WOLLASTONITE[3]
Figures Intro.7, p17; 3.3, p225.

Although Whiting (calcium carbonate) may well be the most common source of pure calcium oxide, its carbonate component creates glaze-surface problems. As early as 1517°F, bubbles of carbon gas erupt from Whiting and flow through the glaze layers to the surface with explosive force. If the glaze magma is not very fluid (high silica, feldspathic magmas of stoneware glazes tend to be viscous), the flight of carbon will leave telltale traces on the glaze surface in the form of tiny pinholes. If this occurs, it is preferable to use the relatively carbon-free, low-iron, scaly structured mineral known as Wollastonite, named after the eighteenth-century British geochemist William Hyde Wollaston (Frye 1981, 743).

Our domestic source of Wollastonite combines calcium oxide with equal amounts of silicon oxide and contains less than 1% carbon. This mineral formed thousands of years ago when hot, silica-rich magmas infiltrated the adjacent limestone rock. This torrid contact metamorphosed the outer layers of the limestone into a calcium silicate mineral with a needlelike crystalline structure. The length and width of the needles determine Wollastonite's industrial function. Wollastonite is manufactured in two forms—powdered, in which the needlelike crystal is relatively short, and acicular, in which the needles are long and narrow. Powdered Wollastonite is used primarily by the ceramic industry in glazes. Acicular Wollastonite, which is more expensive to manufacture, has far-ranging, important potential as a substitute for asbestos and as a filler in the plastics and resin industries. This acicular form of Wollastonite is also used as a filler in ceramic claybodies.

Although Wollastonite is a common accessory mineral in metamorphic rocks, large deposits are rare. It is perhaps for this reason that Wollastonite is a relatively new addition to the ceramic materials list. Commercial exploitation on a large scale in the United States did not begin until the 1950s. At that time, the primary commercial manufacturer was (and still is) NYCO, a division of Processed Minerals Inc., who worked the large deposits of Wollastonite found in New York state. NYCO has recently acquired mineral rights to mines in Mexico, which produces more than half the world's production (Eppler and Robinson, "New Opportunities for Wollastonite in Traditional Whitewares," *Ceramic Industry,* April 1998). Up until 1994, production concentrated on acicular Wollastonite for the asbestos-replacement, plastics, and resin industries. However, the company has now expanded its markets to include both industrial and nonindustrial ceramics markets with emphasis on the use of Wollastonite in glazes and claybodies. The ceramics industry remains a major portion of this company's market.

We see in Wollastonite the relationship between a kiln firing and the formation of the earth's minerals. Crystals of Wollastonite will form within a glaze surface whenever sufficient amounts of calcia and silica are present in a glaze magma fired above 2192°F and cooled slowly. (Below these temperatures, calcia and silica will combine to form crystals of the calcium feldspar known as Anorthite; Hamer and Hamer 1986, 196.) It is the presence of these Wollastonite crystals that create the jade-like, satin-matt surface in the Sanders Wollastonite glaze (p. 160). *Figure 3.3, p225.*

In addition to creating satin-matt, jade-like surfaces, Wollastonite will create glassier surfaces and more brilliant colors than can be obtained with Whiting (calcium carbonate). The reason for this is that Wollastonite, unlike Whiting, is not a pure source of calcia. In Wollastonite, the glassmaker silica joins calcia to provide more powerful melting action and increased surface brilliance. *Figure Intro.7, p17.*

WOLLASTONITE*

TRADE NAME	Wollastonite
GEOGRAPHICAL SOURCE	New York: R. T. Vanderbilt Company, Inc. New York, Mexico: NYCO Minerals Inc.
GEOLOGICAL SOURCE	Wollastonite—Contact metamorphosed limestone of the Grenville Formation of Canada and New York. Precambrian age (1000–880 m.y.)
CHEMICAL STRUCTURE**	$CaSiO_2$

	VANSIL®W	NYAD®325
SiO_2	50.0	51.0
CaO	44.0	47.5
FeO	0.3	0.4
Al_2O_3	1.8	0.2
MnO		0.1
MgO	1.5	0.1
Na_2O	0.2	
TiO_2		0.02
L.O.I.	2.2	0.68

MINERALOGICAL STRUCTURE***
Wollastonite	97.5%
Calcite	2.5
Diopside	Trace

*C. S. Thompson, Manager of Minerals, R. T. Vanderbilt Company, Inc. (Telephone interview 18 January 1992); *Ceramic Industry* 1998, 177; Keeling 1962, pp. 877–894; Power 1986, pp. 19–34.
**R. T. Vanderbilt Company, Inc. 1998; NYCO Minerals, Inc. 1998.
***Sample examined by Geological Dept. of British Research Ceramic Assoc., Keeling 1962, 877.

Glaze Function

Cone 9–10 Reduction

1. In the test series that compared the melting action of 14 different melters to Whiting, the Wollastonite test produced the most brilliant blue-gray color. It was, in addition, slightly shinier than the Whiting test.

2. Twenty percent Wollastonite was added to Sanders Celadon and compared with a test in which 20% Whiting was added to Sanders Celadon. The Wollastonite test produced a slightly shinier, bluer surface than the comparable test with Whiting (see p. 160).

Cone 5–6 Oxidation

1. Tests of 90% soda feldspar and 10% Whiting were compared to 90% soda feldspar and 10% Wollastonite. Once again, a glassier, shinier surface appeared in the Wollastonite tests.

2. A test in which 30% Wollastonite was added to a gloss glaze known as Jacky's Clear (see p. 212) produced a greater transparency and shine compared to the test with Whiting.

3. Note the stonier surface of Portchester Copper glaze (22.7% Whiting) compared to the surface produced by Nepheline Syenite 80%, Wollastonite 20%. *Figures 1.6, p83; 4.7, p271.*

Conclusion

The differences in most cases were subtle; in certain glaze batches, the substitution of Wollastonite for Whiting may make only a negligible difference. Wollastonite is a useful substitute for Whiting whenever it is necessary to:

1. Eliminate pinholes
2. Slightly lower the firing temperature
3. Slightly increase shine and transparency

Such an effect can be accomplished by the substitution of Wollastonite for Whiting, without greatly disturbing the glaze surface. Conversely, a slight increase in firing temperature or a lessened gloss and transparency can be achieved by substituting Whiting for Wollastonite.

Although there were no striking color changes and the gray-blue-green celadon color basically remained the same, some of the Sanders Celadon tests did show a distinctly bluer shade with the substitution of Wollastonite for Whiting. Recent tests by other potters have confirmed these results and point to the general conclusion that celadon surfaces will be bluer, smoother, slightly more melted, and clearly more brilliant with the Wollastonite melter. In this connection, research on the Song Dynasty glazes show that the lime ash used by the ancient Chinese Potters resembles Wollastonite in its chemical structure (Hennessy 1982, 95–96).

The same conclusions with respect to the advantages of

Wollastonite held true in most instances for oxidation cone 9–10 and 5–6 glaze surfaces.

It can thus be seen that Wollastonite is an indispensable ceramic melter material. Except for those cases where the objective is a lowered gloss or a more muted color, Wollastonite constitutes a superior source of calcium oxide.

Claybody Function

Both individual studio potter tests and industrial research indicate that the semifibrous, needlelike structure of Wollastonite makes it a useful and desirable claybody material at all firing temperatures. Claybodies that contain Wollastonite in its fibrous, nongranular form have lower drying and firing shrinkage, lower moisture expansion, lessened weight, increased green and fired strength, and increased resistance to thermal shock (*Ceramic Industry* 1998, 177; Keeling 1962, 878; Power 1986, 24). Wollastonite (10%–15%) is frequently added to sculpture, floor tile, and cone 5–6 claybodies in place of an equivalent proportion of feldspar and flint (Ibid.).

WOOD ASH
Figures 3.4–3.6, pp226–228.

It is an apocryphal part of potters' lore that the first stoneware glaze was born some 1500 years ago in the Far East when the ashes of a wood-fueled kiln accidentally settled on the shoulders and neck of the pots during the kiln firing and melted into a green, glassy surface. Wood ash is the powdery residue of inorganic material left over after the completed combustion of wood. It contains the working trinity of glassmaker, adhesive, and melters, plus iron oxide, and thus constitutes an iron-bearing glaze core for stoneware firing temperatures. For centuries, wood ash has been an invaluable natural glaze material for the potter. One has only to see the Japanese Tokonabe wares of the late Heian period or the "Old Shigaraki" wares of the Muromachi period (1392–1573), to appreciate the extraordinary surfaces created by the ashes of wood (Koyama, 184, 188). *Figure 3.4, p226.* Today, its special appeal comes from its earthbound origins—the trees, shrubs, and plants that make up the natural landscape of our lives. In this modern age of commercial and overpurified ceramic materials, wood ash stands alone and without rival as a rich and obtainable source of unique glaze surfaces.

As a start, it is necessary to examine the terms *wood ash* and *ash glazes* as used by potters. Implicit in these words is the fact that we are now dealing with what is left of a specific kind of organic material after it has been consumed by fire. Although the term *ash glazes* generally means wood ash, ceramic materials contain many different kinds of ash materials. It is obvious that there are as many different kinds of ash as there are materials that burn. Fiery volcanic eruptions produce "volcanic ash." The cremation of animal bones results in "bone ash." Banana ash was produced by firing a banana peel inside a covered pot to cone 9–10 reduction temperatures. Bananas contain considerable amounts of potassium, and the burning of the peel created an ash that singed the interior of the pot in a manner similar to the action of pure potassium (see Chapter 1, pp. 25–26). Thus, a lustrous, glassy, orange-red patch appeared in the interior of the pot in place of the banana peel. In the case of wood, we are dealing with an organic material that turns to powdery ash after burning for a short time in a relatively low-temperature fire. Thus, unlike some of the other kinds of ash mentioned above, it is possible for any ordinary mortal with a fireplace to transform wood into a unique glaze material, which functions as both a melter and a glaze core.

Wood ash is a highly complex mixture of silica, alumina, calcia, magnesia, potash, soda, phosphorus, and iron oxide—the same trinity of glassmaker, adhesive-glue, and melters that is present in the feldspars—and that is the reason wood ash makes such an effective glaze core at the higher stoneware temperatures. The most common kinds of wood ash provide an interesting and varied glaze surface when fired alone at cone 9–10 reduction temperatures. Pine, Beech, and Ash wood ashes each melt into a green, yellow-brown, sugary surface on an iron-bearing stoneware claybody. Beech ash created an exquisite, green-orange-black, mottled, stony-matt interior ringed with orange on a small porcelain bowl.

Despite their complexity, it is possible to make some generalizations about the numerous kinds of wood ashes and their function as glaze materials.

1. Soft Wood Ash

Wood ashes have been classified as soft, medium, or hard depending on the ratio of the melters to the total silica, alumina, and phosphorus content. Bernard Leach states the general rule that if the result of subtracting the total melters from the total of the silica, alumina, and phosphorus content is minus 20 or more, the ash is soft (Leach 1962, 161). Soft ashes function in glazes as an impure source of calcium, containing as much as 40%–50% calcium oxide. Thus, soft wood ash such as pine and apple ash may be substituted for commercial supplies of Whiting and Wollastonite with very interesting results.

Pine ash in amounts of 10%–20% were combined with 90%–80% Cornwall Stone respectively, and fired to cone 9–10 at 2340°F in a reduction atmosphere. The soft blue surfaces that resulted were similar to the surfaces produced with comparable Whiting (calcium carbonate) and Cornwall Stone Tests. Soft Pine ash substituted for Whiting in Sanders Celadon with almost identical color, flow, and surface results. In this test, we eliminated the added iron colorant (Barnard 3%–5%) and found that the iron mineral carried by the Pine ash produced a celadon shade that was close to the original glaze. When we added the iron colorant, the result was a slightly more fluid, glassy surface, with a deeper green color than in the original Sanders Celadon glaze. The difference, however, was very subtle.

Despite the similarity of test results, the chemical structure of soft wood ash differs markedly from the chemistry of Whiting—which makes the results even more remarkable! In addition to 40%–50% calcium oxide, soft wood ash contains varying amounts of silica, potash, phosphorus, soda, magnesia, and iron oxide. The structural difference between Whiting and soft wood ash appears clearly when each are fired alone. As stated above, soft wood ash creates a sugary, green-yellow, stony surface. Whiting does not melt at high stoneware temperatures, because the melting temperature of calcium oxide is well over 3000°F, and the stoneware kiln rarely exceeds temperatures of 2360°F. Thus, most of the calcia remains in the same form—white, stiff, and unmelted. However, a faint, underlying layer creates a green-yellow-brown surface, which is visible wherever the calcium layer is thin. This thin layer of calcia has pulled some silica and alumina from the claybody and produced a melted surface very like the green-yellow-brown surface of Pine ash.

WOOD ASH

TRADE NAME	Wood Ash, Plant Ash
GEOLOGICAL SOURCE	Trees, shrubs, vegetation
CHEMICAL SOURCE	Variable and complex

EXAMPLES*

	Pine	Apple (washed)	White Mahogany	Box (washed)	Wheat Husk
Silica	24.39	2.65	51.51	14.29	68.53
Alumina	9.71	1.98	3.81	10.34	4.44
Iron Oxide	3.41	0.70	4.53	2.74	4.46
Phosphorus	2.78	1.59	2.08	4.73	2.23
Calcia	39.73	54.20	9.49	37.55	7.23
Magnesia	4.45	3.25	4.39	6.12	1.88
Soda	3.77		10.98		
Potash	8.98	0.89	9.49	2.58	8.03
Manganese Oxide	2.74				
Carbonate		34.69		21.49	
Sulphate					8.03
Chloride					8.03

*Sanders 1982,232; Leach 1962,162.

Thus, despite the considerable difference in oxide structure, it is possible to substitute soft wood ash for calcium carbonate in a celadon glaze and achieve strikingly similar results.

2. Medium Wood Ash

Medium wood ash contains higher amounts of silica, alumina, and phosphorus and, correspondingly, lower amounts of calcia. According to the Leach rule, if the total melters are subtracted from the total silica, alumina, and phosphorus content with the result between 40 (hard) and minus 20 (soft), the ash is labeled medium. As is bound to be with all classification systems, overlapping is inevitable. For example, Pine ash lies just at the borderline between the Leach test for soft ash and medium ash. The best way to determine the nature of the ash is to test the ash alone.

Tests of Beech and Ash wood ashes compared to pine ash proved less runny. This indicates that they contain lower melter content than the pine ash. Despite this difference, these ashes functioned very effectively as a substitute for calcium carbonate and produced unique color effects as well. When Beech ash was combined with equal amounts of Cornwall Stone, an opalescent blue opacity appeared that suggested the brilliant blue bowls of the Song Dynasty. Only beech ash produced this surface—none of the other kinds of ashes that were combined with Cornwall Stone in similar tests achieved this brilliant blue surface. A similar ring of milky-blue brilliance appeared when Beech ash substituted for Whiting in Sanders Celadon. A comparable test of Pine ash produced a gray-green surface similar to the original glaze surface. The substitution of Beech ash for Kona F-4 (the soda feldspathic glaze core in Sanders Celadon) changed the glossy, transparent, blue-green-gray surface to a pale, yellow-green, stony surface ringed with a brilliant sea green crackled glass. In a second test, the ash replaced both the feldspar and the melter Whiting (18%) and thus functioned as both a glaze core and an auxiliary melter. The resulting surface was stony, yellow matt, as in the first test, but the green, glassy ring changed once again to the opalescent, milky, blue ring so characteristic of this kind of ash. The reasons for this unique color effect may lie in the higher amounts of silica and phosphorus (another glassmaker), which this kind of ash contains. Glasses of differing refrac-

tion rates suspended inside each other can cause various color and surface effects (see p. 156). We would need further testing of the same batch of ash on larger test pieces in numerous kiln firings in order to substantiate these conclusions. A test of 20% Walnut ash and 80% Nepheline Syenite produced a gray-blue, matt-shine, highly fluid surface that recalled the surface produced with 10% Whiting. *Figure 1.3, p80.* See also p. 42.

The results that appear at this very early stage of testing are but a tantalizing hint of the enormous potential that exists in the combination of wood ash and feldspars. The brilliant, sparkling interior rings of milky blue and sea green, which appeared so consistently on all of our feldspar and wood ash combinations, suggest wonderful possibilities for the decorative interiors of bowls. These are but a few of the many, unexpected rewards for the diligent tester. Though the same results may not repeat, these unique wood ash surfaces are treasured precisely because they do not bear repetition. They are true gemstones—a gift from the gods.

3. Hard Wood Ash

Hard ash contains high amounts of silica, which may reach 70%–80% of the total oxide content. Because the melter content is proportionately low, hard ash does not function as a glaze core, but rather as a natural source of iron and other colorants. Vegetable ash such as Rice ash, Kelp ash, and Wheat ash are primary sources of hard ash. It is believed that fern ash was the colorant in many of the ancient Chinese celadons. The Song Dynasty potters burned layers of Fern ash and limestone and used the resulting ashes in combination with powdered petuntse rock (one-third feldspar, one-third kaolin, and one-third quartz) to create the jade-like surfaces of their celadons (Sanders 1982, 197–198; see also Wood 1978, 22). Our tests produced a rich, gray-green, lustrous surface when 5%–10% Fern ash was added to 90% Nepheline Syenite and 10% Whiting. The refractory nature of hard ash appeared when 90% Cornwall Stone combined with 10% Kelp ash. Unlike the comparable test with Pine ash, which produced a blue-green gloss surface, the Kelp ash test was white, stiff, and unmelted.

Oxidation Atmosphere
Cone 9–10 Oxidation; Cone 5–6 Oxidation

Thus far we have described wood ash as a melter/glaze core for reduction cone 9–10 temperatures. What of the cone 9–10 and cone 5–6 oxidation temperatures? The green-blue colors of the wood ash tests, activated by the iron colorants in the ash, occur only in the smoky, roaring atmosphere of a fuel-burning kiln. In the still, airy atmosphere of oxidation, the iron colorants in the ash mixture create a pale yellow surface. Beech ash fired alone on a porcelain claybody becomes a pale yellow glassy puddle, compared to the reduction test— a brilliant green center ringed with gray-black and singed with orange-red. A 10%–20% beech ash, combined with 80%–90% potash feldspar and fired to cone 9–10 oxidation temperatures melts into a glossy, yellow, brown-speckled surface. Compare this result with the soft, silky blues of the same tests fired in the reduction atmosphere. Although these oxidation tests show that the ash functions as a substitute calcium melter, it is not as powerful a melter in the oxidation atmosphere as it appears to be in the reduction tests. (This, however, is the case with calcium carbonate as well and is evidence of the greater fusion produced in the reduction firing.)

This does not mean that wood ash has no value for oxidation glaze surfaces. On the contrary, the impurities in the wood ashes can bring excitement, life, and depth into the often flat and dead surface of an oxidation glaze. A red-brown surface speckled with yellow-green appeared when 60% Spodumene combined with 40% soft wood ash and 10% Gerstley Borate. Although fired in an oxidation atmosphere, the mottled surface was not unlike the surface of wood-fired clay, licked by the flames and scored by the flying ashes of the fuel. *Figure 1.12, p89*. Even at the lower stoneware temperatures (cone 5–6), it is possible to use this impure material as an exciting source of new kinds of surfaces. Naturally, at these lower temperatures, larger quantities of powerful melters are necessary. Gerstley Borate, a combination of calcium and boron (glassmaker and melter combined, see pp. 188–190) melts alone into a green-yellow glass at these temperatures, and is, therefore, a natural melter for wood ash. A combination of 50% Gerstley Borate and 50% soft wood ash creates a mottled, yellow-green, glassy surface on a white-burning clay. The same combination, plus an additional 10% Flint (silica) on a red-burning claybody produces a mottled, dark, yellow-green, matt-shine surface, which echoes the 9–10 reduction firing of pine ash. *Figure 3.6, p228*. Soft wood ash combined with equal amounts of kaolin plus 10% lithium carbonate creates a soft, speckled, yellow matt surface on the white claybody. Equally interesting surfaces appear when Barnard clay substitutes for kaolin. An extraordinary black, leopard-skin spotted surface appears when 35% soft wood ash is combined with 50% yellow ochre (hydrated form of iron oxide),[4] and 15% Whiting.

If we accept these complexities and limitations and are willing to devote the time for exploration, soft, medium, and even hard wood ashes will prove to be exciting materials for oxidation glazes—at both low and high stoneware temperatures. Once activated by powerful melters, the impurities of wood ash should restore the depth and variation, which is so often missing from the oxidation surface.

Preparation of Wood Ash

Whether or not any of these test surfaces will bear repetition in subsequent tests remains to be seen. Wood ash is far more variable than feldspar. Herein lies both its virtue and its vice. Although it contains the potential for extraordinary "one of a kind" gemstone glaze surfaces, such surfaces may be extremely difficult to repeat. Strictly speaking, there is no such thing as wood ash; there are only the ashes of various kinds of wood. Thus, the actual proportions of each oxide contained in a particular wood ash batch will vary, depending not only on the kind of wood, but also on the particular part of the tree that is burned. The chemical composition of each tree will be different depending on its particular locality and soil. Furthermore, silica content is higher in the older and more mature portions of the tree. In other words, even the various sections of the same tree can have vastly different chemical compositions. For this reason there can be significant variations in different batches of the same kind of wood ash. A test from one batch of Ash wood ash produced a green, glassy interior and a pale yellow, dry matt exterior when it was combined with equal proportions of Kona F-4 feldspar and fired to cone 9–10 reduction temperatures.

The second test, which used a different batch of Ash wood ash, produced a milky blue interior ring, in marked contrast to the surface, gloss, and color of the first batch test! In view of the countless variations that are possible, it is advisable to collect a large amount of ash from one particular source before beginning your tests. Test the ash from this large batch alone and then in simple combinations with feldspars or other glaze cores at various firing temperatures and in both oxidation and reduction atmospheres. This procedure will tell you more about its glaze potential than any scientific chemical analysis. Be sure that you have screened the ashes to eliminate any large particles of unburned carbonate material. The final batch should consist of uniform, powdery particles. Lumps and larger size particles will not combine properly with the rest of the glaze materials and could result in a rough and unmelted surface. **Caution:** Rubber gloves and goggles or eyeglasses should be worn when handling ashes; the alkali materials in wood ash are highly caustic and may irritate the skin and mucus membranes of the eyes.

The washing of the ash removes some of these alkali, water-soluble materials, such as soda and potash. This is accomplished by soaking the ashes in a garbage can filled with water for about three days. Each day, the water is replaced with fresh water, thereby removing the water-soluble materials in the water. The ash is then dried on newspapers and is ready for use. The result is an ash with less melting power due to the loss of the soluble soda and potash melters. A test of washed Pine ash was not as glassy or as mottled as the unwashed test. In addition, it produced a stonier surface.

Artificial Wood Ash

Certainty, rather than the romantic pursuit of natural materials, may be your goal. If so, it is possible to eliminate the unpredictability of natural wood ash by combining ceramic materials that duplicate its chemical structure. Joseph Grebanier (1975, 127–131) provides a complete formula for artificial Pine ash and mixed common ash. The formula, which contains feldspar, whiting, flint, soda ash, and innumerable other ceramic materials, approximates the results obtained with natural soft wood ash. Tests of artificial Pine ash combined with Albany Slip clay produced a rich yellow-brown gloss, which flowed into a black-brown pool flecked with gold spots. The comparable test with natural Pine ash was darker in color, less fluid, and had more depth and complexity than its artificial counterpart. Despite these differences, the overall effect was similar. A test that combined Albany Slip clay with artificial mixed ash (more silica, less calcium, and less melters than artificial Pine ash) produced a rich, brown-black surface flecked with iridescent silvery spots. This surface had some depth, but it was darker and shinier than its natural wood ash counterpart. A simpler approach to the duplication of a wood ash surface is to combine 60% Albany Slip clay, 30% Whiting, and 10% Barium carbonate (*a highly toxic material*). When fired to cone 9–10 reduction temperatures, the result is a mottled, yellow-green surface that strongly suggests the natural wood ash surface. However, the even, ordered pattern of the mottled surface bears witness to its artificial origin and proves again the difficulty of repeating nature's work. Thus, predictability is possible, but at the cost of the very essence of a wood ash glaze surface—surface depth and free-flowing design.

Conclusion

In conclusion, certain basic principles emerged from our tests. Soft and medium ashes will function as substitutes for both the glaze core and the calcium melter with unique and special surface effects. Hard ashes serve best as a source of iron-bearing colorants. For this reason, it would be helpful to know the kind of wood or plants from which your ashes derive, as well as their general chemical composition, but this is only a preliminary starting point. Above all, it is essential to test the particular batch of ash alone and in simple combinations with other glaze cores and melters.

The more tests performed with different kinds of wood ashes, the more we uncovered the hidden personalities of

LAB IV CALCIUM OXIDE

WHITING   WOLLASTONITE   WOOD ASH

1. The first step is to become familiar with the calcium melters when they are fired alone. Take two to four identical test pots with a large enough surface to show clearly the effect of three different materials. Glaze 1/3 of each pot with Whiting, Wollastonite, and a wood ash. If you do glaze the outside of the pot, leave a good margin in case of excess fluidity. There will be no problem with Whiting or Wollastonite, but a wood ash at high stoneware temperatures can be runny. Fire each pot in as many firing temperatures as are available.

2. Mix up a familiar glaze (see Lab. II, #3, p. 94) without the calcium melter.

3. Substitute Whiting, Wollastonite, or the wood ash for the particular calcium melter in your glaze. Thus, if your glaze contains 18% Whiting, mix it up with 18% Wollastonite, and then again with 18% wood ash in place of Whiting.

4. Substitute your wood ash for the particular feldspar or glaze core in each of your two glazes. For example, if your glaze contains 50% potash spar, mix it up instead with 50% wood ash.

5. Add 10%–30% Whiting to your glaze formula. There will be three tests of 10%, 20%, and 30% additions of Whiting. Do the same kind of tests with Wollastonite and your wood ash.

Note: Place the original glaze on 1/2 of each of the tests. This is your control.

---

these materials. Not only did they exhibit unique characteristics, but also the equally unique personality of the feldspathic materials became clearly visible as we compared tests that combined the same wood ash with different feldspars and feldspathic rocks. This was particularly true in connection with the Cornwall Stone and Beech ash tests, which consistently produced a brilliant, milky-blue, opalescent ring of glass.

The rewards of prolonged testing with wood ashes and feldspathic materials are great. If you stay with it, you will eventually produce unique glaze surfaces that contain all the richness, depth, and diversity of unpurified raw materials. The use of wood ashes as a glaze material becomes a wonderful way of restoring the lost connection between the twentieth-century potter and his or her natural environment.

It is from this connection that very special pots are born.

BONE ASH (Calcium Phosphate)

TRADE NAME	Bone Ash: $4Ca_3(PO_4)_2 \cdot CaCO_3$
GEOLOGICAL SOURCE	Animal (cattle) bones calcined.
OTHER SOURCES*	*Apatite Group*: Apatite (CaF) $Ca_4(PO_4)_3$

Crystalline calcium fluorophosphate found in limestones, igneous rocks, and in clays. (Chloride or hydroxyl may substitute for fluoride.)

Calcium Phosphate: $Ca_3(PO_4)_2$:

Precipitate from mixtures of sodium phosphate and calcium chloride solution.

CHEMICAL STRUCTURE (variable)**

	Bone Ash Typical
P_2O_5	40.9%
CaO	54.8
MgO	1.1
CO_2	1.1
SiO_2	1.0
L.O.I.	1.5
Soluble Ash	99.5
Acid Insoluble Ash	0.6

*Bates 1969, 179. *Ceramic Industry* 1998, 65–66,74.
**Murlin Chemical Inc. 1998.

BONE ASH

You will recall that calcia bonds with the glassmaker silica to form the stable melter Wollastonite; it joins with the glassmaker boric oxide to form the powerful melter colemanite (see p. 183). Calcium bonds with yet another glassmaker—this time its mate is phosphorus. This combination produces organic bone ash and the natural minerals of the Apatite Group—materials that are strikingly different from any of the calcium minerals discussed thus far. The Apatite Group is the mineral source of calcium phosphate. Apatite contains more calcium fluoride than organic bone ash and has a lower melting point (2372°F compared to 3038°F for bone ash; Hamer and Hamer 1986, 358). Both are used commercially in the production of bone china. The organic material, bone ash, which is in fact the residue of burned cattle bones, is the common source of calcium phosphate for pottery studios. Hence, the following discussion of calcium phosphate is based on tests with bone ash.

Phosphorus pentoxide (P_2O_5) is the phosphorus material contained in bone ash. Alone, phosphorus pentoxide melts at the low temperature of 1058°F–1076°F but melts at 3038°F after it combines with refractory calcia (in a ratio of 40% to 55% respectively; Hamer and Hamer 1986, 32, 358, 362). Unlike the combinations of calcia with the glassmakers silica and boron, the union with the glassmaker phosphorus produces an opacifier instead of a melter. The special characteristics of this opacifier depend on the glassmaking properties of phosphorus.

In the colemanite section, it is pointed out that glasses of differing refraction rates, suspended inside each other, create the visual effect of opacity. In this way, the glassmakers silica and boric oxide create the milky, opaque or opalescent surfaces that are so characteristic of high-colemanite glazes at the cone 5/6 oxidation temperatures. The glassmaker phosphorus pentoxide, which is contained in bone ash, creates unique opacifying surfaces for the same reason. However, unlike colemanite, which functions both as a melter and an opacifier at the lower firing temperatures, bone ash functions primarily as an opacifier at the higher stoneware cone 9/10 firing temperatures. This difference is reflected in the different melting points of the two materials: colemanite's melting point is 1472°F; bone ash has a melting point of 3038°F, which is more than twice the melting point of colemanite! At the cone 9/10 temperatures, the heat of the fire causes the decomposition of bone ash and the release of highly reactive, liquid tricalcium phosphate, which dissolves the rest of the glaze materials. At the end of the firing, globules of phosphorus glass float inside the melted silica glass; the result is a milky, opaque, glaze surface. The ash of wood is known to contain fair amounts of phosphorus. Thus, the startling, milky blue surface of some of our wood ash and feldspar tests, as well as the incomparable, blue, semi-opaque surface of certain Song Dynasty celadons, is attributed to the presence of phosphorus (Hetherington 1948, 67–68). The color of the opaque surface (white or blue) and the degree of the opacity (opaque or semi-opaque) would depend on the size and density of the globules of phosphorus glass that float within the silica glass.

Glaze Function
Cone 9/10 Reduction

The extraordinary opacifying properties of bone ash at the higher firing temperatures appear in the following tests:

IA

Potash feldspar	90.0%	Blue-gray color.
Whiting	10.0	High gloss.
Red iron oxide	0.5%	Transparent.
		Visible craze lines.

Color of reduction claybody beneath glaze coat is usually gray; hence, a grayish surface color reflects underlying claybody color and is a good indication of transparency. White cast indicates opacity.

Analysis of shards of certain Chinese Song Dynasty celadons revealed the presence of low amounts of phosphorus, possibly derived from iron-phosphate minerals contained in the glaze (Hetherington 1948, Ibid.). This prompted us to add 2% bone ash (calcium-phosphate) to the above glaze batch. The result was a surface that strongly resembled an ancient Chinese blue celadon.
Figure 3.13, p235.

Add 2% bone ash	Lighter, bluer color. Milky surface. Less transparency. Visible craze lines.
Add 4% bone ash	Paler blue color. Milkier surface. Increased opacity. Pinholes. Visible craze lines.
Add 6% bone ash	Milky-white color. Increased opacity. Pinholes. Visible craze lines.

IB

Potash feldspar	90%	Deep, blue-gray color.
Whiting	10	Transparent.
Red iron oxide	1%	Visible crackle.
Add 6% bone ash		Lighter blue-gray color. Opaque. Fat, milky surface. Transparent.

Further tests made with small amounts of iron phosphate in place of bone ash and iron oxide were not as successful.

II COMPARATIVE TESTS OF WHITING AND BONE ASH

Potash feldspar Whiting	75% 25	Gray-blue-green color. Transparent.High gloss. High craze. Pools of green glass in center of test.
Potash feldspar Bone ash	75% 25	White opacity where thick; yellow-brown gloss where thin. No visible craze.
Potash Feldspar Whiting	50% 50	Stony, gray-green matt. A few spots of green glass in center.
Potash Feldspar Bone ash	50% 50	Stony, opaque, white. Rim has sheen.
Potash Feldspar Whiting	25% 75	Stony, yellow-green matt.
Potash Feldspar Bone ash	25% 75	Stony, opaque white (thin). Pinholed, dry, unmelted (thick). Rim has sheen.
Whiting	100%	Small center section stony, green-yellow; remainder of pot bare.
Bone ash	100%	White, glossy, needlelike crystals in center. Stiff, white, dry coat on sides of test pot.

III SAM HAILE GLAZE

Potash Feldspar	44 grams	Gray-blue, milky white.
Cornwall Stone	20	Semi-opaque;
Whiting	20	semigloss.
Kaolin	6	
Calcined Kaolin	8	
Zinc oxide	5	
Add 6% bone ash.		Whiter color; lower gloss. Pinholes

IV PAVELLE GLAZE

Figure 3.7, p229.

Bone ash	42.0%	Orange-yellow color.
Cornwall Stone	17.5	Opaque, smooth,
Nepheline Syenite	10.5	matt.
Dolomite	30.0	
Zinc oxide	4.0	
Rutile	3.0	

Note the high amount of bone ash in this glaze formula. Bone ash functions as a glaze core. The stony, opaque surface results from the combination of bone ash and dolomite (calcium-magnesium melter, see pp. 198–199). This glaze has not been stable. The use of different supplies of bone ash and/or Cornwall Stone produced a shiny and blemished glaze surface.

V PERSIMMON GLAZE

Figure 3.8, p230.

Potash Feldspar	48.6%	Red to red-orange
Whiting	7.2	to brown. Gloss.
Bone ash	10.0	
Talc	6.3	
Flint	21.6	
Kaolin	6.3	
Red iron oxide	11	

Tests that would determine the contribution of bone ash to the color and surface of this glaze need to be performed. See Lab V, p. 182.

Test Conclusions

1. Bone ash does not interact with the underlying claybody as does Whiting. Nor does bone ash behave as a melter. In all of these tests, its sole function is that of an opacifier. This opacifying function is evidenced by the milky, white surface color of the bone ash tests, compared to the blue-green-gray color of the Whiting tests. The blue-green color comes from the interaction of Whiting and feldspar with the iron in the claybody; the gray cast of these tests reflects the underlying gray reduction claybody, and thus indicates a transparency caused by the melting power of Whiting. Bone ash does not have this melting power, nor does it interact with the claybody, as indicated by the milky-white surface of the bone ash tests.

2. Bone ash is a powerful opacifier. Relatively low amounts of bone ash (6%) produce opacity and whiteness. A semi-opaque, milky-blue surface appeared with the addition of only 2% bone ash. It was necessary to add more than 25% Whiting before any opacity appeared. Low amounts of Whiting (up to 25%) increase the fusion of the glaze, whereas similar amounts of bone ash appear to have the opposite effect. In this connection, it is interesting to note that Whiting's melting point exceeds that of bone ash by about 1000°F. Despite the lower melting temperatures of bone ash, a relatively small addition will produce an opaque, dense surface, whereas the same amount of Whiting added to a feldspar will not produce any appreciable increase in opacity. The reason for these differences lies in the different nature of the opacifying properties contained in these two calcium compounds. The opacity caused by 25%–30% additions of Whiting derives from crystalline formations together with some unmelted particles of material, all of which require the presence of substantial amounts of Whiting. Bone ash achieves opacity by the suspension of phosphorus glass globules inside silica glass.

3. General rules with respect to permissible limits of the amount of bone ash are difficult to make and would obviously depend on the composition of the glaze. More than 4% bone ash caused some pinholing and crawling in thickly coated tests of high-feldspar glazes. On the other hand, note that in the Pavelle glaze more than 40% bone ash was used successfully in combination with 30% dolomite.

4. A thin coat of bone ash creates a surface sheen on the claybody. This sheen was visible in all of the bone ash tests; even the stony, unmelted surface of the 75% bone ash test revealed a slight sheen at the edges of the glaze where the layer was thinnest. Hence, it is common practice to spray a

thin coat of bone ash on an unglazed claybody. This is especially effective on sculptural forms.

Cone 5/6 Oxidation

Bone ash fired alone at the cone 5/6 oxidation firing temperature displays a white, dry, and powdery surface similar to Whiting, calcium carbonate. However, bone ash adheres more closely to the claybody than does Whiting, which remains unmelted and powdery at the cone 5/6 temperatures. Despite the dry, stiff surface of bone ash, a professional potter reports that a thin spray of bone ash and water will produce a faint sheen on a white, fine-grained claybody, at the cone 5/6 temperatures.

Bone ash does not appear frequently in glaze formulas at these temperatures. Out of 14 cone 5–8 glazes listed by John Conrad in *Ceramic Formulas: The Complete Compendium* (1973, 164–202), only three cone 8 glazes contained bone ash. And in the list of 229 cone 5–8 glazes reported by Emanuel Cooper in *The Potter's Book of Glaze Recipes*, only four contained bone ash (1980, 49–136). Bone ash was not contained in any of the cone 5/6 glazes tested by my students over the years.

Ceramic chemists do not appear to agree on the role of bone ash at the lower stoneware temperatures. A respected text on ceramic glazes states that bone ash will function as an opacifier up to cone 11.

> "Bone ash and some other phosphates will give opacity up to cone 11 and possibly higher.
>
> "Their use is frequently accompanied by defects associated with the decomposition of these compounds, as well as crawling and beading."
>
> (Parmalee 1973, 13)

Yet later on in the same text, it is stated that bone ash functions as an opacifier only at low temperatures.

> "As a glaze material, bone ash is used to a limited extent as an opacifier at low temperatures. If used in excessive amounts or at too high a temperature, it causes dullness, pinholes and blistering."
>
> (Parmalee 1973, 61)

Ceramic Industry repeats this conclusion:

> "Bone ash is used occasionally in glazes at low temperature to produce opacity, but if used in too large an amount or at too high a temperature, blistering will occur."
>
> (*Ceramic Industry* 1993, 42)

Other authors characterize bone ash primarily as a melter for the stoneware temperatures. The English potter Michael Cardew stated that bone ash "has occasionally been used as an ingredient in glazes as a part-substitute for calcite...." He then went on to add that "It has a well-known opacifying effect in glazes, which may perhaps be traceable to the fact that phosphorus is one of the three main glass-forming elements..." (Cardew 1973, 53). James Chappell describes bone ash as a "flux in higher-fired glazes..." (Chappell 1991, 397).

Because bone ash is not a common ingredient of cone 5/6 oxidation glazes, my students have not tested bone ash in combination with other glaze materials at this firing range. Tests of this nature remain to be done. Consequently, whether bone ash opacifies glaze surfaces or increases fusion at the cone 5/6 firing temperatures are questions that remain to be answered. The few tests that we have performed at this temperature together with our numerous tests at the cone 9/10 temperatures show only that bone ash functions as a powerful opacifier at the higher stoneware temperatures and, if used sparingly, will contribute a luminous opacity to the celadon glaze surface. Larger amounts combined with calcium-magnesium and feldspathic materials help to produce smooth, opalescent, shell-like opacities (see Pavelle Glaze, p. 176; *Figure 3.7, p229*).

Claybody Function

It is possible, through the use of bone ash, to create translucency in claybodies that are porous and not fired to their vitrification point (Grimshaw 1980, 338). Bone ash contains up to 15% calcium carbonate, and even small amounts in a claybody are said to intensify the fusion power of the feldspar (*Ceramic Industry* 1998, 74). Larger amounts of bone ash create translucent claybodies, and this is one of its most important industrial uses.

Bone ash is the important constituent of the English bone-china claybody and makes up 30% to 50% of the formula. It functions in the claybody as a powerful melter and reacts with feldspathic materials, such as Cornwall Stone, to create a translucent claybody at cone 8/9 firing temperatures. The process by which this occurs is fascinating. Once the kiln reaches 1830°F, bone ash begins to decompose and release calcium carbonate. The released calcium carbonate combines with silica and alumina in the claybody to form Anorthite, a calcium feldspar. The remaining material becomes tricalcium phosphate, a powerful, highly reactive fluid that interacts with the balance of the ingredients in the claybody. Tricalcium phosphate dissolves the quartz minerals to form a high-calcium glass. The final result after cooling is a translucent, white body that contains varying proportions of Anorthite feldspar, tricalcium phosphate, apatite minerals, and glass (Grimshaw 1980, 748; *Ceramic Industry* 1998, 74). Once again, the ceramic firing process recalls the recrystallization of various earth minerals from a single magma.

As in the case of glazes, it is not clear whether bone ash imparts translucency to otherwise unvitrified claybodies at lower temperatures or whether it functions in this capacity only at cone 6/9 firing temperatures. Only further testing can provide the answer to this question. John Conrad lists a casting porcelain claybody, with 50% bone ash for cone 6/8 firing temperatures, and a stoneware claybody with 5% bone ash for cone 6 firing temperatures (Conrad 1973, 44, 50). None of the lower claybody formulas contained bone ash.

Claybodies with sizable amounts of bone ash can be used more effectively in industrial factories than in private pottery studios and schools because of the following conditions:

1. Claybodies with sizable amounts of bone ash tend to be unplastic, and difficult to work with—hence the casting function of the Conrad claybody.

2. Cattle bones are essential and in England only oxen bones are used.

3. Organic matter in bone ash decomposes in time, so the claybody must not be aged.

4. The firing range is rigid and allows for little variation in the firing temperature (Tichane 1990, 62).

Conclusion

The lack of agreement among ceramic experts as to the function of bone ash bears witness to its complex, variable, and unpredictable nature. Professor Rex W. Grimshaw, a mineral science expert and professor of ceramics, concludes that "little is known of its structure..." (Grimshaw 1980, 205). Because bone ash results from the calcining of animal bones (preferably cattle), its composition can and often does vary from bag to bag. For this reason suppliers offer an artificial bone ash (synthetic dicalcium phosphate [$CaHPO_4$]), which seeks to duplicate the chemical structure of the calcined animal bones and which would remain stable from bag to bag. The effectiveness of artificial bone ash as a substitute for the natural material is a matter of contention. On the plus side is its overall stability together with its reported increased firing rate and translucency in the bone china claybody. On the downside is its negative effect on claybody plasticity (Bill Hunt, "Bone China," *Studio Potter*, June 1998, 33–35; *Ceramic Industry* 1998, 74). The effect of artificial bone ash in glazes awaits further testing. Interestingly enough, a replacement of the supplier's bone ash (presumably made up of animal bones, judging by the smell of the bag) with home-calcined chicken bones added produced a glaze surface with far more depth and luminosity than the suppliers' bone ash. This suggests that, once again, the price of reliability could well be the loss of surface depth and excitement. Unpredictability and surprise are a normal and expected part of a potter's experience. Even with the most standardized materials, it is usually difficult to predict with complete accuracy the complex reactions that will occur within the kiln. Thus, Professor Grimshaw's final conclusion is one that all potters should take to heart.

"In probably no other industries does the art of controlling ceramic reactions, *whose very nature is largely unknown*, rise so high as in the firing of ceramic wares."
(Grimshaw 1990, 749, emphasis added)

FLUORSPAR
(Fluorite) (Calcium-Fluoride)

TRADE NAME	Fluorspar
GEOGRAPHICAL* **SOURCE**	*Domestic:* Kentucky-Illinois region. Mexico.
GEOLOGICAL* **SOURCE**	Escaping fluorine gases from earth's interior attack surrounding country rock of limestone and dolomite. Result is contact metamorphic limestone and dolomite. Found as accessory mineral with malleable, ductile metals of rare earth group (yttrium, cerium) and many other different minerals such as apatite and tourmaline.

CHEMICAL ANALYSIS*
$4(CaF_2)$

Ca	51.3%
F	48.7
Specialty Grade**	Guaranteed 97.00
	(Typical)
CaF_2	98.00
SiO_2	1.11
CaO_3	0.57
Al_2O_3	0.015
Fe_2O_3	0.03
$BaSO_4$	0.04
P_2O_5	0.0065
As	0.00025
MgO	0.02
Mn	<0.01
S	0.024
Cl	0.003
Zn	0.0040
organics	0.0015
MP	2450°F

*Frye, 1981, 612; *Ceramic Industry* 1998, 105; Seaforth Mineral & Ore Co., Inc. Brochure 1998.
**Seaforth Mineral & Ore Co. Inc. Specification Sheet 1998.

FLUORSPAR[5]

Fluorite, or fluorspar, derives from the Latin "fluere, to flow" (Frye 1981, 611). Fluorspar formed when reactive fluorine gases from the earth's interior penetrated the surrounding limestone rock ($CaCO_3$). This combination of calcium carbonate and fluorine creates an exciting and volatile calcium melter for a wide spectrum of ceramic firing temperatures, ranging from lower temperature enamels to high temperature stoneware (1500°F–2380°F.)

Fluorspar is used industrially as an opacifier and flux in the manufacture of opal glasses and enamels. It is also used to correct color and refraction defects in the optical glass of microscopes, spectroscopes, and telescopes.

At enamel temperatures, *Ceramic Industry* reports that fluorspar functions both as melter and opacifier, depending on the amount used and the rest of the glaze ingredients. (In one enamel test series, 4% additions of fluorspar functioned as melter; 5%–25% additions of fluorspar behaved as an opacifier; and finally, with 26%–50% additions, fluorspar returned to its earlier melting function!)

Fluorspar creates opacity by forming crystals of fluorite, which are scattered throughout the glaze magma. This kind of opacity occurs primarily at the lower firing temperatures of enamel. However, as we shall see in the following tests at cone 9/10 reduction, crystals can sometimes form at the higher temperatures as well. At stoneware temperatures, fluorspar functions mainly as a fusion catalyst and produces a faster melt of the glaze materials.

At temperatures somewhere between 1652°F and 2190°F (provided silica is present, the disassociating temperature of fluorspar in the absence of silica rises to 2300°F–2500°F) fluorspar breaks apart and releases volatile fluorine gas. The liberated gas attracts silica from the glaze and/or claybody to form silicon tetrafluoride (SiF_4), which escapes from the kiln, leaving behind the newly separated calcium. For reasons that are not completely understood (perhaps not all of the fluorine decomposes, and therefore, some remains behind to function as a powerful melter), the final glaze surface is usually shinier and more transparent than it would be with equal amounts of calcium carbonate.

The consequences of using fluorspar can sometimes be injurious to the ceramic surface, the kiln interior, and even the potter. The formation of silicon tetrafluoride robs the glaze and/or claybody of needed silica. In addition, the forcible exit of silicon tetrafluoride is often marked by blisters and pinholes. It is believed that the coarser grades of fluorspar encourage these pinholes. Although there is a difference of opinion as to whether the liberation of this volatile gas injures kiln interiors (see Parmalee 1973, 39), *Ceramic Industry* reports definitively that silicon-tetrafluoride corrodes furnace linings. Our experience confirmed the corrosive nature of silicon fluoride. A test series that added 20%–40% fluorspar to Sanders Celadon glaze not only corroded the kiln shelves, but also attacked the surface of adjoining pots. Even more important is the fact that tetrafluoride gas is a known carcinogenic material.

For these reasons, fluorspar is best used in small amounts. Parmalee advises no more than 6%–7% (Parmalee 1973, 292). When fine-grained fluorspar is used sparingly, it becomes an exciting auxiliary melter. "Fluorspar has been found to offer promise in glazes as a substitute for whiting, tending to promote more fusible glazes" (*Ceramic Industry* 1998, 105).

For the more adventurous, amounts in excess of 7% will produce many surprises, some of which are wonderful, and some of which are not, as the following tests will show.

Glaze Function
Cone 9/10 Reduction

FLUORSPAR TESTS

Fluorspar	100%	*Stoneware claybody.* Yellow-green similar to calcium carbonate test. Matt-shine surface. Rough, pinholes, bubbles.Spotty, uneven, coverage.
		Porcelain claybody. Gray-beige color. Matt-shine surface. Rough, pinholes, bubbles. Spotty, uneven coverage.
Feldspar glaze		*Stoneware claybody.*
Potash feldspar	90.0%	Blue-gray color.
Whiting	10.0	Crackle. Gloss.
Red iron oxide	0.5%	
Bone ash	2.0%	
Add 2% fluorspar		No difference in color. Pinholing where thick

SANDERS CELADON

		Stoneware claybody
Kona F-4	44%	Gray-blue-green.
Whiting	18	Transparent.Gloss.
Flint	28	
Kaolin	10	
Barnard clay	3%	
Add 2% fluorspar		Color greener. Increased gloss. Increased fluidity. White flecks in pool of green glass in interior of test bowl.
Add 4% fluorspar		Similar to 2% test.
Add 6% fluorspar		Color greener.
Add 10% fluorspar		Similar to 6% test.
Add 20% fluorspar		Deep green glass. Interior ringed with blue and beige crystals afloat in green glass. Smashing test!
Add 30% fluorspar		Gray-green glass interior ringed with pools of greener glass flecked with metallic, matt beige crystals afloat on surface of glassy ring. Glaze very fluid. Light gray-green. Stony-matt, where glaze flowed off walls and rim.
Add 40% fluorspar		Similar to 30% test but increased crystalline formation. Increased fluidity leaving larger portions of wall and rim stony-matt and light gray-green in color.

Note: Repeat of 20% and 30% tests did not produce the same results. New tests appeared similar to original glaze without fluorspar! Surface had slightly higher gloss and slightly greener color, but differences were subtle.

Cone 5/6 Oxidation

At the stoneware temperatures of cone 5/6 oxidation, it is possible that fluorspar would function as a catalyst melter in the same manner as it does at the higher stoneware temperatures of cone 9/10. Fluorspar fired alone at cone 5/6 oxidation temperatures produced a surface similar in texture to the cone 9/10 reduction tests. An intricate network of yellow matt crystals float on the pale olive-green, glossy, pinholed, crazed background, producing a complex, highly textured surface. However, my students did not test cone 5/6 glazes that contained fluorspar, and for this reason, tests that would determine the quality of its catalytic performance in glazes at these temperatures remain to be done. According to Parmalee, "calcium fluoride may be used to advantage in glazes up to about 0.05 equivalent for temperatures below cone 7," but the precise nature of this advantage is not described (Parmalee 1973, 312).

Claybody Function

As in the case of bone ash, fluorspar is used in white claybodies as an additional melter to raise the vitrification of the claybody. Here again, industrial research indicates that only low amounts of fine-grained fluorspar (2.5%) produce increased fusion (Ibid.).

In general, industrial research on the function of fluorspar in both claybodies and glazes remains woefully incomplete.

> "Fluorspar has received scattered attention as a constituent in white-ware bodies and glazes but has never been systematically evaluated in sufficient detail to permit fair appraisal of its possibilities....It is probable that fluorspar and other fluorine compounds have usefulness for ceramic glazes, but their successful general application will require the development of additional information relative to advantages and limitations."

(*Ceramic Industry* 1998, 105).

Conclusion

1. When fluorspar is used alone, and is not combined with silica, it remains stable. However, this is an impossibility in ceramics due to the inevitable presence of silica in the claybody. Fluorspar applied alone to the clayform attacks the silica in the claybody. The result of this interaction creates complex, mottled surface effects at stoneware temperatures. (A thin wash of fluorspar and water sprayed on to an iron-bearing claybody and fired to cone 9/10 reduction produced a gray-green-yellow, mottled, roughened, matt-shine surface. Similar results appeared at cone 5/6 oxidation temperatures.)

2. Additions of fluorspar will add excitement and volatility to any glaze surface in which it appears. As with all volatile materials, unpredictability is its keynote.

a. Low amounts of fluorspar (up to 6%) increased fluidity and intensified color.

b. Larger quantities (10%–40%) produced surprises from one kiln firing to the next. In all cases, fluidity and brilliance of color increased.

3. The disadvantages of fluorspar are as follows:

a. The formation of volatile silicon tetrafluoride gas can result in loss of silica, and stony-matt, pinholed, blistered surfaces.

b. Kiln walls and adjoining pots could be damaged.

4. Fluorspar is very different from both calcium carbonate (Whiting) and calcium phosphate (bone ash). It is a more powerful melter than calcium carbonate, creates shinier surfaces, and interacts more visibly with the claybody when fired alone. Unlike bone ash, fluorspar usually induces transparency, rather than opacity. (At times, as seen in our tests with 30%–40% amounts, fluorspar produced crystalline opacities, but the results were not predictable.) Fluorspar interacts forcibly with the claybody and glaze materials; after the exit of the silicon tetrafluoride gas, the remaining calcium functions as a powerful melter (either alone, or in concert with some remaining calcium-fluoride) to increase the fusion of the glaze materials.

Despite the volatility of fluorine, fluorspar is first and foremost a calcium compound, and as such, is a valuable auxiliary melter whenever it becomes necessary to increase the fusion point of a calcium-base glaze. In low amounts it can perform this service without causing drastic color or surface disturbances. Thus, nature provides yet another choice in the infinite list of ceramic materials that are available for the potter.

LAB V CALCIUM OXIDE

BONE ASH FLUORSPAR

1. Take two to four identical test pots. Glaze one-fourth of each pot with Wollastonite, bone ash, Whiting, and Fluorspar. Leave a good margin to account for the volatility and fluidity of fluorspar should you glaze the outside of the pot. Fire each pot in as many firing temperatures as are available to you.

2. Substitute bone ash for the calcium melter in your glaze.

3. Substitute Fluorspar for the calcium melter in your glaze. Glaze only the inside of the test pots.

4. Add 2%, 4%, 6%, 8%, and 10% bone ash to your glaze. Repeat these tests with Fluorspar.

5. Select a glaze with which you are familiar that contains bone ash,* and mix it up without the bone ash.

6. Select a glaze with which you are familiar that contains Fluorspar, and mix it up without the Fluorspar.

7. With the exception of the first test series, each test you make should have a control; that is, one-half of each test pot should be glazed with the original glaze.

8. When testing with Fluorspar, glaze *only the inside* of the test pot.

*If you are not familiar with a bone ash glaze, use the Pavelle or Persimmon glaze (p. 176).

BORATE MINERALS

Figure 3.9, p231.

Calcia combines in nature with many different oxides, and each calcium compound used in ceramics will produce unique effects within a particular glaze magma. We have previously described calcia joined with the glassmakers silica and phosphorus. Calcia bonds with yet another kind of glassmaker, boric oxide (B_2O_3), made up of the elements boron and oxygen.

Neither boric oxide nor boron is found in a pure state in nature; instead they exist in complex combinations with various oxides. There are about 150 different boron compounds, but only the following five are of major commercial importance (Interoffice Correspondence, United Borax & Chemical Corporation, 7 April 1989).

| Colemanite | $2CaO \cdot 3B_2O_3 \cdot 5H_2O$ |
| | 50.81% B_2O_3 |

Ulexite[6]	$2\,CaO \cdot Na_2O \cdot 5B_2O_3 \cdot 16H_2O$
	42.96% B_2O_3
Probertite	$2CaO \cdot Na_2O \cdot 5B_2O_3 \cdot 10H_2O$
	49.6% B_2O_3
Kernite	$Na_2O \cdot 2B_2O_3 \cdot 4H_2O$
	50.96% B_2O_3
Borax (Tincal)	$Na_2O \cdot 2B_2O_3 \cdot 10H_2O$
	36.52% B_2O_3

Gerstley Borate = Ulexite + Bentonite clay.

These five boron compounds contain variable amounts of boric oxide. They are chemically bonded with molecules of water due to their initial formation in ancient lake basins near boron-rich thermal springs. Large-scale deposits of borate minerals are found in southwestern California, Turkey, southern Russia, Argentina, Chile, Peru, and mainland China (Kistler and Smith 1983, 533–560).

CHARACTERISTICS OF BORIC OXIDE

Boric oxide, like silica, functions as a glassmaker in a glaze magma. This means that boric oxide will create transparent glaze surfaces because, as in the case of silica, its atoms do not recombine into an ordered crystalline pattern during the relatively short heating and cooling period of a ceramic kiln fire. Hence, boric oxide's unstructured atoms permit light rays to travel freely through the glaze layers. Like silica, boric oxide has a low expansion and contraction rate, and therefore tends to reduce the crazing pattern induced by the high-shrinking melters of soda and potash. However, unlike silica, boric oxide is a powerful melter and has a melting point alone of 1292°F. The addition of soda (Borax) causes its melting point to lower to 662°F; the further addition of calcia (Ulexite) produces a melting point of 572°F; and finally, the combination of boric oxide and H_2O (Boric acid) results in a melting point of 392°F. The combination of calcium and boric oxides that make up the borate mineral of colemanite raises the melting point to 1472°F. Compare all of this to the melting point of silica, which is 3110°F, or the melting point of calcium oxide, which is 4658°F.

Most important, boric oxide minerals, in addition to their glassmaking properties, contain adhesive powers that enable them to remain on the walls of the pot without the separate addition of alumina minerals. Boron and aluminum are in the same IIIA group in the periodic table of elements; boric oxide, like alumina, has a 2 to 3 ratio of first element atoms to oxygen atoms, and because of common properties, is placed by the chemist in the same R_2O_3 column for the purpose of molecular computations (see Appendix A). In addition, certain boric oxide materials (colemanite and Gerstley Borate) are found combined with alumina minerals (bentonite clay, or shale), and may be marketed with these impurities (Kistler and Smith 1983, 535).

For all of the above reasons, it is easy to understand the important ceramic function of boric oxide as both a glaze core and melter, particularly at the lower firing temperatures. At the earthenware firing range (below 2012°F) and the low stoneware firing temperatures (2150°F–2230°F), boric oxide materials function as both glaze core and melter. At the higher stoneware temperatures (2300°F–2380°F), they function primarily as a catalyst in promoting glaze fusion and brilliance of color.

Just as calcia and soda leave characteristic imprints on the color and surface of glazes, so too does boric oxide. Combinations of boric oxide materials with tin oxide (SnO_2) and/or rutile (TiO_2) and/or zinc oxide (ZnO) produce brilliant pinks.

CONE 5/6 OXIDATION

POWDER BOX PINK	
Boron Frit 3134	50 grams
Flint	25
Whiting	5
Ball clay	15
Tin oxide	7

See also, Nancy's Pink Icing Glaze (Cone 4–6, McWhinnie, Harold J. 1982, "Cone 4–6 oxidation glazes" *Ceramics Monthly*, January 1982, 71) in which a combination of Gerstley Borate, tin oxide, and dark rutile produces boudoir pink in a high-lithium, high-silica, cone 6 glaze.

Boric oxide intensifies the brilliance of cobalt blue, (Transparent Blue glaze, p. 186). Copper green becomes a brilliant, turquoise green in the presence of boric oxide (see tests of R-15 Blue-Green, p. 186).

These effects are best seen in the context of specific boric oxide materials. Of the more than 150 naturally occurring boron compounds (known as borates), only three are commonly used in ceramics: Tincal or Borax ($Na_2O \cdot 2B_2O_3 \cdot 10H_2O$); Colemanite ($2CaO \cdot 3B_2O_3 \cdot 5H_2O$); Ulexite ($2CaO \cdot Na_2O \cdot 5B_2O_3 \cdot 16H_2O$), the primary mineral in Gerstley Borate.

SOURCES OF BORIC OXIDE
COLEMANITE

Boric oxide combines with calcia to form the natural mineral colemanite—the most important ceramic material for lower firing temperatures. The geological formation of colemanite is closely related to ulexite (the primary mineral in Gerstley Borate), which contains both sodium and calcium in its complex chemical structure. It is believed that colemanite formed as a secondary mineral when calcium-rich waters invaded ulexite deposits and replaced the soda content with calcium (Kistler and Smith 1983, 537). Colemanite takes its name from William T. Coleman, the prominent San Francisco businessman who had the vision to recognize the importance of California's borate deposits. As the distributor for the Pacific Coast Borax Company, his scouts discovered, in 1880, rich deposits of a new, calcium-borate mineral, named colemanite, in the vicinity of Death Valley, California. Prior to the discovery of rich mineral deposits of tincal (borax), colemanite was processed with sodium materials to yield borax as an end product (U.S. Borax & Chemical Corporation 1985, 8–9).

These major deposits of colemanite originated in 10–6-million-year-old lake sediments of the Furnace Creek Formation (late Miocene to early Pliocene Epoch), overlaying deposits of ulexite, and were the primary source of the United States' supplies of colemanite until the mines closed in 1927. American Borate Co. mines commercial deposits of colemanite in Death Valley, California, for export only (Technical Services, American Borate Company, August 1998).

"Pure colemanite is $2CaO \cdot 3B_2O_3 \cdot 5H_2O$ and contains 50.8% B_2O_3, but it is marketed as an impure ore. The only current supplier of colemanite ores is Etibank, the borate company owned and managed by the government of Turkey."

(R. J. Brotherton, U.S. Borax Research Corporation, Memo to Dr. H. Steinberg, 5 April 1989)

Turkey's borate districts are said to contain the world's largest deposits of colemanite. Turkey, via its U.S. representative, The American Borate Company, is a primary supplier of Turkish colemanite to domestic glass fiber industries (Phyllis A. Lyday, Boron Commodity Specialist, U.S.G.S., Minerals Information, 1998, 1; Technical Services, American Borate Company, 1998). Turkey produces run of the mine ore and concentrates from the Miocene Epoch (15–20 million years old), which, based on its 38%–48% boric oxide content, contains 75%–86% concentrations of colemanite. The balance consists of shale, clays, limestone, sandstone, volcanic tuff, and marls (R.J. Brotherton, U.S.

Borax Research Corp., March 1989). Argentina also supplies colemanite via its representative, F&S International, Inc. Argentinian colemanite has a somewhat different chemical composition (see p. 185). Note that a chemically precipitated colemanite with a more stable chemical composition is now produced by Fort Cady Minerals Corporation (U.S.G.S. Mineral Industry Surveys, Boron, 1997, Annual Review, 1–2).

In view of the variable nature of this material, together with its solubility and high chemical water content, the use of colemanite in pottery glazes has always been subject to risks. (For a parade of possible horrors such as pinholing, crawling, bloating etc. see pp. 190, 192–193, and Zamek, "Gerstley Borate and Colemanite," *Ceramics Monthly*, July–August 1998, 73–74, 118–120).

Colemanite has had a long history of appearance and reappearance on the nonindustrial ceramic scene. At the time of the writing of the first edition of this book, colemanite was available for potters. At the present time this is no longer the case because of the large minimum order requirement for imported Colemanite. Domestic sources in California are produced primarily for export. Once again, we see how the availability of our ceramic materials depends on current manufacturing practices.

Characteristics of Colemanite

Colemanite is less soluble in water than the major sources of boric oxide, such as borax and boric acid. In addition, colemanite combines adhesive, glassmaking and powerful melting properties; it thus becomes an ideal glaze core for the lower firing temperatures of cone 5–6 and under. A test of colemanite alone on a bowl made from a buff-colored, low-iron, stoneware claybody and fired to cone 5/6 oxidation created a shiny, transparent glaze surface with a decided yellow-olive cast. This test showed how admirably colemanite performed as a natural glaze core at these lower stoneware temperatures.

The range of colemanite is extraordinary. Consider the fact that from 1650°F–2380°F, colemanite in both oxidation and reduction atmospheres, without the help of any other material, creates a glassy, transparent glaze surface. (The color of this surface will depend, of course, on the iron content of the claybody and the temperature and atmosphere of the fire. The color of the earthenware test surface was brown (red-brown claybody); the color of the cone 5–6 stoneware test was yellow (buff stoneware claybody), and the color of the oxidation, cone 9–10 test was greenish-yellow (same buff claybody). The reduction, cone 9–10 stoneware test displayed a glassy, mahogany-brown surface ringed with black.

Colemanite's chemical structure reveals a low alumina content—0.10%. Yet despite this fact, it clings to the walls and surface of the clayform as if it contained considerable amounts of alumina. (Note that colemanite is marketed as "impure ore," and that Turkish colemanite is described as containing "clays, marl, limestone, sandstone, and volcanic tuff" [Ibid.].)

Most important is the fact that colemanite's low melting temperature of 1650°F makes it possible to incorporate refractory materials in a low-fire glaze magma, which otherwise would not melt. A case in point is wood ash, which remains stiff and unmelted at the cone 5–6 oxidation firing temperature. When combined with equal amounts of colemanite at this temperature, pine ash becomes a yellow-green-brown mottled, matt-shine glaze surface, not unlike the surface of a wood ash glaze at cone 9/10 firing temperatures. And in the low-fire earthenware and raku temperatures (1069°F–1940°F), colemanite is an indispensable ingredient of many glaze formulas (see Conrad 1973, 118; Cooper 1980, 29–79).

COLEMANITE

TRADE NAME Colemanite

GEOGRAPHICAL SOURCE American Borate Co., Death Valley, California
Etibank, Turkey, American Borate Co. (Representative)
Ulex S.A. Argentina, F&S International, Inc. (Representative)

GEOLOGICAL SOURCE*
(*domestic*) 10–6-million-year-old lake sediments fed by boron-rich thermal springs in Furnace Creek formation of late Miocene—early Pliocene age. Found as secondary mineral with ulexite or as a replacement mineral (sodium replaced by calcium).

CHEMICAL STRUCTURE** $2CaO \cdot 3B_2O_3 \cdot 5H_2O$

$50.8\% \cdot B_2O_3$ (ideal)

38–43% B_2O_3 (average)

	IDEAL	TURKEY (Typical)***	ARGENTINA (Ulex)†
B_2O_3	50%	42.00%	41.50–43.00%
CaO	26	26.40	23.50–24.50
Na_2O		0.10	0.01–0.02
SiO_2		4.40	
Fe_2O_3		0.10	0.40–0.55
Al_2O_3		0.30	2.30–2.60
MgO		1.80	1.40–1.70
K_2O			0.03–0.06
As_2O_3			200ppm–500ppm
SrO		1.00	Cl 0.05–0.07
SO_3		0.20	0.04–0.60
L.O.I.		23.70	

*Kistler and Smith 1983, 533–560.
**U.S. Borax & Chemical Corporation, Interoffice correspondence, 7 April 1989. U.S. Borax Research Corporation, Memo, 5 April 1989. Hammill & Gillespie, Inc. Data Sheet.
***Technical Data, American Borate Company 1998
†Technical Data, F&S International Inc. 1998.

Color Effects of Colemanite

Colemanite has characteristic color effects on a glaze surface in both oxidation and reduction atmospheres.

Cone 5–6 Oxidation

Over a high-iron oxide slip or claybody, colemanite colors a feldspathic glaze surface a deep, purple blue. (*See Figure 3.10, p232*. Boron Frit produces a similar effect.)

BLUE HAZE SEEDS	
	Purple-blue gloss over iron slip.
Kona A-3 Feldspar	42.00%
Colemanite	24.00
Zinc oxide	9.25
Flint	19.50
Kaolin	4.75

In combination with volcanic ash, colemanite produces a soft yellow surface.

VOLCANIC ASH GLAZES		
Figure 1.16, p93.		
Volcanic ash	64.2%	62.3%
Colemanite or	35.8	28.3
Gerstley Borate		
EPK kaolin		9.4
Bentonite	2.0	

The intensity and hue of the glaze color in the following glazes depend primarily on the fusion powers of colemanite.

TRANSPARENT BLUE (Crafts Students League)	
	Brilliant blue gloss.
Potash feldspar	44 grams
Colemanite	20
Whiting	1
Zinc oxide	3
Kaolin	1
Flint	24
Cobalt oxide	0.5%
Copper oxide	4%

RANDY RED	
Figure 3.11, p233.	Persimmon, green, gloss.
Kona F-4	20%
Colemanite	32
Talc	14
Flint	30
Kaolin	5
Red iron oxide	15
Omit Colemanite	Dry brown matt.

R-15 BLUE-GREEN SATIN-MATT (Crafts Students League)	
Nepheline Syenite	41.6%
Colemanite	11.9
Dolomite	7.6
Talc	14.6
Kaolin	4.4
Flint	19.9
Cobalt oxide	2
Rutile	6
Blue-Green *Under Glaze*	0.5
Omit Colemanite	Green-blue dry matt.

JACKY'S CLEAR	
Figure Intro.6, p16.	

Cone 6–9 Oxidation and Reduction

Transparent, gray, gloss (reduction).
Transparent, beige, gloss (oxidation).
Color derives from underlying claybody.

Nepheline Syenite	50%
Wollastonite	10
Zinc oxide	10
Colemanite or	5
Gerstley Borate	
Flint	20
Ball clay	5

Cone 5/6 Oxidation

Omit Colemanite: Surface changes to soft, white, satin-matt.

Note that in both Transparent Blue and Randy Red, colemanite has the dual function of melter and glaze core.

The low alumina content of all four glazes is made possible by the considerable adhesive powers contained in colemanite.

The amount of colemanite included in the glaze batch is important for the resulting color and surface. Transparent Blue glaze with 20 grams colemanite, 1/2% cobalt, and 4% copper turned a dull green when colemanite was reduced from 20 to 5. The persimmon red-brown gloss of Randy Red (32% colemanite and 15% iron oxide) became a green-brown satin-matt when the colemanite was reduced by half. Hence, not only is colemanite largely responsible for the surface gloss of these glazes, but the proportion of colemanite is also crucial for the achievement of desired color effects.

Cone 9/10 Reduction

Colemanite functions primarily as a fusion catalyst at the high stoneware temperatures. At these higher temperatures, colemanite is used sparingly because of its powerful effect on the flow and gloss of the glaze surface (see Jacky's Clear).

PAM'S SATURATED IRON		
		Black-brown, gloss.
Custer Feldspar	52.4%	
Flint	17.0	
Whiting	10.1	
Colemanite or		
Gerstley Borate	6.2	
EPK kaolin	5.6	
Red iron oxide	8.6	

A thin coat of colemanite and water can be sprayed on a bisque pot in order to give a sheen to the unglazed claybody. At cone 9/10 reduction, this produces a greenish, brown sheen on the fired claybody surface.

The increased fusion produced by colemanite intensifies and deepens the color of the cone 9/10 reduction copper red glaze. Hence, copper red glazes frequently contain a small amount of colemanite or comparable borate materials. (See Conrad 1973, 243–245, G446–G450, 5 copper red glazes, all of which contain low amounts of colemanite.) Similarly, the increased fusion produced with low amounts of colemanite intensifies the green color of the celadon glaze surface.

In order to compare the melting powers of colemanite with the calcium melters, we combined 25% to 75% colemanite with 75% to 25% potash feldspar respectively and compared the results with comparable Whiting tests. The

Whiting tests progressed from a blue-green, glassy, crackled surface to a yellow-green, noncrackled, matt surface. The colemanite tests were greener in color throughout and never lost their glossy shine, even though the surface became increasingly mottled as more and more colemanite was added.

Interestingly enough, and despite the greater melting power of colemanite compared to the calcium melters, the substitution of low amounts of colemanite for Whiting did not always produce a greater melt at the higher stoneware temperatures. The test of 5% colemanite and 5% Whiting combined with 90% potash feldspar produced a stiffer, less-melted surface than the comparable test with 10% Whiting. When 18% colemanite substituted for 18% Whiting in Sanders Celadon the color of the colemanite test was again lighter, and the surface, although glassier, seemed stiffer and less evenly covered than the comparable test with Whiting. It was only when we doubled the percentage of melter to 36% that powerful melting and color changes occurred in the colemanite tests. The color of the surface with 36% Whiting was a pale, gray-green. The glaze flowed off the rim of the pot, leaving a dull, matt, gray-white edge. The colemanite substitution was deeper green with an overall, glassy surface. Similarly, when we added from 10%–40% colemanite to Sanders Celadon, the color of the surface became greener and shinier with each addition. The 30% and 40% additions produced a strong green color with an overall glassy surface that was in marked contrast to the semimatt, yellow-green, satin surface of the comparable Whiting tests.

Note that the trial-and-error substitutions that we performed in the foregoing tests were not exact measurements. The actual weights of colemanite and Whiting differ markedly; colemanite is four times as heavy as Whiting. Consequently, when 18 grams of colemanite substitute for 18 grams of Whiting, there is probably less melter in the colemanite glaze mixture than in the Whiting tests, and this would account for the less-melted appearance of the colemanite substitutions in some of these tests. A more exact test would result from a molecular substitution of colemanite for the amount of calcium oxide required in the molecular formula of Sanders Celadon. (See Appendix A1.)

Opalescence and Opacity

One of the most interesting effects of the calcium borate mineral colemanite is its ability to create opaque and opalescent glaze surfaces at the cone 5–6 firing temperature. Boric oxide is a glassmaker. Silica is also a glassmaker. The presence of boric oxide globules of glass suspended inside the solution of silica glass traps, refracts, and/or reflects the light rays and prevents them from traveling through the glaze layers. The visual result is a glossy opalescence or a milky-blue opacity, depending on the density of the boric oxide globules in relation to the silica glass. Heavy masses of boric oxide tend to reflect the light and create opacity. Thin particles of boric oxide tend to refract or break up the light and create the visual effect of opalescence.

A high-colemanite glaze that contains both boric oxide and calcium oxide may also become opaque due to the formation of calcium borate crystals. These crystals can refract or obstruct light rays and will produce the effect of milky, opaque, or opalescent surfaces. In any event, and for whatever reason, oxidation cone 5–6 glazes that are high in colemanite characteristically present a cloudy, milky-blue, glossy surface. This milky-blue opacity is usually a good indication of the presence of colemanite.

Cone 5–6 Oxidation

MILKY COLEMANITE	
	Cloudy, milky-blue gloss. Opaque.
Potash Feldspar	43.9 grams
Flint	24.7
Colemanite	20.6
Whiting	1.5
Zinc oxide	3.2
Barium Carbonate	6.4
Kaolin	1.0
Copper oxide	0.75%
Rutile	0.50%
Granular Rutile	0.50%

I have not seen the same result at the cone 9/10 firing temperature. At this higher stoneware temperature, the primary effect of combined boric oxide, silica, and calcia seems to be that of greater transparency and shine.

GERSTLEY BORATE

TRADE NAME	Gerstley Borate
GEOGRAPHICAL SOURCE	Death Valley, California, U.S. Borax Inc.

Mine closed in Jan. 2000 with a one-year supply left.

GEOLOGICAL SOURCE* Principal borate mineral is ulexite; a sodium-calcium borate formed in marsh or dried-up lake basins, interlayered with sediments of clays, mudstones, tuffs, and limestones. Gerstley formation, part of Furnace Creek formation of Miocene age (10 million years old).

CHEMICAL STRUCTURE**

B_2O_3	25.0–32.0
Na_2O	3.6–5.3
SiO_2	9.7–10.7
Fe_2O_3	0.3–0.35
Al_2O_3	1.1–1.3
MgO	2.9–3.59
CaO	16.2–20.6
K_2O	.14
SO_3	.14
Cl	.008
P_2O_5	.066
As_2O_3	.006
undetermined	.24
L.O.I.	21.59

MINERALOGICAL STRUCTURE*

Principal borate mineral:	Ulexite
Also contains:	Bentonite clay gangue (clay, marl, limestone, sandstone, volcanic tuff) 18.7%
Small amounts of:	Colemanite, Probertite

Water of Crystallization 25%
Free Water 0.03%

*Kistler and Smith 1983, 539-542; Memoranda, U.S. Borax Research Corporation, 5 April 1989; Interoffice correspondence, U.S. Borax & Chemical Corporation, 7 April 1989; Zamek, *Ceramics Monthly*, August 1998, 73–74.
**Memorandum U.S. Borax Research Corporation, 5 April 1989; Technical Data: Laguna Clay Company 1992; U.S. Borax Inc. 1998.

GERSTLEY BORATE—AN OBITUARY

Two kinds of calcium borate ores have been available for the potter. One of these was colemanite, discussed previously. The other calcium borate material, known as Gerstley Borate, has a lower boric oxide and higher carbonate content than colemanite and is not a natural mineral.

> "As noted above Gerstley Borate is not a specific compound or mineral and varies in B_2O_3 content. Material from the Gerstley mine contains approximately 25% B_2O_3. There are no refining steps involved in processing this material other than grinding to -3 mesh for customers that cannot use the ore as mined."
>
> (Memorandum U.S. Borax Research Corporation, April 5, 1989)

The name Gerstley Borate does not identify a specific mineral; it refers to a mine "located just north of Shoshone, California, or about 24 miles south of Death Valley Junction on Rt. 127" (Interoffice correspondence, U.S. Borax & Chemical Corporation, 7 April 1989). It was operated by U.S. Borax Company under the direction of the late James Gerstley, former Managing Director of Borax Consolidated Ltd. The Gerstley mine opened in 1923, closed in 1962, reopened in 1988, and finally, as of January 2000, has closed once again due to unfavorable economic conditions (see Zamek, March 2000, *Ceramics Monthly* 49–50). The mine's closure left only a one-year supply available to potters. See pp. 193–195 for substitutes.

The ore extracted from the mine consists of a sodium-calcium borate mineral known as ulexite, together with small amounts of colemanite and probertite (same structure as ulexite, except for six fewer molecules of H_2O) and a bentonite clay gangue of claystone, sandstone, and volcanic rocks of basalt and andesite (Ibid.). The less than 1000 tons per year mined by U.S. Borax Inc. reached the nonindustrial ceramic market via Hammill & Gillespie and The Laguna Clay Company. They were the exclusive suppliers of Gerstley Borate for the smaller ceramic suppliers and potters. Although they ground the ore to a 200 mesh size for their customers, they did not guarantee overall quality; the material was sold to them as run of the mill only; hence, some shipments contained 100 mesh size particles (Technical Representative, U.S. Borax Inc., letters to author, April 1989, August 1998; Zamek, "Gerstley Borate and Colemanite," *Ceramics Monthly*, August 1998, 73–74, 118–120; Gerstley Borate, Notes, http://www.ceramicssoftware.com/education/material/gerstley:htm, July 1998).

Ulexite takes its name from the nineteenth-century German chemist Georg Ludwig Ulex (1811–1883). The ulexite ore is of Miocene age and is 10 million years old. The mineral combination of ulexite and bentonite clay made Gerstley Borate a useful material in the aerial control of forest and brush fires. It was marketed as a fire retardant under the name of Firebrake. At the present time, the primary industrial market is the ceramic industry (Ibid.).

> "Gerstley Borate never has been widely used in any of the glass applications, except in pottery glazes—a very small segment of the general market. For many years, our only customer for Gerstley Borate has been Hammill and Gillespie."
>
> (Charles D. Frame, Sales Manager, U.S. Borax & Chemical Corporation, letter to author, 26 April 1989)

Regardless of a very different chemical structure, during colemanite's periodic disappearances, Gerstley Borate often substituted for colemanite. (See pp. 194–195. The colemanite tests described on pp. 186–188 could also be made with Gerstley Borate and would probably yield similar results.) The receptacle bin of a major pottery school was labeled "colemanite," despite the fact that it actually contained Gerstley Borate. Whether or not colemanite or Gerstley Borate was the material used in any of these glazes often depended on which material was available to the supplier or whether it was suspected that a particular supply of colemanite or Gerstley Borate had caused troublesome glaze defects. Although my own experience with Colemanite and Gerstley Borate has not been particularly troublesome, potential problems are inherent in their chemical and mineralogical structures (see Zamek, "Gerstley Borate and Colemanite, *Ceramics Monthly*, August1998, 73–74, 118–120; Gerstley Borate, Notes, http://www.ceramicssoftware.com/educationmaterial/gerstley.htm, July 1998). For example, the fast release of the water in their chemical structure could cause crawling and glaze-flaking. Their solubility, even though slight, could change the glaze solution if the physical water in the glaze bucket is poured off or altered through evaporation. Then too, overfiring of these powerful, low-fire melters is always a possibility, leaving the glaze surface with unsightly pinholes and scars.

Despite the common practice of interchanging these two materials, they do contain significant differences. More variability is possible with Gerstley Borate than with colemanite, which is, after all, a natural mineral with a definite composition. Gerstley Borate's higher carbonate content could cause even more pinholing and scarred surface effects than would colemanite. In addition to this increased carbonate content, the ulexite mineral in Gerstley Borate also contains soda, which is normally not present in colemanite. In certain glazes, this free and easy substitution may radically change the color and quality of the glaze surface. For example, a student reported that the substitution of Gerstley Borate for colemanite in a cobalt blue, glossy, cone 5/6 oxidation glaze resulted in a weaker blue color and a less-shiny surface. These color and surface changes created serious problems for this production potter. In such a case, a molecular substitution of the materials would be necessary.

By and large, however, notwithstanding their considerable chemical, mineral, and weight differences, it is surprising how many glazes accept the substitution of one for the other without a major surface change (see pp. 194–195). In this connection, tests of Gerstley Borate fired alone at cone 5/6 and cone 9/10 oxidation temperatures did not differ from comparable colemanite tests. Both were glassy, yellow-green surfaces, and it was not possible to distinguish one from the other. All of this simply proves the flexibility and large margin for error that exists in most glaze mixtures. It thus enables the potter to experiment on a trial-and-error basis, free from the time-consuming mathematical calculations that are required for molecular substitution in the absence of expensive glaze-calculation software programs. This trial-and-error substitution approach is especially valuable for adventuresome potters who wish to explore the potential of various ceramic materials. As long as the differences produced by the substitution of materials are within an acceptable standard (and in so many instances this is in fact the case), there is no limit to the opportunities for new and exciting glaze surfaces that the trial-and-error substitution method provides. There is always time at some later point to perform those molecular computation gymnastics.

TINCAL (BORAX)	
TRADE NAME	Borax
GEOGRAPHICAL SOURCE (*domestic*)	Boron, CA. U.S. Borax, Inc. Searles Lake, CA. North American Chemical Co. (Harris Chemical North American, Inc.)
GEOLOGICAL SOURCE	Evaporite of chemical precipitate deposits, which formed in sediments around or in ancient lake basins fed by volcanic boron springs, during and after mid-Tertiary period (26 million years ago) (Kistler and Smith 1983, 537–538).
CHEMICAL STRUCTURE	$Na_2O \cdot 2B_2O_3 \cdot 10H_2O$ (*Ceramic Industry* 1998, 74)
Sodium oxide	16.25%
Boric acid	36.51
Water of crystallization	47.24

Gerstley Borate and colemanite are but slightly soluble in water. We come now to a sodium-borate mineral that is highly soluble in water, hence its commercial importance. Borax is the most commercially important of all the borate minerals precisely because of its high rate of solubility and reactivity with other chemicals.

The name "borax" is derived from the Arabic "buraq" (Frye 1981, 564). The mineral is known as "tincal." After refining, it is called "borax." The chemical structures of both tincal and borax are identical.

Because it is both glassmaker and melter, and, in addition

has a low thermal expansion rate, borax plays an important industrial role in the production of heat-resisting glass, fiberglass, and other glasses. In the manufacture of these glasses, relatively large amounts of borax contribute the properties of reducing thermal expansion and increasing corrosion resistance (*Ceramic Industry* 1998, 76).

Southwestern California contains our total domestic source of tincal (borax), as well as all of the other borate minerals, and contains 50% of the world's borate deposits. A major domestic producer of borax is U.S. Borax Inc. This company mines large-scale mineral deposits of tincal in Boron, California. The deposits formed about 13–7 million years ago, during the middle to upper Miocene Epoch from chemical precipitation in a dried lake basin located near a boron-rich volcanic spring in an arid climate. Additional deposits of tincal formed less than 5000 years ago in another continental basin from the evaporation of lake waters. These deposits are found in the salt brines of central salt flats in Searles Lake, California. Harris Chemical North American, Inc.[7] is the sole producer of this source of borax (Grimshaw 1980, 332; *Ceramic Industry* 1998, 74; Kistler and Smith 1983, 533–542; USGS Mineral Industry Surveys, Boron, 1997, 2).

Characteristics of Borax

Like colemanite and Gerstley Borate, borax is both a glassmaker and a melter. However, borax is a more powerful melter than either. Colemanite, which links boric oxide with calcia, has a melting temperature of 1472°F. Borax melts at 662°F—an indication of the higher melting power of soda as compared with calcia.

Borax is highly soluble in water. It is possible, however, to obtain an insoluble form of borax, known as anhydrous, or fused borax (Dehybor®), which contains no water molecules. Although this form of borax is used for glazes that contain more than 5% borax, our tests with soluble borax had more depth and brilliance than those with the insoluble form. When substituting anhydrous Borax, use approximately one-half the amount (*Ceramic Industry* 1998, 76).

Borax is a very interesting glaze material—especially for the low-stoneware, oxidation firing range. Low amounts of borax (2%–4%) will increase the fluidity of the glaze, heal pinholes left by escaping gases, and deepen the brilliance of its surface. All of this may be accomplished without substantially increasing the thermal expansion rate of the glaze. The addition of 2% borax to a glaze magma toughens the raw glaze coat and prevents flaking and chipping during the stacking of the kiln.

Although an excess of borax is said to produce crazing, blistering, and destruction of underglaze colors (*Ceramic Industry* 1998, 76), we have tested large amounts of borax in combination with feldspars and produced interesting glaze surfaces.

Combinations of 10%–30% borax and feldspars produced some fascinating results at cone 5/6 oxidation. When 10% borax was combined with 90% potash feldspar (Kingman, oxide structure, p. 47), it produced a stiff, white, bubbled surface. The surface changed to a warm, red-brown, glassy, crackled gloss with the tests of 20% borax and 80% potash feldspar. A pale green, glassy surface with a deep, crackled pattern appeared when 30% borax was joined with 70% feldspar. The depth and gemstone quality of this last test was clear proof of the potential in borax for the creation of sparkling, decorative surfaces on the insides of bowls. (It would probably be too fluid to risk on the outer walls.) The edges of all the tests were ringed with bright orange flashing—an indication of the presence of the soda component of borax. In this connection, it is worthwhile to compare sodium carbonate and feldspar tests. A 10%–30% soda ash was combined with 70%–90% Kingman feldspar. The surfaces in the 10%–20% tests remained white and pinholed. Only in the 30% soda ash and feldspar test did a deep-red-brown gloss finally appear, but it could hardly compare in depth or quality to the sparkling green glass of the borax tests—a tribute to the greater melting power of soda combined with boric oxide. However, the edges of the soda ash and feldspar tests were ringed with deep orange that was brighter in color than the comparable borax tests. This, once again, was proof of the flashing power of pure soda.

The 9/10 reduction tests again showed the greater melting power of borax compared with soda. The 10% borax combined with 90% Kingman feldspar produced a glossy, pale, milky, gray surface, flecked with the black iron spots from the claybody. The comparable soda test exhibited a pale, blue-gray gloss. No iron spots from the claybody showed through the surface. The 20% borax produced a deep, brown-green, highly crackled gloss, flecked with deeper spots of iron. The comparable soda ash test was again a pale, blue-gray gloss of somewhat deeper hue than the 20% test. The 30% test displayed a lustrous, black-brown surface in both the borax and the soda ash tests, but the soda test contained more visible pitting and less even coverage. The borax test had a deeper, blacker, and more lustrous surface with greater flow, more even coverage and less pinholing than the comparable soda ash test.

Similar results were obtained with 90% Rotten Stone, 5% Whiting, and 5% Borax at the cone 9/10 oxidation temperatures. These tests (repeated in two separate firings) produced a rich, reddish-brown-black gloss, studded with oilspots that contained more depth and excitement than the comparable test with 10% Whiting. The presence of Borax created an incomparable gemstone surface. The substitution of 5% soda ash for 5% Borax resulted in an unmelted, bubbled, red-brown surface. Once again, the superior melting power of Borax compared to soda ash appeared in this test.

Borax is clearly a material that we should explore further and in greater depth. The few tests performed so far point to its rich potential for the creation of complex, exciting glaze surfaces, especially at the oxidation cone 5/6 and cone 9/10 firing temperatures.

BORIC ACID[8]

$B_2O_3 \cdot 3H_2O$

$56.5\%\ B_2O_3$

Pure boric oxide is found in boric acid, which consists of pure boric oxide and water. This material has one of the lowest melting points of the ceramic melters—it melts at 392°F. The bulk of boric acid is produced by processing kernite (sodium borate with six fewer molecules of water than borax, see p. 182) with sulfuric acid (U.S. Borax 1985, 54). Rare deposits of the natural mineral form of boric acid, known as Sassolite, are found in volcanic waters near the Tuscan volcano in Sasso, Italy (Frye 1981, 709; Grimshaw 1980, 332, *Ceramic Industry* 1998, 76). Boric acid, like borax, is highly soluble in water and would present all of the same problems arising from the use of water-soluble materials, such as soda ash, in a glaze magma (see p. 26). Irrespective of its high water-solubility rate, it remains the purest available source of boric oxide for the potter.

It is important to remember that colemanite, Gerstley Borate, borax, and boric acid are hydrates. Colemanite, for example, contains five molecules of water that are chemically bonded to the calcium and boric oxide molecules. As the temperature rises, the water molecules break their bonds and escape as steam. If the temperature rise occurs too fast during the early period of the kiln firing, the energetically escaping water molecules may also carry off sections of the glaze materials that have not yet melted sufficiently to bond with the claybody. I will never forget the sight of a hastily fired kiln

	SiO2	Al2O3	B2O3	CaO	Na2O	K2O	Melting Point (°F)
BORON FRITS[9]							
Ferro 3134 (Pemco 54) (Hommel 90)	46.5		23.1	20.1	10.3		1450* 1600**
Ferro 3124 (Pemco 311) (Hommel 90)	55.3	9.9	13.7	14.1	6.3	0.7	1600* 1750**
Ferro 3195	48.6	12.1	22.5	11.3	5.6		1450* 1600**
Ferro 3110	69.8	3.7	2.6	6.3	15.3	2.3	1400* 1700**

*fuses, **flows

that contained many pots glazed with a high-colemanite glaze. When the kiln was opened, a group of half-naked pots huddled on the shelves, with pools of colored glass at their feet. This catastrophe might have been avoided had the kiln been fired more slowly during the early stage of the firing.

The substitution of fritted borate materials can eliminate such problems; it is these materials that are the subject of the following section.

BORON FRITS

The boric oxide-soda-calcia combinations discussed previously are problematic glaze materials. Borax and boric acid are highly water soluble. An excess of either material in a glaze magma may blister and pit with disruptive surface effects. Colemanite and Gerstley Borate, although not as water-soluble, are also hydrates, and the release of their water molecules is often responsible for a pinholed and scarred glaze surface. All these borates are highly powerful melters; however, the brilliance and excitement that they bring to the glaze surface are often obtained at the price of disastrous fluidity and overfiring. To mitigate these problems, pure oxides of boron, calcium, and sodium are combined with silica and alumina in artificially made combinations known as "frits." (For a detailed explanation of frits, with tables of chemical structures, see Hunt 1978, 48–54.) Pure oxides of silica, alumina, boron, calcium, and sodium are combined by the chemist, fired to glass-forming temperatures, plunged into cold water to shatter the glass, and ground into powder. This is a process known as "smelting." The result is an artificially made glaze core containing the trinity of glassmaker (silica and boric oxide), adhesive-glue (alumina), and melters (calcia, soda, and boric oxide) without the troublesome water molecules, or the solubility property, both of which are inherent in the natural minerals. Because of their low fusion points (1400°F–1750°F), these frits are most useful at the lower firing temperatures. At both earthenware and cone 5/6 firing temperatures, they create smooth and bright glaze surfaces in combination with feldspars and other refractory materials, which would otherwise not enter the melt at these lower firing temperatures.

Note, however, that high frit glazes are similar to feldspathic glazes in that they require the presence of a suspension agent, such as Epsom Salts, in order to facilitate glaze application (see p. 23).

As is the case with the feldspars, we have many available frits, some of which are replaced by slightly different combinations from time to time (45 unleaded frits from the Pemco company are listed in the article by Hunt 1978, 51–52). Once again, an in-depth exploration of one or two gives better results than a superficial acquaintance with many. The following tests focus on four boron frits that have been frequently used in earthenware and cone 5/6 oxidation glazes.

Our tests show, that despite the considerable variation in oxide structure, the first three frits produce fairly similar surfaces when fired alone. At the cone 5/6 firing temperature, all the tests were yellow-brown to beige in color, transparent, glassy, and lightly crackled, and all were marked by a few pinholes. The underlying claybody of these tests was a low-iron stoneware reduction claybody, which fired to a pale buff at cone 5/6 oxidation. Thus, the yellow-beige color of the glaze surface reflected the underlying body color—an indication of transparency.

A thicker layer produced a milky, white interior ring in frit 3134 and frit 3124 tests. The test of frit 3195 did not show this milky white opacity, possibly due to an over-thin application.

At the earthenware firing temperature, the surface of frit 3134 was similar to its appearance at stoneware firing temperatures. The stability and long firing range of these fritted materials proves remarkable indeed!

The greatest differences between the frits themselves appeared in the color tests that were fired to cone 4 in the oxidation atmosphere. Various percentages of 10 coloring oxides were added to frit 3124 and frit 3134. Frit 3124 produced a milky, dense, opaque surface throughout the entire range of color tests. Frit 3134, which combines a higher amount of boric oxide and soda with no alumina, produced a glassy, brilliant, color series with a highly crackled surface. In this connection, it was interesting to compare the specific color results. Frit 3134 plus 3% copper oxide became a deep glassy green color ringed with opaque turquoise where the glaze layer was thin. The comparable test with frit 3124 showed just the opposite color effect. The thick ring in the center of the test bowl, caused by the flow of the glaze, was an opaque turquoise; the thinner layer on the sides of the wall was a crackled, glassy green. Unlike the other boron frits described above, frit 3134 contains no alumina, and this, plus the higher boric oxide and soda content, could well account for its different color effect.

DIFFERENCES BETWEEN FRITS, GERSTLEY BORATE, AND COLEMANITE

It is important to remember when comparing the pure, laboratory produced frits with the natural borate materials, such as Gerstley Borate and colemanite, that these minerals are marketed as impure ores, are not processed or refined, and contain clays, marl, limestone, sandstone, and volcanic tuff. In addition, the incorporation of water molecules in the chemical structure of natural borate minerals should make a difference with respect to the smoothness of the fired surface. Some of these differences in structure are reflected in the tests, but overall, the differences were not dramatic.

Comparative Tests of Frit 3134, 3124, 3195, Colemanite, and Gerstley Borate

Cone 5/6 Oxidation The substitution of each of these three frits for colemanite and Gerstley Borate in a volcanic ash glaze produced the following subtle color differences.

VOLCANIC ASH GLAZE		
Porcelain claybody		Soft, yellow gloss.
Volcanic ash	62.3%	
Colemanite	28.3	
EPK kaolin	9.4	
Omit Colemanite and substitute:		
Gerstley Borate		Soft, yellow gloss.
Frit 3195		Soft, gray-green gloss.
Frit 3124		Soft, gray gloss. Crackle.
Frit 3134 *Figure 1.16, p93.*		Opalescent blue-gray-white gloss. Collects in fat rolls of opalescent blue-gray-green glass at base. Distinct crackle network. Beautiful test!

Tests of 100% frits 3134, 3124, 3195, Gerstley Borate, and colemanite fired alone were all transparent, glassy, lightly crackled, and yellow-brown or beige in color; the variations in color reflected the shade of the underlying claybody; all were marked by some pinholes. The main differences shown were the following:

1. Greater number of pinholes in the colemanite test.

2. White, milky rings appeared where the glaze was thicker in many of the tests of the frits, in both earthenware and cone 5/6 firing temperatures. They were not visible in the colemanite/Gerstley Borate tests.

Cone 9/10 Oxidation

1. Pinholing in the colemanite/Gerstley Borate tests was even more pronounced at this higher temperature. The surface of the frit tests was much smoother than the colemanite tests and resembled their cone 5/6 test.

2. The color of the colemanite/Gerstley Borate surface was slightly more greenish than the frit tests, which again appeared yellow-brown in color.

Cone 9/10 Reduction

The surface color of both the colemanite/Gerstley Borate and the frit tests was orange-brown ringed with black. The main differences were as follows:

1. The colemanite/Gerstley Borate tests were blacker and less glossy than the frit tests, which were browner and greener in shade.

2. Frit 3195 (higher alumina) was the least like the colemanite/Gerstley Borate test—it was browner and glassier.

3. Frit 3134 (highest amount of boric oxide and no alumina) showed the closest resemblance to colemanite, which theoretically has the highest amount of boric oxide and no alumina. In actuality, colemanite does contain some alumina due to the impure content of this mineral.

4. The surface of the colemanite test again contained more pinholes than did the frit tests. Thus, smoothness of surface is probably the most important difference between colemanite and the frits.

Summary

l. Frits 3134, 3124, and 3195 did not show any dramatic differences when fired alone at the low and high stoneware firing temperatures.

2. Like colemanite and Gerstley Borate, these frits are most successful as lower temperature glaze cores. At the higher stoneware temperatures, their powerful boron and soda melters will cause the glaze surface to be overfired and dangerously fluid. On the other hand, the sparing use of these frits (on the insides of the pots only) can sometimes produce a dramatic and complex glaze surface at the higher stoneware temperatures. Like colemanite and Gerstley Borate, the frits can also be combined with more refractory materials such as kaolin and Cornwall Stone; in small amounts they will function as a fusion catalyst in the same manner as colemanite. Similarly, a thin coat of these frits can also be applied to the outer walls of a form to create a clay-body sheen.

3. Frits proved fairly similar to colemanite when fired alone at the low and high firing temperatures. The main differences appear at the high stoneware temperatures and are as follows:

   a. Greater pinholing in colemanite surface

   b. Greener cast of color in colemanite surface

   c. Reduction cone 9/10 colemanite test produces lower gloss and blacker shade of color when compared to frits

   d. Frit 3134 most closely resembles colemanite and would be the preferred substitute for the purpose of correcting pinholes and other problems of a high colemanite glaze surface.

The tests of colemanite and the three boron frits suggest that it would be possible to substitute any one for the other in most cases, especially at the lower firing temperatures, without causing dramatic changes in the glaze surface.[10] The most likely result of substituting a frit for colemanite should be a smoother and less scarred surface, and of course, this would be the whole point of the substitution. The mere fact that this kind of substitution is a possibility at all firing ranges is a tribute to the tremendous melting power, stability, and long firing range of these boric oxide frits. Despite the fact that they are artificially made and not part of nature's bounty (and therefore more expensive), they are an invaluable aid to the potter who works in the lower firing temperatures. Their reliable uniformity compared to the impure nature of the natural borate minerals such as colemanite and Gerstley Borate makes them well worth the extra expense.

All of these borate materials, no matter how impure, contain that many-sided oxide, boric oxide, a natural blend of glassmaker, adhesive, and melter functions all rolled into one. Boric oxide is surely the most versatile of all the oxides discussed thus far. It is an unparalleled example of the extraordinary materials that make up our natural environment.

SODIUM BORON FRIT: FERRO # 3110			
SiO_2	69.8	CaO	6.3
Al_2O_3	3.7	Na_2O	15.3
B_2O_3	2.6	K_2O	2.3

It is not possible to leave the subject of boron frits without mentioning the high-soda frit, Ferro 3110. This frit combines high amounts of silica and soda, with low amounts of boric oxide and alumina. At cone 04 oxidation temperatures, frit 3110 creates a glassy, transparent surface marked by a distinctive crackle pattern. Its surface is ringed with orange, which is the characteristic marking of soda. At the cone 9/10 reduction firing temperatures, on a porcelain clay, the frit produces a lustrous, black-gray-white opalescent surface that once again is edged in orange, interspersed with pools of yellow glass. A color test with 2% copper oxide produced a magenta-purple, blue-ringed surface. These test results indicate frit 3110's potential for creating dramatic interior surfaces with exciting color and surface effects at both the low and high oxidation and reduction firing temperatures. In this connection, a complete color test series would surely produce some intriguing and brilliant color results (see Lab VI, p. 196). As in the case of the other boron frits discussed above, this frit would be too fluid to use on the outer walls of a clay form at high stoneware temperatures.

LAB VI BORIC OXIDE

1. a. Take at least two identical test pots with a surface area large enough to apply the following four materials: Colemanite, Gerstley Borate, Borax, and Boric acid. Glaze one-fourth of each pot with one of the above materials.

b. Take six identical test pots and glaze each pot with one of the following materials: Colemanite, Gerstley Borate, Frits 3134, 3124, 3195, and 3110.

2. Take a glaze with which you are familiar that contains one of the eight boric oxide materials listed above. Mix up your glaze *without* this material, and place this mixture on one-half of your test pot. Place the original glaze, *as your control*, on the other half of your test pot. *Do this for all tests.*

3. Substitute colemanite and/or any of the other boric oxide materials for the calcium melter in your standard test glaze, and place this mixture on one-half of the test pot.

4. Substitute colemanite and/or any other of the boric oxide materials for the glaze core in your standard test glaze and place this mixture on one-half of your test pot.

5. Add 10%, 20%, 30% Colemanite and/or Gerstley Borate and/or boric oxide frits to your standard test glaze, and glaze one-half of your test pot with each of these test mixture. Glaze one-half of another test pot with each of these test mixtures and the other half with comparable test mixtures of Whiting and/or Wollastonite additions. (See Lab. IV #5, p. 173, in which these test series were performed. Hopefully you have retained these calcium test mixtures.)

6. *Color Tests*: Weigh out 1000 grams of any one of the boric oxide materials listed in #1. Add enough water to make a standard glaze consistency similar to light cream. Mix well, and pour identical levels of this mixture into 10 identical clear plastic cups. Add the following percentages of colorants to 9 of the 10 cups of the boric oxide mixture. (The 10th cup will contain the boric oxide material without the colorant.)

Cobalt carbonate or oxide	0.5%
Copper carbonate or oxide	3.0%
Cobalt carbonate or oxide ⎱	0.25%
Copper carbonate or oxide ⎰	1.0%
Red iron oxide	6.0%
Rutile	3.0%
Ilmenite	3.0%
Tin oxide	5.0%
Nickel oxide	2.0%
Black mix[11]	4.0–7.0%

If you have poured identical levels into the 10 identical clear plastic cups, each cup should contain approximately 100 grams of the boric oxide mixture, plus the percentage of the coloring oxide (except for the 10th cup, which contains only the 100 grams of the boric oxide mixture).

Mix each cup well, and glaze identical test pots with each one of the 10 glaze mixtures. These test results should give you a fair indication of the color range of the boric oxide materials.

> **Note:** It is presumed that you will have at least two firing temperatures available to you: bisque and/or earthenware and/or cone 5/6 oxidation or reduction and/or cone 9/10 oxidation or reduction. Fire each test in each of the available firing temperatures. Thus, always make at least two tests of each test mixture.

Caution: All of these boric oxide materials are powerful melters, especially at the higher temperatures, and will increase the fluidity of the surface. In addition, certain colorants, such as red iron oxide, will increase the melting action, particularly in the reduction atmosphere. Hence, if you glaze the outside of your test pot, be sure to leave a good margin to account for excess fluidity.

MAGNESIA MINERALS

Figure 3.12, p234.

Magnesium oxide, or magnesia, increases the melt of a stoneware glaze if used in small amounts, and for this reason it is classified here as an auxiliary melter. At temperatures below 2138°F, it functions as a refractory opacifier (Hamer and Hamer 1986, 204). Hence, in the metals industry, magnesia functions as an important refractory material (Frye 1981, 442). Although magnesia performs the dual role of both melter and refractory, most important is its unique effect on glaze texture and color. These are the qualities that will receive primary attention in the following description of magnesium materials.

Magnesium is the seventh most abundant element in the earth's crust, and together with calcium and iron, is the first mineral to crystallize from the earth's magma (Craig, Vaughan and Skinner, 1988, 366–367; Tarbuck and Lutgens 1984, 55). Minerals that are first to crystallize are heavier, and consequently sink to the bottom of the melt; they are also the last to melt. The boiling and bubbling movements from the later melting of magnesia beneath layers of earlier fused materials cause uneven, streaked effects. This is the phenomenon that produces exciting, variegated glaze surfaces (Hamer and Hamer 1986, 204).

Magnesium belongs to the same family of metals in the periodic table as calcium. Both are classified as alkaline earth metals; each possesses two more electrons than their nearest noble gas neighbor. From this shared trait flow a host of similar characteristics, many of which directly relate to glaze surfaces.

At stoneware firing temperatures:

1. Magnesia and calcia add hardness and durability to a glaze surface. The melting points of magnesia and calcia reach well over 4000°F. (Calcia melts at 4658°F, and magnesia melts at 5072°F.)

2. The amount of calcium and magnesium oxides that appears in a glaze determines their behavior. Low amounts (up to 10% for magnesia and up 25% for calcia) increase fusion (depending on the rest of the materials in the mixture). Larger amounts will opacify and finally dry up the glaze surface.

3. Calcia and magnesia form crystalline bonds with silica to create the semigloss, opaque surface known as a satin-matt.

4. Magnesia rarely appears in a glaze without calcia. The natural affinity of magnesia for calcia appears in the mineral dolomite and dolomitic limestone. These naturally formed calcium-magnesium carbonate minerals are the most common ceramic source of magnesia. Without the steadying presence of calcia, high-magnesia glazes often reach the extremes of a super-gloss opacity, or a dry, crawled surface. According to industrial research, a 3:2 ratio of calcium to magnesium, respectively, creates the strongest fusion in the production of glass. Further additions of magnesia (for example, 50 calcia to 50 magnesia) will decrease fusion and increase opacity. Ohio dolomites naturally contain this 3:2 ratio of calcium to magnesium and are of major importance in the glass manufacturing industry.

(*Ceramic Industry* 1998, 100).

Despite its dependency on calcia, magnesia has its own unique contribution to make to ceramic surfaces, not the least of which is, as mentioned above, the satin-matt surface—the very essence of magnesium oxide. A silky, sensuous, satin-matt surface is a prime characteristic of magnesia. *Figures 3.12, p234; 3.14–3.15, pp236–237.* (Note that Rotten Stone, which produces those extraordinary satin surfaces at cone 9/10 oxidation, contains 10% magnesia, see pp. 68–75.) In this respect, magnesia is more effective than calcia, possibly due to the high surface tension that is characteristic of magnesia materials. High surface tension tends to round edges and fattens the appearance of the glaze surface. On the other hand, this same property results in crawling. Magnesia materials characteristically produce intentional (and sometimes unintentional) crawled surfaces. (See Test IB, p. 207; *Figure 3.16, p238.*)

Secondly, magnesia has a low rate of expansion, lowest of the melters after boric oxide (Hamer and Hamer 1986, 155). The presence of a magnesia in a glaze will lessen the craze network produced by the high-shrinking sodium and potassium melters of the feldspars.

Thirdly, its high melting temperature makes it a

significant refractory, which adds hardness and durability to the fired glaze surface. In certain glaze mixtures, even the smallest amounts will stiffen the glaze surface.

Perhaps magnesia's most striking contribution is its impact on glaze surface color. Magnesia materials enhance certain colors and completely destroy others. A cobalt blue glaze could turn purple in oxidation and red-blue in reduction with the substitution or addition of magnesia. The combination of tin and magnesia in an oxidation cone 5–6 glaze produced a pink colored surface. A brilliant copper red reduction glaze will turn liver-colored if magnesia is added to the calcia melter. A low-iron, celadon glaze changed from gray-blue-green to yellow-green and, finally, to dark-brown with additions of magnesia (see tests, pp. 206–213).

Note that additions of magnesia to a glaze often opacify the surface. The color of the glaze in such cases would naturally become more muted and pastel-like.

This description of magnesium oxide does not tell the real story because magnesia is found not in a pure form, but only in various compounds. Each material in which magnesia appears has its own unique characteristics, which temper the behavior of magnesium oxide. Thus, in order to understand the actual workings of magnesia in ceramic surfaces, it is necessary to consider our three major sources of magnesia—dolomite, $(CaMg[CO_3]_2)$, talc, $(3MgO \cdot 4SiO_2 \cdot H_2O)$, and magnesium carbonate, magnesite $(MgCO_3)$.

SOURCES OF MAGNESIA

DOLOMITE

The name dolomite refers to a mineral, a rock, and a mountain range in the Alps of northern Italy. To avoid the inevitable confusion, some geologists refer to the rock as dolostone (Tarbuck and Lutgens 1989, 138).

The pure mineral dolomite forms from the interaction of magnesium carbonate and calcium carbonate in two ways: first is a pure, but rare process of formation that occurs when magnesium and calcium carbonate materials of weathered rocks and soils are carried to the seas and lakes to be dissolved by the action of the water. Eventually, some of these dissolved carbonate materials precipitate out to form the

magnesium-calcium carbonate material known as dolomite. A secondary primary source of dolomite formed millions of years ago, when waters enriched with magnesium carbonate ions reacted with high calcium limestone rock.[12] Some of the calcium carbonate ions in the limestone were replaced with magnesium carbonate ions to form the replacement mineral dolomite. This process took millions of years to complete; the fact that most of our supplies of dolomite are known to be of ancient origin is evidence of this second method of formation (Ibid.).

The firing cycle of dolomite is more complex than that of calcium carbonate because it consists of both magnesium and calcium minerals, which, when heated, create two possible stages of decomposition (Grimshaw 1980, 713). (Magnesium carbonate decomposes at above 1450°F and produces carbon dioxide, with a loss of weight of 52.4%. Later, at about 1650°F calcium carbonate decomposes into calcia and carbon dioxide with a weight loss of 44%.) Dolomite is said to lose its carbonate material and decompose into calcium and magnesium oxides at about 1472°F (Hamer and Hamer 1986, 105). However, as in the case of calcium carbonate, the true fusion point of dolomite would be much later. It would take place somewhere between the melting points of calcia and magnesia, which are 4658°F and 5072°F, respectively. Once again, as with calcium carbonate, the presence of additional melters in the glaze mixture lowers the melting temperature of these otherwise refractory oxides. Thus, dolomite usually appears in conjunction with the sodium, potassium, and calcium melters in the feldspars, together with additional auxiliary melters of calcium, boron, zinc, and/or lithium minerals.

Tests of dolomite fired alone at cone 9/10 reduction temperatures display a stony, yellow-green, white surface that closely resembles the tests of calcium carbonate. In the oxidation atmosphere, the green cast gives way to yellow. This similarity to calcium carbonate that appeared in our tests was corroborated by industrial research.

> "In general, dolomite may be substituted for nearly any other type of lime flux in pottery bodies and glazes when other batch constituents are properly compensated."
> (*Ceramic Industry* 1998, 100)

Claybody Function

Dolomite has an extraordinary range as a claybody melter—it functions as an auxiliary claybody flux with feldspar in cones 1–12. According to industrial research, the substitution of dolomite for Whiting as an auxiliary body flux in conjunction with feldspar increases the firing range of the claybody by one or two cones (*Ceramic Industry* 1998, 101). We created a successful, cone 5–6, white claybody with 8% dolomite, 15% Nepheline Syenite, 7% potash feldspar, 50% Grolleg clay, and 20% Flint. At the cone 9/10 reduction temperatures, there was some deformation of shape and a self-glazed, translucent, glossy surface, which suggested that the original cone 5/6 firing range of this dolomite claybody could be extended to cone 7–8. Dolomite is said to be a useful auxiliary flux in earthenware bodies as well, and thus its range extends throughout many firing temperatures (Ibid.).

Glaze Function

The glaze function of dolomite will be discussed on pp. 202–213, in comparison with the other magnesium minerals of talc and magnesium carbonate.

TALC

More than 600 million years ago, hot, reactive, high-silica solutions from the earth's interior erupted onto the marbleized, metadolomites of what is now the northwestern Adirondack Mountains. Violent metamorphic tectonic movements folded and refolded the rock formations. The intense heat and pressure on the metadolomites during these turbulent stages of metamorphic activity formed layers of rock that contained abundant deposits of hydrated magnesium-silicate minerals, such as talc, tremolite, and serpentine (antigorite). These minerals together make up our commercial talc deposits. The minerals were formed in complex stages of formation; intense metamorphic folding and refolding earth movements released high-silica hydrothermal solutions, which removed magnesium, calcium, and silica ions from surrounding beds of quartzite and metadolomite to form the mineral tremolite ($2CaO \cdot 5MgO \cdot 8SiO_2 \cdot H_2O$).

The serpentine minerals ($3MgO \cdot 2SiO_2 \cdot 2H_2O$) appeared in a later sequence, due to the increased magnesium and lower temperature of these later hydrothermal solutions. And finally the last stage—a resurgence of magnesium-rich hydrothermal solutions without any further folding or violent earth upheavals—produced the pure mineral talc ($3MgO \cdot 4SiO_2 \cdot H_2O$) (Bates 1969, 328–332).

The tremolitic talc deposits of New York are said to be the most productive in the country, and the ceramic industry is one of their most important markets. The New York talc deposits manufactured by R. T. Vanderbilt Company for sale to the ceramic industry (described as "tremolitic talc," owing to the high concentration of tremolite) contain 40%–60% tremolite, 20%–30% antigorite, and but 20%–30% pure talc. The calcium in the tremolite talc (7%–8%) is said to be a valuable addition to claybodies because it increases the fired strength and lowers the absorption, shivering, and moisture expansion of the claybody (*Ceramic Industry* 1993, 110, 1998, 170; R. T. Vanderbilt Company, Technical Data, 1998). Note that the pure mineral talc is actually a very small part of these commercial talc deposits (20%–30%), and in some commercial deposits, talc is not present at all.

> "Large minable bodies of the pure mineral are rare, and in commercial usage the term talc has quite a different application. It refers to various rocks that are composed of magnesium silicates, in which talc, the mineral, may be dominant, abundant, minor, or entirely absent."
>
> (Bates 1969, 328)

Herein lies the problem of asbestos contamination. Asbestos is the term that describes fibrous crystalline formation in magnesium-silicate minerals. Prior to 1992, OSHA included the minerals of tremolite, actinolite, and anthophyllite in the dreaded asbestos classification. However, these minerals can grow fibrous (asbestiform) or nonfibrous (nonasbestiform) crystals. According to R. T. Vanderbilt Company, fibrous formation (asbestiform) is a rare occurrence in tremolite, actinolite, and anthophyllite, and their nonfibrous mineral form should not have been classified as asbestos.

"When the Asbestos Standard was published in the summer of 1972, by OSHA, the standard defined the minerals tremolite, anthophyllite and actinolite, as asbestos minerals without further qualification. In reality, these minerals are common rock forming minerals which only rarely and under special circumstances grow and crystallize in the form of fibers...."

(Rieger, 38)

The Material Safety Sheet of R. T. Vanderbilt Company describes its tremolitic talc as nonasbestiform and noncarcinogenic with no asbestos fibers.

"New York State talc has not shown any carcinogenicity in animal studies. Epidemiologic studies in humans have been interpreted in conflicting ways with no clear evidence of an increased risk of lung tumors in association with exposure."

According to K. C. Rieger, Manager of the Ceramic Laboratory, R. T. Vanderbilt Company, mineralogists from the U.S. Bureau of Mines and U.S. Geological Survey "have demonstrated the differences between tremolite present in New York State, and the uncommon rare variety of asbestiform tremolite" (Rieger, 38).

"In 1986, when OSHA revised its 1972 asbestos standard, the agency redefined 'asbestos' to include only the six asbestiform minerals chrysotile, crocidolite, amosite, tremolite asbestos, anthophyllite asbestos and actinolite asbestos.

...OSHA also added a separate definition for tremolite, anthophyllite and actinolite—referring to only the nonasbestiform forms of these three minerals. *However, the nonasbestiform minerals were included in the scope of the standards.*" [emphasis added]

(NEWS, United States Department of Labor, Office of Information, Washington, D.C., May 29, 1992)

In 1992, OSHA changed its standards to exempt nonfibrous forms of these minerals.

"The U. S. Department of Labor's Occupational Safety and Health Administration (OSHA) has determined that the evidence is insufficient to conclude that nonasbestiform varieties of asbestos minerals present the same type or magnitude of health risk as asbestos.

[...] *As a result, OSHA has decided to exclude nonasbestiform tremolite, anthophyllite and actinolite from coverage under its asbestos standards.*"

(Ibid.)

"These nonasbestiform amphiboles, which comprise upwards of fifty percent of our talc, are now considered as 'nuisance dust.'...Of course, there never was any asbestos in our talcs...."

(Paul Vanderbilt, R. T. Vanderbilt Company, Inc., letter to author 17 March 1993)

Each shipment of R. T. Vanderbilt Company's talc deposits are regularly subjected to testing prior to shipment as part of the company's quality control procedures (K. C. Rieger, R. T. Vanderbilt Company, 1998).

For those purists who are willing to pay more, there are sources of pure talc on hand. California, Montana, and France produce steatite deposits that contain the pure mineral talc, primarily for use in the manufacture of electrical insulators and kiln furniture (*Ceramic Industry* 1998, 170–171). Then too, there is always the option of substituting separate magnesium and silica materials for talc. Bear in mind that none of these substitutions will produce the same results as the calcium-rich, tremolitic talc, and that some of them will prove to be far more expensive.

Claybody Function

Talc is a valuable claybody melter at both low and high firing temperatures.

1. Talc forms the mineral enstatite at the low temperature of earthenware bodies. Enstatite has a high thermal expansion and contraction rate, which causes the glaze to be put in compression upon cooling, and thus helps to eliminate crazing. Talc reduces the postfired moisture expansion of the often porous earthenware claybodies and thus significantly

decreases the amount of moisture-induced body expansion that produces delayed crazing. In this connection, it is interesting to note that lime-bearing tremolitic talc proved superior to other talc minerals in reducing moisture expansion. This form of talc also produced stronger claybodies, fast firing schedules (1 hr.!), stable shrinkage and absorption rates, and lower warpage. Thus, tremolitic lime-bearing talc constitutes the predominant talc mineral for commercial wall tiles bodies of cone 02–2 (*Ceramic Industry* 1993, 110; 1998, 170).

2. Talc increases thermal shock resistance of high-temperature claybodies by the production of cordierite and mullite minerals that have low thermal expansion rates (Ibid.). As an auxiliary flux with feldspar and/or Nepheline Syenite, small additions (3%–5%) increase the tightness and strength of the claybody, lower firing temperature, and reduce crazing of the glaze (R. T. Vanderbilt Company, Inc., *The Use of Talc in Ceramic Whitewares*).

3. High talc bodies fire white, thus increasing the brilliance of glaze colors (Ibid.).

Because of all of these advantages, talc minerals have long been an important ingredient of low-fire, cone 5–6 "porcelain" and high-fire, thermal, flameware claybodies. (See Claybody Formulas, pp. 113, 139.)

> **Note:** The manufacturer's safety data sheet warns that prolonged inhalation of talc dust could cause lung injury. Thus, proper ventilation and frequent cleaning of the workroom is essential when working with a fine-grained, talc claybody.

MAGNESIUM CARBONATE

Magnesium carbonate (magnesite) and light magnesium carbonate (hydromagnesite) are the potter's source of pure magnesium oxide. In contrast to magnesium carbonate of which there are worldwide deposits, the mineral form of pure magnesium oxide, periclase, is a rare geological occurrence.

Light magnesium carbonate results from a boiled solution of magnesium sulfate and sodium carbonate. The raw, powdered form of light magnesium carbonate is exactly as its name implies—light and fluffy. A 20-pound bag of this material will go a long way!

The Potter's Dictionary states that light magnesium carbonate is the preferable glaze material because it "mixes better in the glaze slop" (Hamer and Hamer 1986, 205). Although both forms of magnesium carbonate are soluble in acidic water, light magnesium carbonate is described as less soluble than the heavier magnesite (Ibid.). Whatever the form, magnesium carbonate ends up as magnesium oxide at the end of the kiln firing and is, therefore, the ceramic source of pure magnesium oxide. In this sense, it is comparable to Whiting (calcium carbonate), which is the source of pure calcium oxide. As is true of calcium carbonate, industrial research describes magnesium carbonate as a strong melter at high temperatures and an opacifier only at low temperatures.

> "In glazes, magnesium carbonate acts as a refractory up to a relatively high temperature when it then becomes an active flux."
>
> (*Ceramic Industry* 1998, 130)

However, we have not found this to be the case at high stoneware temperatures. Unlike calcium carbonate, magnesium carbonate functions in our stoneware tests as a most refractory material. As demonstrated in the following cone 5–6, and cone 9–10 tests, (pp. 206–213), magnesium carbonate functioned in most cases more effectively as an opacifier than as a melter. Magnesium carbonate, more than any of the other magnesium materials tested by us, required the presence of additional melters—primarily calcium, in order to increase fusion at high stoneware temperatures. Another startling characteristic of magnesium carbonate was its effect on the glaze texture. When used in fairly large amounts (30% or more) it created a uniquely patterned, crawled glaze surface. (See Test IB, p. 207; *Figure 3.16, p238.*)

Claybody Function

We have not tested magnesium carbonate as a claybody material. According to industrial research, many manufacturers of porcelain and semiporcelain wares use small amounts of magnesium carbonate for the whiteness it produces. Industrial test results of magnesium carbonate as a claybody melter are confusing and obviously depend on the amount used and the rest of the claybody ingredients. In one case as little as 0.1% raised the maturation temperature and prevented blistering; 0.4% increased translucency. In a feldspathic claybody, 5% magnesium carbonate lowered the vitrification point of the claybody. Once again, magnesium carbonate proved most effective as a claybody melter when used in conjunction with feldspar and other auxiliary melters (*Ceramic Industry* 1998, 130).

The fact that magnesium carbonate is somewhat soluble could have adverse effects on the claybody.

"For this reason it cannot be used in clays as a catalyst because it upsets workability. Talc is used instead for this purpose."

(Hamer and Hamer 1986, 205)

GLAZE FUNCTION OF DOLOMITE, TALC, AND MAGNESIUM CARBONATE

The three magnesium materials discussed—dolomite, talc, and magnesium carbonate—share the general characteristics of magnesium. On the other hand, they each have their own distinct personality and will leave their particular mark on a glaze surface. In order to understand which magnesium mineral will best achieve the desired result, a comparative analysis of test results obtained in actual glaze surfaces is helpful. In making this analysis, the important questions are as follows:

How do dolomite, talc, and magnesium carbonate compare with each other and with calcium minerals in terms of *melting power, opacity, texture,* and *color effects?*

The following summary of our test results provides some answers to these questions.

Melting Power
Cone 9–10 Reduction

1. In general, our tests showed that dolomite is the least eccentric of the magnesium trio and would cause the least change in the glaze surface when substituted for calcium carbonate. The dolomite substitution in Sanders Celadon (Test IX, p. 210) produced only a slightly deeper blue-green color. And again, the test with 5% dolomite and

5% Whiting did not display any substantial difference from the test with 10% Whiting (Test IIA, p. 207). However, the satin-matt and stony-matt glazes all showed less transparency and lowered gloss when 25%–30% dolomite substituted for equivalent amounts of Whiting (see Tests III, IV, VI, and VII, pp. 208–209). It may be possible, therefore, to substitute lower amounts of dolomite (up to 20%) for equivalent amounts of calcium carbonate without causing dramatic changes in the glaze surface, particularly in cases where the glaze contains additional melters of zinc oxide or colemanite. The presence of colemanite may facilitate the melting power of magnesium, and for this reason, equivalent substitutions of dolomite for calcium carbonate should be more successful in glazes that contain colemanite.

2. Both talc and magnesium carbonate are less powerful melters than dolomite. The replacement of 5% Whiting with 5% talc and, alternately, with 5% magnesium carbonate in a glaze batch of 90% feldspar and 10% Whiting caused the original gray, glossy, transparent surface to opacify and whiten in color. A corresponding test with dolomite did not change the surface (Test IIA). On the other hand, although whiter in color and, hence, less transparent, the gloss of the surface increased with the talc and magnesium carbonate substitutions.

3. Magnesium carbonate is the most refractory of all three magnesium minerals and causes the greatest change in the glaze surface. The presence of silica in talc and calcium carbonate in dolomite modify the refractory property of magnesium. Despite its refractory nature, low amounts of magnesium carbonate increase the fusion of a feldspar at high stoneware temperature. (Test IB, p. 207). In this test series, 10% magnesium carbonate caused the original opaque-white surface of Nepheline Syenite to become shinier and transparent. However, 20% magnesium carbonate caused the surface to opacify and whiten; glaze coverage became uneven and slightly crawled. This result contrasted sharply with the gray-green, matt-shine, even coverage of the comparable test with calcium carbonate (Test IB). The replacement of 18% Whiting with 18% magnesium carbonate in Sanders Celadon (13% yellow ochre) caused the silky, deep green surface to change to a leathery, brown, crawled lizard skin (Test IX #5, p. 210).

4. Because of the low expansion rate of magnesium minerals, we expect less crazing to occur when they replaced calcium minerals, and this result did at times occur (Test IB).

To summarize, only dolomite, of the three magnesium minerals, can substitute for calcium minerals in some glazes without changing the fusion point of the glaze (Tests IIA–B, IX, pp. 207–208, 210). However, in most of the glazes that we tested, large-scale dolomite substitutions caused the surface to appear whiter and, therefore, less melted (Tests III, IV, VII, X, XI, pp. 208–210). The color of a reduction clay-body beneath the glaze is gray. The change of glaze surface from gray to white indicates lessened transparency and therefore less melt.

As a general rule, even in low amounts, any substitution of dolomite for other calcium minerals is not without the risk of some surface change. For this reason, all substitutions should first be tested to determine if they do in fact alter a particular glaze surface.

Cone 5–6 Oxidation

1. As with cone 9–10 firing temperature, the magnesium minerals appear to be less-powerful melters than their calcium counterparts (Tests I, VI, VIII, IX, pp. 211–213).

2. Again, as in the case of the 9–10 firing temperature, the dolomite substitution for Whiting causes the least change in the glaze surface. (Tests I, VI, IX)

3. The presence of colemanite or Gerstley Borate activates the melting power of talc and magnesium carbonate (Tests VI, VII).

Opacity and Texture

Magnesium minerals create opaque glaze surfaces in the same manner and for many of the same reasons as calcium minerals. Thus, a satin-matt opacity will result from magnesium-silicate crystalline formations. A stony-matt opacity results from unmelted particles of magnesium that lie suspended in the melt. The main differences between calcium and magnesium in this respect are as follows:

1. Magnesium minerals create a fatter, satin surface than comparable amounts of calcium.

2. A lesser amount of magnesium minerals will create an opaque surface.

3. The purest source of magnesia, magnesite, or magnesium carbonate creates unique crawled surface effects, which find no parallel in calcium materials.

Cone 9–10 Reduction

The following tests revealed the unique opacifying properties of each of the three magnesium minerals.

1. Dolomite favors both a stony, opaque surface and the formation of yellow, matt crystals (Tests III–VII, IX, X, XIII, pp. 208–211). *Figure 3.14, p236.*

2. Talc is most effective in the production of satin-matt surfaces. Combinations of talc and potassium feldspar produced a more even, satin surface than did comparable tests with Whiting (Test IIC). *Figure 3.12, p234.*

3. As in the case of the calcium minerals, changes in the firing and cooling cycle can disturb the magnesium satin-matt surface. The soft, satin surface produced by dolomite and talc in the Charlie D glaze changed into a glossy, transparent surface flecked with opaque, matt, yellow, flower-shaped crystals when the firing temperature increased from cone 9–10, to cone 10–11 (Test X). These yellow, opaque crystals were similar to the crystalline formations resulting from additions of dolomite to a celadon glaze (Test IX). *Figure 3.14, p236.*

4. Magnesium carbonate exhibits peculiar physical characteristics that directly affect the way in which it opacifies a glaze surface. The powdery particles of magnesium carbonate are light and fluffy and possess a high surface tension. This high surface tension rounds the edges of the glaze surface to create the visual effect of fatness, which is so characteristic of the satin-matt surface produced by magnesium minerals. Even more extraordinary textures occur as a result of this property. The surface tension of the high-magnesium carbonate glaze forms a crawled pattern on the bisque ware, which remains on the surface of the fired ware. A fascinating, crawled texture is achieved with combinations of 30%–40% magnesium carbonate and 70%—60% Nepheline Syenite, respectively (Test IB). This combination of materials produced a glossy, white, opaque, evenly

crawled surface with a distinctly cellular pattern. Wonderful surface effects were achieved by applying this combination over or under a darker colored glaze. *Figure 3.16, p238.*

Cone 5–6 Oxidation

The strength and stability of magnesium materials is reflected in the fact that they reveal similar opacifying properties at the lower, cone 5–6 temperatures. Despite the lower 100-degree difference in firing temperature and the clear, oxygenated atmosphere, we observed many comparable results.

1. Once again, dolomite leans toward the creation of stony opacities (Tests VI, VIII, IX).

2. Talc again produces the silkier, more satin surface (Tests VI and IX).

3. Magnesium carbonate creates the same kind of glossy, controlled crawled surface that was observed at the higher firing temperatures (Test VI).

All three kinds of opacities resulted from 30% additions of magnesium minerals to a high-gloss glaze (Test VI). Note that Whiting (calcium carbonate) required an addition of 40% to achieve comparable opacity, and that additions of 30% Wollastonite resulted in a transparent, shiny surface (Test VI).

Color Effects

Magnesium minerals strongly influence the color of the glaze surface.

Cone 9–10 Reduction

1. Calcium minerals, such as Whiting and Wollastonite, favor the gray-blue, gray-green, celadon color in a low-iron (1/2–3%), high feldspathic glaze. The substitution of magnesium minerals (dolomite, talc, or magnesium carbonate) will cause these colors to become yellower and/or whiter (Tests I–X, pp. 206–211). Calcium oxide minerals appear to be more powerful melters than their magnesium counterparts and therefore increase the fusion and transparency of the surface. Increased transparency permits the color of the claybody beneath the glaze to strongly influence the final surface color. The underlying color of the reduced claybody is usually gray—hence the gray cast of the transparent surface. Magnesium minerals, on the other hand, create a whiter, less-transparent surface and therefore appear to be more refractory than the calcium materials. In this connection, note that the stony-white background of Rhodes 32 relies on Dolomite (22.4%) and Kaolin (25.1%) for this color (Test V). The same is true of the stony, orange-white surface of the Pavelle glaze (Test XIII, *Figure 3.7, p229*) and the Cascade glaze (Test IV). The substitution of dolomite for Whiting in a blue-gray, semimatt glaze changed the surface to a stony, yellow-white, similar to the surface of Rhodes 32 (Test III).

2. Dolomite encourages a yellow-colored glaze surface, even more than calcium carbonate minerals (Tests I–IX). Note the yellow, stony, crystalline surface (30% dolomite) of the Pavelle glaze (Test XIII, *Figure 3.7, p229*). And again, note that the predominantly yellow-white-brown, orange flecked surface of Rhodes 32 (22.4% dolomite) became less orange and more yellow without whiting (3.5%) (Test V). Note that talc also can encourage a yellow surface color. The addition of only 5% talc to a Nepheline Syenite-Whiting glaze increased the yellow color of the glaze. The yellow colored surface of a high-dolomite glaze often consists of visible yellow, crystal. (Tests V, VI, IX). Although it is true that large additions of calcium carbonate will often produce a yellow surface (see Test VII), these clearly visible, bright yellow, flowerlike crystalline formations occurred most consistently with additions of dolomite. Thus, the foregoing tests identify a primary characteristic of dolomite—the production of a yellow, crystalline, stony surface.

3. Magnesium minerals do not show the same property of claybody-glaze interaction that is shown by the calcium minerals. Note that claybody-glaze interaction is not necessarily a concomitant of melting power. Boric oxide, a more powerful melter than calcium, does not exhibit this trait. Calcium minerals in a glaze interact with the claybody and pull the iron-specks of the claybody into the glaze surface. This interaction is visible in small tests of high feldspar glazes that do not contain any iron colorant, such as iron oxide or Barnard clay.[13] In many of these tests, the surface color is the blue-green hue of a celadon glaze, which typically contains

low amounts of iron colorants (iron oxide or iron-bearing clay, such as Barnard) in combination with feldspar, silica, and calcium melters (Whiting or Wollastonite). Comparable tests with magnesium minerals (with the possible exception of dolomite; see *Figure 3.13, p235*) do not produce this same celadon color. The substitution of dolomite for Whiting in the Rhodes 32 stony-matt glaze caused the surface to lose the orange spots that flecked the stony-white surface of the original glaze. The original glaze appeared on one-half of the test pot as a control, and it clearly showed the presence of orange speckling on a stony-white background. The other half of the test, which contained the dolomite substitution, did not contain these orange speckles. We concluded that the orange flecks were in fact reoxidized red iron spots from the clay-body, which had disappeared with the dolomite substitution because of the loss of the calcium melter. (This was a most surprising result in view of the fact that the amount of Whiting that was replaced by dolomite was less than 4% of the total glaze batch. Repeated testing in various kiln firings would be necessary to substantiate this conclusion.)

4. In the presence of higher amounts of iron oxide (6%–15%) magnesium minerals bring out a red or purple-mahogany-brown color; calcium carbonate minerals favor the deep green or black-brown color (Tests IX, XI, XII). In this connection, note the presence of talc (6.3%) in the Persimmon glaze, which achieves its remarkable shade of red-orange-brown through the combination of bone ash, talc, and 15% red iron oxide (Test XII; *Figure 3.8, p230*).

5. Magnesium minerals have a pronounced effect on the blue color produced so consistently with cobalt. The usually stable and predictable cobalt blue turns purple or mauve when magnesium minerals are present (Test IIC).

6. The brilliant copper red of a high feldspar glaze with calcium melters changes to a dreary liver-color or dull pink when magnesia appears in the glaze. Temple White (Test VII, 19.6% dolomite) turned an unattractive, liverish pink color with the addition of 1/2% copper oxide. On the other hand, combinations of 1/4% cobalt oxide and 1/2% copper oxide create a brilliant turquoise blue-green color in the presence of dolomite and talc (Test X).

These are but a few examples of the remarkable color effects produced by magnesium minerals at the cone 9/10 reduction firing temperatures. Once again, we see the importance of the melters in the production of glaze color.

Cone 5/6 Oxidation

Because of the refractory nature of magnesium minerals, one would expect additions of magnesia at the lower stoneware temperatures to lighten the color of the glaze. However, even at these lower temperatures, additions of magnesia produce unexpected color changes.

1. A soft, buttery white surface results from additions of 30% talc to a high-gloss, transparent glaze. Substitution of 30% dolomite produces a yellower, stonier, white surface. Substitution of 30% magnesium carbonate creates a white opacity, but the surface is glossy and crawled (Test VI).

2. As in the cone 9/10 reduction tests, magnesia minerals play a crucial role in creating the red color of the glaze. The pattern is surprisingly similar—magnesia minerals destroy certain reds and pinks and help to produce others.

Iron oxide depends on the presence of magnesia to achieve the red color of the Randy Red glaze (Test VII, *Figure 3.11, p233*). The glossy red-green surface of this glaze changed to blue-yellow without the original 14% talc. The red remained with the substitution of magnesium carbonate for the talc, but disappeared with the substitution of dolomite and Wollastonite, which both produced a yellow-brown color. (Wollastonite produced a shiny, yellow brown surface in contrast to dolomite, which resulted in a yellow-brown surface of lower gloss—evidence of the greater fusion power of Wollastonite.)

3. As in the cone 9/10 reduction tests, the presence of magnesia tended to make purple the blue of cobalt (Tests IV and VI). However, when cobalt is combined with additional colorants, such as rutile, for example, we can expect different results. Cobalt and rutile went through a series of color transformations with the addition of magnesia. An increase of dolomite in a glaze with cobalt, rutile, Gerstley Borate, talc, and dolomite caused the blue-gray satin-matt to become increasingly greener (Test VIII). The substitution of talc for Whiting in a rutile glaze caused the original yellow-gray-white mottled surface to change to an even blue-gray-white

(Test IX). (It also increased the satin surface of the glaze and lowered its fusion point.)

4. A blue-green color is produced with copper, feldspar, and dolomite (Test III). On the other hand, a combination of feldspar and talc over a copper oxide stripe produced a green-black color (Test IV).

The foregoing tests of magnesia minerals underscore the significant contribution that the melters (separate and apart from their fusion properties) make to glaze texture and color. Melter materials bring to life certain colors and destroy others. By understanding these properties, we can use them to create unique and exciting glaze surfaces.

COMPARATIVE TESTS OF MAGNESIA AND CALCIA
CONE 9/10 REDUCTION

			IA			
Nepheline Syenite	Whiting	Wollastonite	Dolomite	Talc	Magnesium Carbonate	Results
90	10					*Dark gray-blue matt. Flecks of gloss. Black matt. Flecks of gloss.
		10				Gray-green gloss.
			10			Yellow-gray-green. High gloss.
				10		White-gray; high gloss. Gold luster where thin.
					10	White, high gloss.

Note: Whiting test traps carbon (black matt test). It interacts with the claybody and creates more transparency than magnesium melters. (Dark gray surface color reflects underlying dark-gray claybody.) Wollastonite test creates both transparency and gloss (adds more silica and increases fusion), but does not trap carbon as does Whiting test. It does not appear to interact with claybody as much as Whiting, as dark-gray color of underlying claybody has not influenced surface color to same extent as in Whiting test. Dolomite test is not as transparent (yellower color), but creates a higher gloss. Talc test creates even less transparency—the surface with 10 Talc is the most like the 100% Nepheline Syenite without additional melter. See IB.

*Different kiln firings.

IB

Figure 3.16, p238.

Nepheline Syenite	Whiting	Magnesium Carbonate	Results
100%			White. Gloss. Craze. Gold luster where thin.
90	10		Blue-gray-green. Gloss. Craze.
90		10	White-gray. Gloss. No craze.
80	20		Blue-green glass. Fluid. Craze.
80		20	Yellow-white. Beginning crawl. No craze.
70	30		Yellow-green. Matt and glass. Fluid. Craze.
70		30	Yellow-white. Satin-matt. Increased crawl. No craze.
60	40		Increased matt.
60		40	Whiter color. Increased crawl.
50	50		Yellow-green matt.
50		50	Yellow-white. Eggshell texture. Claybody visible. Matt. No craze.
40	60		Yellow-green-white. Dry matt.
40		60	Increased white. Larger eggshell fragments. More clay body exposed. Dry matt. No craze.

Note: Unfired magnesium carbonate is visibly light and fluffy. This material has a high surface tension. A high-magnesium carbonate glaze will form a distinctive cellular pattern; this pattern will appear even prior to firing.

IIA

Figure 3.2, p224.

		Results
Potash Feldspar	90%	Transparent, Gloss.
Whiting	10	Blue-gray color. Distinct craze pattern.
1. Replace 5 Whiting with 5 Dolomite.		No visible change. Slight increase of craze and flow. Less gray, bluer color.
2. Replace 5 Whiting with 5 Talc.		Higher gloss. Larger craze pattern. Whiter color.
3. Replace 5 Whiting with 5 Magnesium Carbonate.		Higher gloss. Whiter color.

Note: A large craze pattern is an indication of a lower expansion rate. A fine network of craze lines indicates a higher expansion rate. The lower expansion rate of magnesium oxide, compared to calcium oxide, is visible in test 2.

IIB

Figures Intro.7, p17.; 3.2, p224; 3.13, p235.

G-200 feldspar   90 grams
Bone ash   2
Red iron oxide   1/2

Add

Whiting	Wollastonite	Dolomite	Results
10			*Stoneware:* Blue-gray. Gloss. Overall craze network. *Porcelain:* Bright robin's egg blue. Gloss. Overall craze network.
	10		*Porcelain and Stoneware:* Brighter blue than Whiting tests. Craze network same.
		10	*Stoneware:* Lighter gray-blue-brown. Gold luster. Craze network same. *Porcelain:* Least blue of all tests. Gray-white color edged in brown.

Conclusion: Tests show Whiting and Wollastonite producing bluer color surfaces than Dolomite tests. Dolomite tests display tendency to produce yellow-brown hues. However, later tests with Dolomite on porcelain definitely produced a blue color. *See Figures Intro.7, p17; 3.13; p235.* Wollastonite consistently displayed the brightest blue color of all the melters.

IIC

| G-200 Feldspar | 50% | Semi-opaque; Semigloss. |
| Whiting | 50 | Gray-green-yellow color. |

| G-200 Feldspar | 50% | Satin surface. Gray-white. Semi-opaque. |
| Talc | 50 | Semigloss. |

G-200 Feldspar	50%	Satin-matt. White color.
Talc	40	Stripe of cobalt oxide results in purple-blue color.
Whiting	10	

Note: Talc encourages satin-matt opacities.

III LORD'S MATT

Potash Feldspar	54.3%	Gray.
Whiting	24.7	Semi-matt.
Ball clay	21.0	

Replace Whiting with Dolomite.   Yellow-white. Stony surface.

IV CASCADE

Potash Feldspar	30%	Orange red (thin).
Spodumene	20	White (thick).
Dolomite	22	Opaque. Matt.
Whiting	2	
EPK kaolin	22	
Tin oxide	6	

Replace Dolomite with Whiting.   Bluer-gray color. Less orange. Increased transparency.

V RHODES 32

Figure 4.2, p266.

Potash Feldspar	48.9%	Orange-brown speckled
Dolomite	22.4	(thin). White (thick).
Whiting	3.5	Stony-matt.
Kaolin	25.1	

I. Omit Whiting	Blue-gray-yellow. No orange speckles. Orange speckles appear on control.
2. Omit Kaolin	Yellow-brown. Opaque Satin-matt. Interior center shows yellow-green matt crystals in puddle of green glass.

Note similarity of test 2 to our other tests with high dolomite. The consequence of omitting 25% kaolin produces a high dolomite glaze, complete with its characteristic yellow, matt crystals.

VI CORNWALL STONE

Cornwall Stone	80%	Blue-gray.
Whiting	20	Gloss. Large crackle.
Red iron oxide	0.5%	

I. Replace one-half Whiting (10%) with Dolomite.	Greener cast. Increased flow.
2. Replace Whiting (20%) with Dolomite.	Blue-green flecked with yellow-white dots.
3. Replace one-half Whiting (10%) with Dolomite (15%)	Gray-green glass flecked with yellow-white crystals (thin). Yellow-white crystalline surface (thick).

Note: Low amounts of dolomite combined with equally low amounts of Whiting (10%) increase fluidity. Larger amounts of dolomite begin to opacify and stiffen the glaze.

The surface color changes from blue-gray to green-gray and finally to yellow-white, which reflects the characteristic opacifying tendencies of dolomite. Once again, the yellow-white, crystalline opacity in test 3 is characteristic of a high-dolomite glaze surface.

VII TEMPLE WHITE

Figure 4.8, p272.

Potash Feldspar	34.7%	Gray-white.
Dolomite	19.6	Semigloss.
Whiting	3.1	Semi-opaque.
Flint	18.9	
Kaolin	23.6	

I. Substitute Whiting for Dolomite.	Blue-gray. Increased shine. Increased transparency.
2. Add 20 Whiting	Blue-gray flecked with tiny, pale, yellow crystals. Stonier surface.
3. Add 10 Dolomite	Whiter color. Stonier surface. Increased opacity.
4. Add 5 Magnesium Carbonate	Gray with white flecks. Increased white opacity. Increased gloss. Increased satin surface.
5. Add 10 Magnesium Carbonate	Separation and crawling. Increased opacity.
6. Add 15 Magnesium Carbonate	Crawling increases. Shine increases.
7. Add 20 Magnesium Carbonate	Whiter color. Crawling increases. Increased opacity. Less shine.

VIII SUSAN SHAPIRO'S SAM HAILE

Nepheline Syenite	62.0%	White-brown-gray-
Whiting	19.4	yellow. Mottled, opaque,
EPK kaolin	5.8	stony-matt.
Calcined kaolin	7.8	
Zinc oxide	4.8	
Add Talc	5.0	Yellow-green-lime. Slight increase in sheen. Silkier surface.

IX SANDERS CELADON

Figure 1.5, p82.

Kona F-4 Feldspar	44%	Blue-green gray.
Whiting	18	Transparent.
Flint	28	Gloss.
Kaolin	10	
Barnard clay	5%	

I. Replace Whiting with Dolomite	18	Deeper blue-green. Slight increase in flow. (Control omitted. Changes could be due to firing variables.)
2. Add Whiting	18	Gray-green; puddles of green glass. Increased opacity and flow.
3. Add Dolomite	18	Pale gray shiny surface ringed with yellow-white crystalline formations. Visible yellow crystals in portion of glaze, which has flowed to base in thick rolls.
4. Add Dolomite to Test 3	18	Pale gray, shiny surface where glaze is thin. Main body of glaze is opaque. Yellow crystalline formations in interior center rimmed with stony-white.
5. Replace 5% Barnard clay with 13% Yellow Ochre		Deep green. Reduced transparency.

and

a. Replace 18% Whiting with Magnesium Carbonate.	Leathery, brown, lizard skin surface.
b. Replace 9% of Whiting with Magnesium Carbonate.	Lustrous, silky, purple, red-brown surface.

X CHARLIE D SATIN-MATT

Figure 4.8, p272.

Kona F-4 Feldspar	20%	Gray-white.
Potash Feldspar	20	Satin surface.
Dolomite	15	Semigloss.
Talc	13	
Whiting	2	
Flint	20	
Ball clay	10	

I. Replace Talc and Whiting with Dolomite	Whiter color. Less transparent.
2. Replace Talc and Dolomite with Whiting	Blue-gray. White speckles. Increased shine. Increased transparency.
3. Replace Dolomite and Whiting with Talc	Whiter and stiffer than both control and test 1.
4. Add: 1/2% Copper oxide 1/4% Cobalt oxide	Turquoise blue-green.
5. Add 4% Manganese oxide.	Pale, yellow-brown. *Figure 3.14, p236.*
Increase firing temperature (cone 11)	Shiny, yellow-brown background flecked with large, yellow crystals.

Note: The yellow crystalline formations that appear in test 5 with the 4% addition of Manganese oxide are similar to the crystalline patterns of previously discussed high-dolomite tests (see p. 209).

XI PAM'S SATURATED IRON GLAZE

Potash Feldspar	52.4 grams	Black gloss edged in brown where thin.
Whiting	10.1	
Colemanite	6.2	
Flint	17.0	
Kaolin	5.6	
Red iron oxide	8.6	

Replace Whiting with Dolomite	Metallic brown. Lowered gloss.

XII PERSIMMON

Figure 3.8, p230.

Potash Feldspar	48.6%	Red to red-orange to
Whiting	7.2	brown. Gloss.
Bone ash	10.0	
Talc	6.3	
Flint	21.6	
Kaolin	6.3	
Red iron oxide	11.0%	

XIII PAVELLE

Figure 3.7, p229.

Bone ash	42.0%	Speckled yellow-orange.
Dolomite	30.0	White where thick.
Cornwall Stone	17.5	Stony-matt.
Nepheline Syenite	10.5	
Zinc oxide	4.0	
Rutile	3.0%	

CONE 5-6 OXIDATION

I

Nepheline Syenite	Whiting	Dolomite	Talc	Results
100%				Opaque, stony surface.
90	10			Semi-opaque. Satin surface. Even coverage. High craze.
90		10		Similar to above.
90			10	High gloss. Crawl. Pinholes. No craze.

II

Nepheline Syenite	Whiting	Colemanite	Magnesium Carbonate	Results
80%	20			Semigloss. Semi-transparent. Satin surface.
80		20		High gloss. Transparent.
80 grams		20	5	High gloss. Transparent. Puddle of green, crackled glass.
80		20	10	Lower transparency. Cloudy green glass. Larger craze pattern.
Add 1/2% Copper Carbonate				Deep blue-green.

Note that surface is still transparent with addition of 10% Magnesium Carbonate. No visible crawling pattern has begun. This test indicates the strong eutectic power of the colemanite (calcium boron) magnesium combination.

III HILDA'S BLUE-GREEN

Portchester 5–6 claybody, with Gold Art instead of Jordan. (See p. 140.)

Kona F-4	82%	Soft blue-green.
Dolomite	18	Satin-matt.
Zinc oxide	1.0%	
Copper carbonate	3.5%	

IV

Potash Feldspar	80%	White, gloss, opaque.
Talc	20	Pinholed.
Copper oxide stripe under glaze.		Green-black.
Cobalt oxide stripe under glaze.		Blue-purple.

V

Potash Feldspar	50%	White, dry,
Talc	50	unmelted.

Note: Although 20% Talc created a melt with 80% Feldspar, the 50% addition with 50% Feldspar did not; note the cone 9/10 reduction test, in which the 50–50 combination produced a soft, semi-opaque gray-white surface (Test IIc, p. 208).

VI JACKY'S CLEAR II

Figure Intro.6, p16; 4.7, p271.

Nepheline Syenite	50%	Tranparent. Gloss.
Colemanite	10	
Wollastonite	10	
Flint	20	
Zinc oxide	5	
Tenn-5 Ball	5	
Bentonite	2%	

	ADD				
	5%	**10%**	**20%**	**30%**	**40%**
Whiting		Most flow and transparency of all other tests, except for Wollastonite.	Still shiny and transparent where thin. Thicker layer shows small, white, opaque spots. Opacity beginning.	Lowered gloss. Cloudy-white, semi-transparent surface.	Similar to 30% dolomite test. Yellow, stony, opaque center where thick. Mottled, white, satin vertical surface.
Wollastonite		The most flow and transparency of all tests, including control.	Similar to 10% test but thicker layer shows beginning of white opacity (Less than Whiting test).	Similar to 20% test. Slightly more transparent than control. White flecks in thicker layer.	
Dolomite		Slightly more flow and transparency than control.	White, cloudy opacity. Mottled, high-gloss, satin surface.	Opaque, stony surface. Mottled, yellow-beige-white color.	
Talc		Less flow, less gloss, less transparency than control. Satin-white where thick.	White where thick. Satin surface. Uneven coverage. Mottled white.	Silky, opaque, even-white, satin surface. High gloss. transparent where thin. Addition of 1% Cobalt oxide results in purple-blue color.	
Magnesium Carbonate	Similar to control, but slightly less transparent. Higher gloss.	More flow than control, but less transparency.	Whiter and more satin surface.	Increased white. Opaque. Controlled crawl.	

VII RANDY RED

Crafts Students League Claybody (p. 140)
Figure 3.11, p233.

Kona F-4	20%	Red-green breaking
Gerstley Borate	32	to brown where thin.
Talc	14	High gloss.
Flint	30	
Kaolin	5	
Red iron oxide	15%	

Test 1.
Omit Talc — High gloss. Blue-yellow.

Test 2.
Replace Talc with Dolomite. — Lower gloss. Yellow-brown color.

Test 3.
Replace Talc with Magnesium Carbonate — Brighter red.

Repeat test: — Dense, dark brown.

Test 4.
Replace Talc with Wollastonite — High gloss. Yellow-brown color.

VIII R-15 BLUE-GREEN

(Crafts Students League claybody and glaze)

Nepheline Synenite	41.6%	
Gerstley Borate	11.9	Satin-matt.
Dolomite	7.6	Blue-green color.
Talc	14.6	
Kaolin	4.4	
Flint	19.9	
Cobalt Carbonate	2.0	
Rutile	6.0	
Blue-Green u.g.[14]	0.5	

Test 1. Omit Dolomite — High gloss. Bright blue color.

Test 2. Omit Talc — High gloss. Brighter blue color.

Test 3. Add Dolomite

	5%	Gray-green color.
	15%	Green satin-matt. Blue earthworm markings. Beautiful!
	20%	Green color. Dry, matt surface.

Dolomite and Talc function together as opacifiers in this glaze. Note the presence of Gerstley Borate, which would activate remaining melters, and contributes to the shiny surface in tests 1 and 2.

CONE 8 OXIDATION

IX ELENA'S SATIN-MATT

Potash feldspar	56 grams	Gray-white-yellow.
Whiting	20	Satin-matt.
Flint	6	
Kaolin	18	
Zinc oxide	9	
Rutile	7	

Test 1. — Substitute Dolomite for Whiting — Increased opacity; satin-matt surface unchanged. Color is white, edged in yellow; blue purple center. Increased flow.

Test 2. — Substitute Talc for Whiting. — Increased opacity. Increased satin surface. Color is bluer and less yellow. Lower fusion.

Talc produces a softer, more satin surface and less fusion than either dolomite or whiting. Talc causes more color changes than dolomite substitution.

DOLOMITE
Dolomitic Limestone

GEOGRAPHICAL*
SOURCE (*Domestic*)

Mill Creek OK; UNIMIN Corporation (Dolomite)
Canaan, CT; Specialty Minerals Inc. (Dolomitic Limestone)
Other deposits widely distributed throughout United States.

GEOLOGICAL**
SOURCE

Pure Dolomites:
Precipitation from fluids rich in calcium and magnesium carbonates produces mixture of these two carbonates. Magnesium carbonate content is at least 44%.

Dolomitic Limestones:
Magnesium-rich fluids react with limestone rocks; calcium ions replaced by magnesium ions to form dolomite of variable chemical composition with less than 44% magnesium carbonate.

CHEMICAL STRUCTURE

$CaMg(CO_3)_2$

Pure Dolomite**
56% $CaCO_3$
44% $MgCO_3$
P.C.E. 1650°F

	Dolomite*** Glass Grade		Dolomitic Limestone† Dolocron®	
CaO	30.70%		$CaCO_3$	55.0%
MgO	20.92		$MgCO_3$	43.0
Al_2O_3	0.20			
Fe_2O_3	0.12		Fe_2O_3	0.3
SiO_2	0.90		H_2O	0.1
Na_2O	0.03			
K_2O	0.10			
L.O.I.	46.66			
(CO_2)				

*Ceramic Industry 1998, 100.
**Parmalee, Harman, *Ceramic Glazes*, 3rd ed., 43.
***Technical Data 1998, UNIMIN Corporation
†Technical Data 1998, Specialty Minerals Inc.; Courtesy of Hammill & Gillespie Inc.

MINERALOGICAL STRUCTURE‡

Dolomite	99%
Crystalline Silica	1%

Detected Substance (amount less than quantifiable limit)

Nonasbestos Tremolite	less than 1%
Arsenic	less than 2 ppm
Cadmium	less than 2 ppm
Chromium	less than 0.1 ppm
Lead	less than 4 ppm

‡Pfizer Inc., Material Safety Data 1991 (July).

TALC*

TRADE NAME	TALC
GEOGRAPHICAL SOURCE	Gouverneur, St. Lawrence County, New York, R. T. Vanderbilt Company, Inc. Also, California, Vermont.
GEOLOGICAL SOURCE	Metamorphic, magnesium-silicate rocks of Precambrian age (more than 600 million years ago). NY talc consists primarily of mineral tremolite ($2CaO \cdot 5MgO \cdot 8SiO_2 \cdot H_2O$). Pure mineral talc is often but a minor constituent of commercial talc deposits. It formed in several stages by reaction of hot, silica-rich solutions with beds of quartzite and metamorphic dolomite (marble).
CHEMICAL STRUCTURE	$3MgO \cdot 4SiO_2 \cdot H_2O$ (pure)

	NYTAL 100HR 200 MESH	CERAMITALC N0. 1 325 MESH
SiO_2	55.20	56.80
MgO	30.00	29.00
CaO	8.42	6.94
Fe_2O_3	0.16	0.20
Al_2O_3	0.31	0.29
MnO	0.17	0.13
Na_2O	0.34	0.18
L.O.I.	5.41	6.46

MINERALOGICAL STRUCTURE

Tremolite	40%–60%
Serpentine (Antigorite)	20%–30%
Talc	20%–30%

*Bates 1969, 328–333; Rieger, 37–39; C. S. Thompson, Manager of Minerals R. T. Vanderbilt Company, telephone interview April 1992; Technical Data R. T. Vanderbilt Company, Inc. 1998; Frye 1981, 10, 546.

MAGNESIUM CARBONATE*

GEOGRAPHICAL SOURCE	Washington, Nevada, California.
GEOLOGICAL SOURCE	Precipitation from seawater. Alteration of magnesium-rich rocks; replacement mineral formed by hydrothermal magmatic solutions reacting with metamorphosed dolomite; calcium replaced by magnesium.
CHEMICAL STRUCTURE	$MgCO_3$ Light Magnesium Carbonate (Washington) $3MgCO_3 \cdot Mg(OH)_2 \cdot 3H_2O$

MgO	46.2%
CaO	0.3
CO_2	50.3
SiO_2	0.5

*Ceramic Industry 1998, 100; Hamer and Hamer 1986, 205.

ZINC MINERALS [15]

ZINC OXIDE	
SOURCE	*French Process:* Vaporized zinc metal burned in air to zinc oxide powder.
	American Process: Smelting of oxidized zinc sulfide ore mixed with coal, to metallic zinc. Vaporization, burning and condensation to zinc oxide.
CHEMICAL ANALYSIS	French Process Ceramic Grade Cerox-506 Zinc Oxide (Zinc Corporation of America). Highly calcined specifically for ceramic industry, which accounts for less than 1% of ZCA's market.)

ZnO	99.8%
PbO	0.09
CdO	0.04
Fe_2O_3	0.009
Acid Insoluble	0.01
H_2O Absorption (cc/30grams)	22.0%
Melting Point	3587°F

A complete account of zinc oxide appears in *The Potter's Dictionary* by Frank and Janet Hamer (1986), and the reader is urged to read this excellent summary. In this connection, note that there appears to be some inconsistency between various ceramic texts as to whether or not zinc oxide is effective in the stoneware reduction atmosphere. According to *The Potter's Dictionary*, at temperatures above 1742°F, the carbon gases in the reduction atmosphere remove oxygen from zinc oxide and transform it into metallic zinc, which has a low melting point of 768°F. At temperatures above 1742°F, the now boiling zinc metal leaves the glaze magma in the form of a poisonous gaseous vapor. The toxic property of hot zinc oxide makes it imper-

ative that kiln fumes be properly ventilated.[16] Hence, zinc oxide could make no contribution to the stoneware reduction glaze other than to promote early glaze fusion (Hamer and Hamer 1986, 343).

Some ceramic texts do not make a clear-cut distinction between the function of zinc oxide in oxidation and reduction atmospheres at temperatures above 1742°F.

"Zinc is a useful flux from mid to high temperatures. It is very active, producing smooth, trouble-free glazes when used in small amounts."

(Hopper 1984, 56)

"It (zinc oxide) also acts as an opacifier in both oxidation and reduction."

(Ibid., 85)

"...It (zinc oxide) sublimes (melts) at 1800°C (3269°F) and is readily reduced to the metal....However, it rarely is used in high-fire porcelain glazes. Small amounts introduced into high-fire porcelain glazes are useful in starting fusion and forming a more perfect surface."

(Parmalee 1973, 45–46)

Similarly, *Ceramic Industry Materials Handbook* makes no mention of this difference in its detailed description of how zinc oxide functions in ceramic glazes (*Ceramic Industry* 1998, 180).

In point of fact, as the following tests show, various amounts of zinc oxide (some of which are fairly sizable) frequently do appear in both oxidation and reduction high-fire glazes.

Cone 9–10

PAVELLE
Figure 3.7, p229.

		Reduction	Oxidation
		Reduction	**Oxidation**
Bone ash	42.0%	Orange-yellow.	Yellow-white-
Cornwall Stone	17.5	Smooth-matt.	pink
Nepheline Syenite	10.5	Smooth-matt.	
Dolomite	30.0		
Zinc oxide	4.0		
Rutile	3%		

JACKY'S CLEAR
Figure Intro.6, p16.

		Reduction	Oxidation
		Reduction	**Oxidation**
Nepheline Syenite	50%	Transparent.	Transparent.
Wollastonite	10	Gloss.	Gloss.
Colemanite*	5	Craze.	
Zinc oxide	10		
Flint	20		
Ball clay	5		*or Gerstley Borate*

SAM HAILE

		Reduction	Oxidation
		Reduction	**Oxidation**
Potash feldspar	44 grams	Gray-blue-white.	White.
Cornwall Stone	20	Semi-opaque.	Opaque.
Whiting	20	Semigloss.	Stony.
Zinc oxide	5		
EPK kaolin	6		
Calcined kaolin	8		

If, as suggested by Hamer and Hamer (1986), a reduction atmosphere at high stoneware temperatures destroys zinc oxide, then its only function in these glazes would be to promote early glaze fusion. Note that these glazes are frequently used in oxidation and that the glaze color that results from the oxidation firing differs markedly from its reduction counterpart. However, despite the active presence of zinc oxide in the oxidation firing, compared to the reduction firing, the glaze surface itself does not always reflect this difference. See Pavelle glaze, for example, in which both oxidation and reduction surfaces appear stony and opaque. In view of the inconsistency that appears in the ceramic texts as to the

exact function of zinc oxide in a reduction atmosphere, it would be interesting to see what effect, if any, the omission of zinc oxide has on the glaze surface in the reduction firing and compare this result to the same test in the oxidation firing (see Lab VII, Problem 4, p. 221). Because of the volatilization of zinc oxide from the Pavelle glaze at stoneware reduction temperatures, zinc oxide may not contribute enough to the glaze surface to warrant its inclusion. Hence, the presence of zinc oxide in this stoneware glaze formula may simply be a carryover from an oxidation firing. *Note that zinc oxide is a toxic material and care should be taken in handling and inhaling fumes and dust.* All the more reason to determine the necessity for its inclusion in glazes fired in the reduction atmosphere.

CHARACTERISTICS OF ZINC OXIDE

In the oxidation atmosphere, there is general agreement among the various ceramic texts as to the valuable properties of zinc oxide for ceramic surfaces.

1. At temperatures of 2000°F and upwards, in an oxidation atmosphere, zinc oxide combines with other melters, such as soda, potash, and calcia, to perform a dual role of melter and opacifier.

a. Low amounts of zinc oxide create a powerful accessory melter (Tests, pp. 218–219). The strength of zinc oxide as melter compared to calcia (Whiting) is indicated by its yellow-white, smooth surface when fired alone to cone 5/6 oxidation temperatures. Although the test surface appeared dry and matt, zinc oxide firmly adhered to the claybody, in contrast to the flaky, and powdery Whiting test.

b. Larger additions of zinc oxide perform the role of both melter and opacifier at stoneware temperatures. Zinc oxide can perform this dual role simultaneously; it vigorously promotes the fusion of the glaze and, at the same time, seeks out the silica in the glaze to form crystals of the mineral Willemite (Zn_2SiO_4). Zinc oxide's extraordinary ability to enter into the melt and yet still retain a crystalline identity produces white, frosty, opalescent opacities; herein lies the unique value of zinc oxide for oxidation ceramic surfaces at stoneware temperatures (*Ceramic Industry* 1998, 180).

Zinc oxide combines with silica in a 2:1 ratio to produce

large crystalline formations of the mineral Willemite in a high-alkaline, low-alumina fluid glaze. The glaze must be cooled slowly at the specific crystallization temperature of the glaze. (See Behrens Fantasy Crystalline Glaze, p. 220. See also Chapter 4, p. 246; Sanders 1974, 21, 25–27.)

> **Note** that the presence of boric oxide in the form of either Gerstley Borate or colemanite will inhibit the opacifying property of zinc oxide and cause it to behave solely as a melter (Jacky's Clear, p. 220; *Ceramic Industry*, Ibid.).

2. Zinc oxide has the lowest coefficient of expansion next to magnesia and reduces the crazing of the high-expansion and contraction soda and potash melters (Katherine Choy, p. 219).

3. As in the case of calcia, zinc oxide produces a hard, durable glaze surface that will resist the dissolving force of water.

4. The high shrinkage of zinc oxide in both the prefired and fired glaze may produce a crawled and beaded glaze surface (Cornwall Stone, p. 219). In the presence of water, zinc oxide readily combines with H_2O to become zinc hydroxide. Calcining the raw zinc oxide increases the density of the particles, inhibits the formation of zinc hydroxide and thereby minimizes shrinkage and crawling problems. Hence, calcined zinc oxide is the preferred form of zinc oxide for ceramic use (*Ceramic Industry*, Ibid.).

5. Zinc oxide has a strong effect on the color of a glaze surface. Making 5%–10% additions of zinc oxide in combination with substantial amounts of calcia, soda, and potash melters produces a white opacity. This whiteness often has a pinkish cast, particularly in an oxidation atmosphere. (See Pavelle glaze p. 217.) Lower additions will increase the brilliance of the copper and cobalt colorants. The presence of even low amounts of zinc oxide will destroy the clarity of the iron colored surface and produce a muddy, mustardy color (see Hopper 1984, 181).

6. A glaze with 5% or more zinc oxide may exhibit peculiar and unreliable behavior after remaining in the glaze bucket over a period of time. We found that after a few months, the Sam Haile glaze (5% zinc oxide) gave us prob-

lems in application that frequently resulted in a blistered and crawled surface. The first glaze layer resisted the application of a second top layer, with the result that the top layer peeled off after the surface had dried. (See p. 217.).

GLAZE FUNCTION

We focus now on specific tests of stoneware glazes that contain zinc oxide. These tests describe far better than any theoretical analysis the effect of zinc oxide on specific ceramic surfaces.

Low amounts of zinc oxide contribute to an uncrazed gloss surface and to the brilliance of cobalt and copper glazes.

Cone 5–6 Oxidation

CLAUDETTE'S TRANSPARENT BLUE		
(Crafts Students League Stoneware Glaze)		
Custer Feldspar	44 grams	Bright blue gloss.
Colemanite	20	No visible craze.
Whiting	1	
Zinc oxide	3	
Flint	24	
EPK kaolin	1	
Cobalt Carbonate	1/2%	
Copper Carbonate	4%	

Low amounts of zinc oxide will help to achieve a smooth, nonpinholed glaze surface.

K F-4 SATIN-MATT		
Kona F-4	70 grams	White, satin-matt.
Wollastonite	30	Semi-opaque,
Ball clay	10	semigloss.
Zinc oxide	2	
Bentonite	3	

At cone 5–6 oxidation, 8%–14% additions of zinc oxide combine with soda, potash, and calcia melters to produce stony- or satin-matt, white opacities.

VIVIKA HEINO

Nepheline Syenite	54.55%	White, stony-matt.
Whiting	17.5	
EPK kaolin	14.0	
Flint	5.2	
Zinc oxide	9.5	

RON'S WHITE MATT #5

Figure 4.7, p271.

Kona F-4	55%	White, stony-matt.
Whiting	15	
EPK	16	
Zinc oxide	14	

See also White Matt glaze p. 259.

KATHERINE CHOY

Figure 4.6, p270.

Nepheline Syenite	53.9%	White-yellow color.
Whiting	11.7	Satin-matt surface
Zinc oxide	10.3	streaked with spots
Lithium Carbonate	4.7	of matt and shine.
Flint	1.5	Clear gloss on
EPK kaolin	17.9	porcelain claybody.

Omit zinc oxide:	Lower gloss. Semi-opaque. Satin-matt surface. Visible craze.
Omit Whiting:	High gloss. Opaque. Loss of satin surface.
Gerstley Borate replaces zinc oxide:	Lower gloss. Transparent surface. Loss of whiteness.

Zinc oxide contributes the properties of opacity, whiteness, gloss, and reduced crazing to this glaze surface. Whiting produces the satin-matt surface.

CORNWALL STONE (V. EDWARDS)

(Crafts Students League Stoneware and Glaze)

Custer feldspar	44%	White, semigloss.
Cornwall Stone	22	Semi-transparent.
Whiting	18	
Zinc oxide	8	
Titanium dioxide	4	
EPK kaolin	5	

Omit Whiting:	Increased gloss. Crawled, leathery surface.

The omission of Whiting leaves zinc oxide as the primary auxiliary melter. Once again zinc oxide increases the gloss and the opacity of the surface. Note the crawled surface with the increase in the zinc mineral.

Cone 8 Oxidation

ELENA'S SATIN-MATT

Custer Feldspar	56 grams	Gray-white-yellow.
Whiting	20	Satin-matt.
Zinc oxide	9	
Flint	6	
EPK kaolin	18	
Rutile	7	

Omit zinc oxide:	White-brown. Stony-matt.

Without zinc oxide, the surface has lost shine, satin surface, and yellow color.

In combination with high amounts of calcium-magnesium melters and alumina, 4%–5% zinc oxide helps to produce stony, smooth opacities at cone 9/10 temperatures.

Pavelle Glaze: (p. 217)

Sam Haile: (p. 217)

This ability of zinc oxide to produce a stony- or satin-matt surface will be inhibited by the presence of boric oxide in the glaze.

Cones 0–4 to 5–6 Oxidation

MIKE'S HB 8		
Nepheline Syenite	15.00 grams	White, gloss.
Frit 3195	30.00	
Colemanite	15.00	
Zinc oxide	9.00	
Borax	4.65	
Ball clay	15.00	
Tin oxide	8.75	
1/2% Cobalt Carbonate		Blue-violet.
2% Red iron oxide		Brown.
2% Copper oxide		Robin's egg blue.
4% Copper oxide and		Jade-green.
1% Iron oxide		

Cones 5–10 Oxidation

JACKY'S CLEAR
Figure Intro.6, p16; pp. 212, 217

Transparent, glossy surface at both cone 9–10 oxidation and cone 5–6 oxidation. Contains both zinc oxide and colemanite (or Gerstley Borate).

Zinc oxide combines with silica to form macrocrystalline formations within a fluid, high-gloss, low-alumina glaze when cooled slowly at the temperature of crystalline formations.

Cone 4 Oxidation

BEHRENS FANTASY CRYSTALLINE GLAZE	
	Shiny, green.
Nepheline Syenite	60%
Frit 3134	20
Whiting	19
Bentonite	2
Copper Carbonate	3

Application of brush of zinc-silica seeding agent (60 zinc, 25 flint) underglaze. (See p. 246.)

Temperatures of 1800°F–2000°F held for approximately five hours after reaching cone 4 temperature (2134°F–2167°F) during the cooling period.

Result: White, matt, flowerlike crystals sprinkled on a green, shiny background.

DOLOMITE
TALC
MAGNESIUM CARBONATE
ZINC OXIDE

l. Take 2 to 4 identical, large test pots. Glaze one-sixth of each pot with Whiting, Wollastonite, Dolomite, Talc, Magnesium carbonate, and zinc oxide.

2. Substitute Dolomite, then Talc, Magnesium carbonate, and finally, zinc oxide for the calcium melter in your glaze in four separate tests.

Apply to test pots that each have a stripe of cobalt, iron, and copper oxides on their outer walls.

3. If your glaze contains Dolomite and/or Talc and/or Magnesium carbonate and/or zinc oxide:

a. Mix it without these materials.

b. Replace omitted Magnesia or zinc oxide material with one of the materials listed above in step 1.

c. Add to your original glaze: 10%, 20%, 30%, 40% increments of:

Dolomite
Talc
Magnesium carbonate
Zinc oxide

Compare these test results to the calcium melter tests (Lab IV, #5, p. 173).

d. Apply the above test mixtures (3a, 3b, 3c, 3d) to test pots that have stripes of cobalt, iron, and copper oxides on their outer or inner walls.

Fire tests to: Cone 9–10 oxidation and reduction temperatures and compare the resulting surfaces.

4. If your glaze does not contain magnesium minerals or zinc oxide, perform the following tests.

Persimmon Glaze: (p. 211)
1. Omit Talc
2. Replace Talc with Wollastonite

Pavelle Glaze: (p. 211)
1. Omit Dolomite
2. Replace Dolomite with Whiting
3. Omit Zinc Oxide
4. Replace Zinc Oxide with Whiting

Fire tests to: Cone 9–10 reduction and oxidation temperatures. Compare the resulting surfaces.

Jacky's Clear Glaze (p. 217)
Omit Zinc 10

Jacky's Clear II Glaze (p. 212)
Omit Zinc 5

Compare surfaces of Jacky's Clear and Jacky's Clear II by applying each to one-half of same test.

Fire tests to: Cone 5–6 oxidation temperatures; cone 9–10 oxidation and reduction temperatures; compare the resulting surfaces.

> **Note:** One-half of each and every test *(except test 1)* should contain the original glaze as a control.

Notes

[1] Source: Kingery and Vandiver 1986, 211–217; Lawrence 1972, 17. See also pp. 5–6.

[2] For a clear presentation of the scientific phenomena that underlie this role, see Kingery and Vandiver 1986, 214.

[3] Geological, chemical, and industrial information: Courtesy of Konrad C. Rieger, Manager, Ceramics Lab; C. S. Thompson, Manager of Minerals, R. T. Vanderbilt Company, Inc.; Keeling 1962; Power 1986, 19–34; Rieger, "Talc, pyrophyllite, and Wollastonite," 39–40, 44; *Ceramic Industry* 1998, 117; NYCO Minerals, Inc., Sara Robinson, 1998.

[4] See Hamer and Hamer 1986, 222.

[5] Source: *Ceramic Industry* 1998, 105; Frye 1981, 612; Parmalee 1973, 39, 292.

[6] Gerstley Borate = Ulexite + Bentonite Clay

[7] IMC Global Inc. is planning to acquire Harris Chemical North America Inc.

[8] Hamer and Hamer 1986, 33, 359; *Ceramic Industry* 1998, 76; Frye 1981, 709.

[9] Ferro Corporation, Frit and Color Division, Technical Data, 1998.

[10] Whenever substitutions of any kind of material are introduced into a glaze, changes are likely to occur, no matter how similar the chemical structures may be. The question is whether or not the changes fall within an acceptable range or whether they cause substantial differences in the glaze surface. Just as certain human bodies reject the implant of substitute organs, so too, certain glazes may be too sensitive to accept the substitute of any different material.

[11] *Black Mix*

Red iron oxide	3.0
Cobalt oxide	2.0
Chrome oxide	2.0

[12] "The term limestone includes those sedimentary rocks made up of 50% or more of the minerals calcite and dolomite, in which calcite is more abundant than dolomite....Limestone with more than 10 percent of the mineral dolomite is termed dolomitic limestone, and that with 5 to 10 percent is called magnesium limestone. A rock in much industrial demand is high-calcium limestone, which contains 95 percent or more of calcite" (Bates 1969, 156).

[13] See characteristics of Calcium melter, Feldspar-Whiting, Tests, p. 159; *Figure 3.2; p224*.

[14] Underglaze. Formula unavailable.

[15] *Ceramic Industry* 1998, 178, 180–181; Hamer and Hamer 1986, 342–344; Zinc Corporation of America, Technical Data 1998.

[16] See Health Hazards of Raw Materials, p. 3.

Figure 3.1

Background: Nephrite carved bowl (Courtesy of Robina Simpson).

Foreground: Porcelain bowl with glaze of Sanders Celadon by Yien Koo Wang King. Cone 9 reduction (p. 160).

Figure 3.2

Rock: Calcite: Calcium Carbonate (Collection of Department of Earth and Environmental Sciences, Columbia University, New York).

Tests: Feldspar and Whiting. Stoneware. Cone 9–10 reduction.

Left: Potash feldspar 100%.

Center: Potash feldspar 90%, Whiting 10%.

Right: Whiting (Calcium Carbonate) 100%.

224

Figure 3.3 Porcelain shell by Sue Browdy.

Cone 9–10 reduction. Sanders Wollastonite glaze (Wollastonite 38.5%) p. 160.

Figure 3.4 Storage jar with natural ash glaze. (The Metropolitan Museum of Art, The Harry G. C. Packard Collection of Asian Art, Gift of Harry G. C. Packard and Purchase, Fletcher, Rogers, Harris Brisbane Dick and Louis V. Bell Funds, Joseph Pulitzer Bequest and The Annenberg Fund, Inc., Gift, 1975 [1975.268.428]. Photograph ©1999 The Metropolitan Museum of Art.)

Japanese Ware, Muromachi period, fourteenth to fifteenth century. Shigaraki ware.

Figure 3.5 Wood-fired jar by Sue Browdy.

Stoneware, cone 9–10 reduction. Unglazed.

Figure 3.6 Wood ash test pots.

Cone 5–6 oxidation.

Left: Brooklyn Red claybody. Glaze: Mixed wood ash 50%, Gerstley Borate 50%, Flint 10%.

Back: center: Same glaze and claybody.

Right: front: Grolleg Porcelain claybody; Mixed wood ash 100%.

Figure 3.7 Covered jar by Sue Browdy.

Stoneware, cone 9–10 reduction. Pavelle glaze (bone ash 42%, Dolomite 30%) p. 176.

Figure 3.8 (top) Footed bowl by author.

Stoneware, cone 9–10 oxidation. Persimmon glaze (bone ash 10%, Talc 6.3%, Whiting 7.2%). Interior brush of Rotten Stone 90%, Whiting 10%.

Foreground: Pillow tests, stoneware, cone 9–10 oxidation. Persimmon glaze. Placed on different shelves in kiln. Bottom shelf (cooler), blacker color. Middle and top (hotter), progressively redder color.

Figure 3.8 (bottom) Bowl by Sue Browdy.

Stoneware cone 9–10 reduction. Persimmon glaze.

Figure 3.9 Test bowls.

Clockwise from left:
Cone 5–6 oxidation, #308 Brooklyn Red claybody; glaze: Borax alone.
Cone 9–10 reduction, stoneware; glaze: Gerstley Borate alone.
Cone 9–10 reduction, T-1 claybody; glaze: Gerstley Borate alone.
Cone 5–6 oxidation, #308 Brooklyn Red claybody; glaze: Gerstley Borate alone.

Boron Rocks.(Collection of Department of Earth and Environmental Sciences, Columbia University, New York.)
Foreground, left to right: Borax and Kernite.

Background, left to right: Colemanite and Ulexite.

Figure 3.10 Purple Bear and spoon holder by Peggy Bloomer. (Photos by Chris Dube)

Cone 5–6 oxidation. Sheffield Red c/6 claybody. Glaze: Volcanic ash 68% and Boron Frit 3134 32% p. 67 . Note purple color, which results when high-boron glaze is applied over high-iron claybody. See *Figure 1.16, p93* for dramatic contrast of surface color on low-iron claybody.

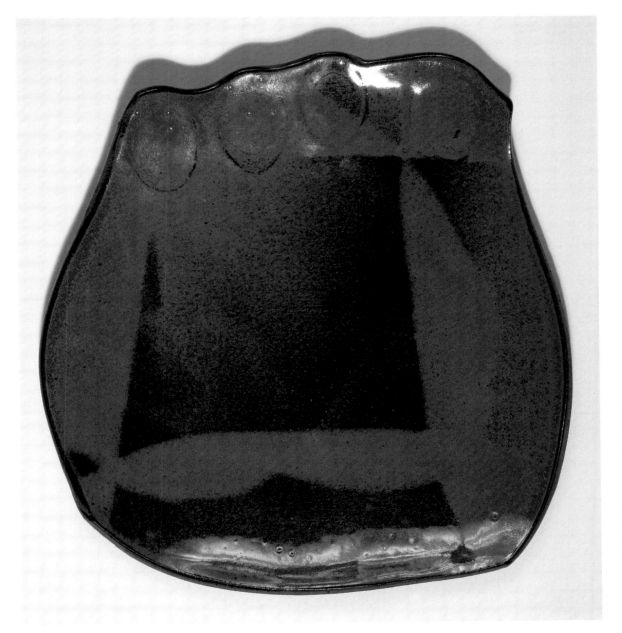

Figure 3.11 Platter by Jacqueline Wilder.

Cone 5–6 oxidation. #308 Brooklyn Red claybody. Glaze: Randy Red (Gerstley Borate 31.7% Talc 13.9%) p. 186.

Figure 3.12 Covered jar by author.

Porcelain, cone 9–10 oxidation. Glaze: Charlie D (Dolomite 15%, Talc 13%) p. 158.

Rocks, *left to right*: Magnesite, Talc, and Dolomite (Collection of Department of Earth and Environmental Sciences, Columbia University, New York).

234

Figure 3.13 Porcelain bowl by Sue Browdy.

Cone 9–10 reduction. Glaze: G-200 Feldspar 90%, Dolomite 10%, Red iron oxide 1/2%, Bone ash 2%.

Figure 3.14 Plate by author; **Cylinder** by Emily Tribich.

Stoneware, cone 9–10 reduction. Glaze: Charlie D with 3% manganese (p. 210). Note crystalline formations and gloss surface of Charlie D glaze on the cylinder caused by higher heat and slower cooling conditions.

Figure 3.15 Porcelain footed bowl by Jane Hartsook.

Hand-built. Cone 9–10 oxidation. Glaze: Temple White (Dolomite 19.6%) p. 209.

Figure 3.16

Stoneware, cone 9–10 reduction. Glaze: Nepheline Syenite 70%, Magnesium Carbonate 30% over Rotten Stone 90%, Whiting 10%.

Chapter 4 | AUXILIARY SILICA AND ALUMINA MINERALS

SILICA

We view an artifact from 2,000 years ago with great awe and reverence. Yet, a material whose age is 400 million years or more is handled without thought or caring. Such is the fate of our ceramic materials whose antiquity reaches far back to the beginnings of our earth. Our irreverence is caused by the changed appearance of the earth materials when used by the potter. They are white, buff, or brown powders, which are housed in mundane paper sacks; they are identified only by a trade name and the supplier's name and address. There is nothing to even faintly suggest their ancient, awe-inspiring origins. A case in point is the ceramic material Potters Flint, or Silica, which is more than 400 million years old and is the primary source for the free silica of our claybodies and glazes.

By way of background, it is helpful to place silica within a geological framework. Although the geological processes described on the following pages may seem irrelevant to our work as a potter, they underlie the very foundation of our craft. As we open our minds to them, the connection between ceramics and the universal processes of the earth's creation will become real and immediate.

Silica consists of the two most abundant elements on the earth—silicon and oxygen. Hence, silica is the most omnipresent of all of our earth materials; it is a presence in 95% of the rocks of the earth. Pure and compound deposits of silica are found throughout the world. The mineral quartz, which is silica, is the most prevalent of all the earth's minerals after feldspar. Silica combines with many minerals, to form a vast geological family of minerals known as the silicates, which includes feldspars and clays. Silica constitutes

60%–70% of feldspars and clays, and thus makes up 50%–60% of our glaze cores and claybodies. The importance of silica in feldspars, clay minerals, and claybodies has already been described in Chapters 1 and 2. Silica combines with calcium and magnesium to form the important melter oxides of Wollastonite and talc (pp. 166–168; 199–201).

Silica constitutes a large portion of the earth's magma. Magma, which contains high amounts of silica and low amounts of iron, is known as granitic magma because it solidifies into the rock granite. Granite comprises 75% of the earth's crust. Melted silica is heavy, dense, and slow flowing. Consequently, the high silica content of granitic magma causes this magma to be heavier and less fluid than low-silica, high-iron, high-magnesium basaltic magma, which flows easily onto the earth's surface and thus cools quickly into the fine-grained rocks known as plateau basalt. Sluggish, silica-rich, granitic magma never reaches the top layers of the earth's surface. Instead, granitic magma cools slowly beneath the protective, outer layers of fast-cooled basalt. Slow cooling fosters large crystal formation; hence, the crystalline structure of silica-rich granitic rock is coarser-grained than fast-cooled basaltic rock. Weathering and hydrothermal forces eventually destroy the protective outer layers of basaltic rock and expose the coarser-grained granitic rocks to the earth's surface. Once exposed to the earth's environment, the coarse-grained, granitic rocks are pounded and broken up into the materials that make up our ceramic claybodies and glazes.

Silica is the hardest and most indestructible of all the minerals that are present in granite. Unlike the softer minerals, such as soda and potash, powerful weathering and hydrothermal forces cannot transform granitic silica into different materials. Instead, these forces only free silica from its parent rock of granite and pound it into smaller and smaller particles. Each tiny grain of silica retains its integrity—microscopic inspection of silica sand reveals intact, massive quartz particles that are wonderful to behold! Water carries these smaller particles, along with the other materials, to various parts of the earth. The heavier silica particles fall to the bottom of oceans, lakes, and streams. In this way, large deposits of relatively pure silica sands formed more than 400 million years ago on the shores and beds of ancient oceans and lakes.

The chemical structure of all forms of pure silica, whether crystalline (tridymite, cristobalite, quartz), glass (obsidian, pumice), or gel is the same. Silica consists of four oxygen atoms attached to one silicon atom. Each silicon atom shares two of its four oxygen atoms with another silicon atom, thus creating the characteristic silica tetrahedron and the familiar chemical formula of SiO_2. Silica glass and silica gel are amorphous forms of silica; their silica tetrahedrons occur in a random pattern, which is comparatively irregular and unpredictable.

The Hall of Gems at the American Museum of Natural History, New York, exhibits magnificent samples of naturally formed quartz, tridymite, cristobalite, obsidian, pumice, and silica glass. Despite this glittering array of silica formations, most of the pure silica deposits that are found on the earth's surface, appear in the form of the mineral quartz— the most stable form of crystalline silica under the atmospheric pressure and temperature conditions that exists on the earth's surface. As previously stated, silica-rich magma tends to flow slowly. It does not penetrate easily to the outer crust of the earth. For this reason, when high-silica magma (the source of the major deposits of pure silica) cools slowly beneath the earth's surface to form granite, the silica it contains is in crystalline form—namely, quartz, tridymite, and cristobalite. Silica glass (obsidian, pumice) occurs only under conditions of extremely fast cooling. Once weathering and erosional forces expose crystalline silica to the cooler temperatures at the earth's surface, tridymite and cristobalite convert to the only stable form of crystalline silica that exists at the earth's temperatures, namely alpha-quartz. *Figure 4.1, p265.*

All of this holds a special meaning for the potter, because these universal processes reoccur during the kiln firing. The high heat of the stoneware kiln converts free quartz in the claybody to tridymite or cristobalite. The comparatively short cooling period of the kiln firing causes the melted quartz in the glaze (Potters Flint) to cool quickly into silica glass. In this way, the pressure, temperature, and cooling conditions of the stoneware kiln recreate silica minerals that are rarely found in our natural deposits of silica. Over a long period of time, most of these forms of silica will once again revert to quartz. However, for the incredibly short span of human life, they remain in their fired condition. Herein lies the bridge between ceramics and the geological processes that have formed the minerals of our earth.

PROPERTIES OF SILICA

INTRODUCTORY NOTE

We have already discussed in Chapter 2, "Clays and Claybodies," the function of silica in a claybody. This chapter describes the effect of silica on the color, surface texture, melting temperature, and glaze fit of the stoneware glaze. Hence, we are now dealing with the properties of silica glass; although silica is added to a glaze in crystalline form as Potters Flint, it emerges as silica glass at the end of the kiln firing.

1. HARDNESS

Stoneware glazes and claybodies are long lasting and strong—Song Dynasty celadons have endured for thousands of years. The high amount of silica contained in this ware is largely responsible for its endurance. Silica is a refractory, high-temperature material, which is impervious to acid and insoluble in water. When fired alone, silica requires a melting temperature of 3110°F to change into a liquid phase. Hence, high amounts of silica create a hard and durable ceramic surface.

2. GLASSMAKER

Silica is a glassmaker. When silica in crystalline form (quartz) is melted and cooled in the short firing and cooling cycle of the potter's kiln, it solidifies into a monocrystalline form (glass). Although it is now believed that glass does in fact contain some crystalline formations, the absence of a predictable pattern creates a transparent surface.

As stated previously, most of the silica found in nature exists in the crystalline form of the mineral quartz. As in the case of all silica minerals, quartz is made up of the silica tetrahedron—a four-sided pyramid of oxygen atoms that encircle a silicon atom. In the quartz mineral, the silica tetrahedrons group together to form a regular, characteristic pattern. The heat of the potter's kiln, accelerated by melters in the glaze and/or claybody materials, breaks apart this pattern. The relatively short cooling period of the potter's kiln is not long enough to permit the silica tetrahedrons to regroup into their original formation; their random flow permits light rays to travel freely, without obstruction, through the layers of the surface, thus creating a transparent surface.

In a glaze magma, the ratio of the glassmaker to the rest of the materials determines whether or not the fired surface will be transparent, semitransparent, opaque, or semiopaque. Large amounts of silica minerals, combined with sufficient melter materials and heat, will create a transparent surface. On the other hand, if the melter materials are not sufficient to melt all the silica into glass, or if they should combine with silica to form crystals, or, lastly, and most importantly, should the glaze contain a disproportionate amount of alumina, the result will be an opaque surface. *Figure 4.8, p272*. The proportion of silica to the alumina is of the utmost significance in determining the final surface of the glaze. This proportion is expressed as the ratio of a single molecule of alumina to the number of silica molecules (A/S) and functions as one method of predicting and classifying glaze surfaces. For example, at stoneware temperatures the A/S ratio of 1 molecule of alumina to 4–6 molecules of silica forecasts a transparent, clear, glossy surface. The A/S ratio of one molecule of alumina to 3–5 molecules of silica predicts a satin-matt surface. Lastly, the A/S ratio of 1 molecule of alumina to 1–3 molecules of silica suggests a stony, matt surface. However, there are exceptions to this kind of prediction. For example, the presence of certain melter oxides such as the alkaline earth minerals of calcium, barium, magnesium, and zinc can change the resulting surface into a satin-matt opacity even though the A/S ratio predicts transparency (see p. 258; *Figures 3.12, p234; 3.14, p236; 4.7, 4.8, pp271–272*).

Subject to these exceptions, the A/S ratio reveals an important clue with respect to the fired glaze surface, and highlights the significant glassmaking property of silica.

3. CRYSTALLINE FORMATIONS

Although melted silica alone will not solidify into its characteristic crystalline pattern, silica combines readily with certain other minerals to form visible, large crystals. We have already seen how silica combines with calcia to create a unique, satin-matt surface, and how it produces a showy, macrocrystalline glaze in partnership with zinc oxide (Chapter 3).

4. EFFECT ON FLUIDITY AND MELT

Siliceous magma is viscous, heavy, and slow flowing. Consequently, large additions of silica reduce the fluidity of a glaze by slowing down the fast, sudden action of soda and potash melters in the glaze magma. Many macrocrystalline glazes contain ten times more silica than alumina because alumina, which is the usual stiffener in a glaze, inhibits large crystal formation. Large amounts of silica slow down the otherwise uncontrollable flow of the low-alumina glaze magma; in these glazes, silica functions both as stiffening agent and as crystal former (see Hamer and Hamer 1986, 344). Low amounts of silica may have just the opposite effect; in a certain proportion, silica enters into eutectic relationships with other materials in the glaze, and thus increases the fluidity and melt of the glaze. (In this connection, note that wollastonite, which unites silica with calcia, is a more powerful melter than whiting [calcium carbonate]. Similarly, talc, which combines silica and magnesia, has more effect upon the fluidity of a glaze than does magnesium carbonate.) Once again, we see the importance of the proportion of the glaze materials.

SILICA

TRADE NAME	SIL-CO-SIL®, Potters Flint, Flint
GEOGRAPHICAL SOURCE*	Central Pennsylvania; northeastern West Virginia; New Jersey; Missouri; South Carolina; Oklahoma; Illinois; U.S. Silica Company
GEOLOGICAL SOURCE*	Oriskany sandstone (Eastern U.S.) deposited in Appalachian Mountain region during early Devonian times (400 million years ago)

CHEMICAL STRUCTURE** SiO_2

SIL-CO-SIL #52 (325 mesh Supersil)
(Berkeley Springs, WV)

SiO_2	99.670
Al_2O_3	0.100
Fe_2O_3	0.025
TiO_2	0.020
CaO	< .015
MgO	< .010
L.O.I.	0.200

Mineralogical Analysis of Oriskany sandstone*

Quartz	97%–99.5%

Minor grains of:
Partially altered Feldspar
Hornblende
Tourmaline
Limonite
Manganese with Cobalt traces

*Bates, 1969, 106, 107. *Ceramic Industry* 1998, 153–154.
**Technical data: U.S. Silica Company, 1998.

SOURCES OF SILICA

Figure 4.1, p265.

Over the course of millions of years, weathering forces release silica minerals from the parent granitic rocks and pulverize them into finer and finer particles. Limestone (calcite), iron minerals, and fine silica particles cement these grains of silica into rocks of quartz, sandstone, quartzite, flint, and chert. Although all of these rocks contain high amounts of silica, some are purer than others. Quartz is the purest form of silica; it usually contains 99% silica. Quartzite and sandstones are also quartz particles that have been cemented together by one of the above bonding agents; quartzite is metamorphosed sandstone; metamorphic forces of extreme heat and pressure melt some of the silica in the sandstone. The resulting silica glass binds the crystalline silica grains together to form quartzite.

Our domestic ceramic industry customarily seeks the purest and least reactive form of silica. For this reason, silica sold to domestic potters is usually crushed quartz rock, quartzite, or sandstone, which contains at least 99% pure silica. This pure form of silica is sold by ceramic suppliers in various mesh sizes under the general name of Potters Flint or Silica and should not be confused with the mineral flint. The confusion in terms between the mineral flint and Potters Flint arose because European potters used the mineral flint as their prime source of silica. Nodules of flint are formed when fine-grained, microcrystalline grains of silica are dissolved by the flowing action of water; eventually they are cemented together into nodules of amorphous silica. Often found in limestone beds, the mineral flint usually contains at least 5% calcium impurities. The calcium content of the mineral flint together with its microcrystalline particles, makes it more reactive in both a claybody and a glaze, than the same amount of medium grade Potters Flint (pure quartzite, sandstone, or quartz). Thus, in a claybody, the mineral flint will increase the production of cristobalite. Jasper and opal are microcrystalline forms of silica, similar to the mineral flint with respect to their reactivity.

The trade names of the various mesh sizes of Potters Flint change from time to time, due to the continual corporate reshuffling, which is so characteristic of our domestic companies. More than six years ago Pennsylvania Glass Sand Corporation merged with Ottawa Silica Company to form U.S. Silica Company. The trade name SIL-CO-SIL was retained as the name of its ground silica sands (U.S. Silica Company, letter to author, March 1, 1989; Technical Data 1998).

According to U.S. Silica Company brochures, the impurities content of SIL-CO-SIL is not more than 0.5%. It is available in at least 6 mesh sizes; 120 mesh (coarse) to 400 (fine). Numbers, ranging from 250 to 37, identify each of the various SIL-CO-SIL products. The listed number before a specific mesh size, such as, for example #52 (325 mesh size) means that 98% or more of the ground silica will pass a 52 μm screen. The higher numbers represent the larger screen sizes, and hence, the coarser grind. U.S. Silica Company also supplies fine ground silica sand of at least 99.2% purity under the name of MIN-U-SIL, which is also available in various particle sizes (U.S. Silica Company, Technical Data, 1998).

The mesh size of silica has vital consequences for both claybodies and glazes. The coarser the grind, the less reactive the silica, and the more it will remain in crystalline form in the melt instead of forming reactive amorphous silica or silica glass. The particular mesh size of the silica has important consequences for the strength and glaze fit of the fired ware (see Chapter 2, p. 104). The silica mesh size of #52 (325 mesh) is a standard grind that is commonly used by potters for both claybodies and glazes. However, for porcelain bodies, it has been found that the coarser grind #75 (200 mesh) contributes the "tooth and thixotropy (properties of gels becoming fluid disturbed, as by shaking) missing with the use of 325-mesh flint. The larger particle size also decreases cracking (Beumée 1994, 39).

The mesh sizes of the silica can have equally dramatic effects on the glaze. For example, during a workshop, it was reported that a glaze with 4.8% lithium carbonate and 19.25% Moon Flint silica (#52-325 mesh) cracked and shivered the ware. When a finer mesh silica (400 mesh) was substituted for the Moon Flint silica, the cracking lessened. It was hypothesized that the increased reactivity of the silica created a lattice network that surrounded the lithium molecules, and thus insulated the claybody from the stressful effects of the lithium melter.

Exactly how and why reactive silica lessened the stress of the complex, lithium glaze can be determined only by a microscopic analysis such as described by W. David Kingery and Pamela B. Vandiver in *Ceramic Masterpieces*. This work analyzes the different visual effects of crystalline and monocrystalline silica in some ancient Chinese celadons.

Extraordinary microscopic photographs of claybody and glaze cross-sections vividly depict these differences (Kingery and Vandiver 1986, 21–29, 82–84, 88, 99, 102–105). Without this kind of analysis, all that is clear is the fact that different consequences resulted from the substitution of a finer mesh size of silica. It is to be expected that the finer-grained silica will be more reactive and will enter into the melt more vigorously than coarser-grained silica. Thus, if you wish to increase the fluidity and glassy surface of the glaze, decrease the mesh size of the silica. Conversely, if the fluidity or the gloss of the glaze is too great, perhaps the substitution of a coarser-grained, less-reactive silica will achieve the desired result.

Always check the mesh size of the silica that appears on each new supply. If the bag does not bear the trade name of SIL-CO-SIL, check with the supplier as to the purity of the material. In addition, be aware of what mesh sizes are in fact available for you—they may not all be listed on the order sheet. This kind of information will have a direct bearing on the way in which silica will behave for you in your claybody and glazes. Once again, the more you know about your materials, the more control you will have over the final result.

GLAZE FUNCTION

The following tests show the surface changes caused by additions of 325 mesh silica to the glaze magma.[1]

1. *Additions of silica minimize surface crazing caused by high-shrinking soda and potash melters in the feldspathic glaze core.*

Cone 9–10 Reduction

Potash Feldspar	90%	Transparent gloss.
Whiting	10	Visible craze network.
Add Flint	20	Transparent gloss. Craze network not visible.

SANDERS CELADON (BASE)

Kona F-4	44%	Transparent gloss.
Whiting	18	Crazing not visible.
Flint	28	
EPK kaolin	10	
Remove Flint	14	Transparent gloss.
		Crazing visible.
Omit Flint		Semitransparent,
		semigloss.
		Crazing not visible.

New batch formula without Flint.

Kona F-4	47%
Whiting	30
EPK kaolin	22

Note: The high-calcia, crystalline, satin-matt surface has obscured the craze pattern, which was not visible on one test and but faintly visible on another. However, after a period of extensive use, some of these satin-matt surfaces revealed a highly visible craze pattern.

Cone 5–6 Oxidation

JACKY'S CLEAR

Figure Intro.6, p16.

Nepheline Syenite	50%	Transparent gloss.
Colemanite	10	Crazing not visible.
Wollastonite	10	
Zinc oxide	5	
Flint	20	
Ball clay	5	
Omit Flint		Transparent gloss.
		Crazing highly visible.

BLUE-GREEN CELADON

(Crafts Students League Stoneware)

Custer Feldspar	46.80 grams	Semigloss.
Gerstley Borate	13.70	Semitransparent.
Whiting	8.30	
Zinc oxide	4.00	
Dolomite	6.00	
Flint	20.00	
EPK kaolin	2.50	
Copper Carbonate	0.05	
Vanadium Pentoxide	0.05	
Omit Flint		Uneven coverage.
		Runny, excess fluidity.
		Highly crazed.

Note: The increased fluidity with the absence of Flint. This is an example of how heavy, slow-moving, silica magma inhibits glaze fluidity.

2. *The ratio of the glassmaker, silica, to the adhesive, alumina, generally determines if the surface will be transparent or opaque, or shiny or matt.* Hence, silica is an indispensable ingredient for the achievement of a transparent and/or glossy surface. Conversely, the reduction of silica helps to achieve an opaque, stony surface.

Cone 9–10 Reduction

RHODES 32

Figure 4.2, p266.

Potash feldspar	48%	Stony, opaque-matt.
Dolomite	22	Yellow-orange-brown.
Whiting	3	White (thick).
EPK kaolin	25	No visible craze network.
Bentonite	2	Delayed crazing.
Add Flint	10	Spots of shine. Stony opaque. Pale yellow-brown (thin). White-gray (thick).
Add Flint	20	Semigloss, semiopaque. White (thick).Gray (thin).
Add Flint	30	High gloss, transparent. Blue-gray color.
Add Flint	40	Gloss, semitransparent. Blue-gray color (thin). Milky blue-white (thick).

TEMPLE WHITE

Figures 4.5, 4.8; pp269, 272.

Custer feldspar	34.7%	Semigloss. Semitransparent.
Dolomite	19.6	Blue-gray where thin.
Whiting	3.1	Gray-white where thick.
Flint	18.9	
EPK kaolin	23.6	
Omit Flint		Stony, opaque-matt. Yellow-orange-brown (thin). White (thick).

Note: The absence of free silica from Rhodes 32 and its presence in Temple White is most significant. It is apparent from these two test series that the basic structure of both glazes is the same—the difference lies in the addition or omission of free silica. Note that in the Rhodes 32 test series the addition of 20% Flint creates a surface similar to the Temple White glaze; conversely, the omission of Flint in the Temple White series creates a surface similar to Rhodes 32. These tests show the significance of the presence or absence of free silica for the glaze surface. *They also underscore the value of glaze dissection tests, which vividly demonstrate the contribution of each of the various materials to the glaze surface.*

RHODES PORCELAIN

Figure 4.3, p267.

Custer feldspar	33%	Transparent gloss.
Whiting	20	Blue-green.
Flint	32	
EPK kaolin	15	
Omit Flint		Soft-matt. Opaque. Yellow-brown color.

New batch recipe without Flint.

Custer feldspar	48.5%
Whiting	29.0
EPK kaolin	22.0

The lowering of the glass content through the omission of the free silica opacified the transparent surface and changed the gloss to a soft matt. The silica omission increased the calcium content and further increased the alumina content. Once again, with the omission of Flint from the glaze formula, note the resemblance to Rhodes 32.

SANDERS CELADON

Figure 1.5, p82.

Kona F-4	44%	Transparent. Gloss.
Whiting	18	Gray-green color.
Flint	28	No craze.
EPK Kaolin	10	
Barnard clay	5%	
Omit Flint	14	Semitransparent. Semigloss. Milky-blue-gray color. Craze visible.
Omit Flint	28	Opaque-matt. Gray color.
Add Flint	20	Semitransparent. Gloss. Milky-blue-green-gray color. Fatter. More opaque. Deep, intense surface.

Cone 8 Oxidation

ELENA'S SATIN-MATT

Custer feldspar	56 grams	Satin-matt.
Whiting	20	Gray-white-yellow.
Flint	6	
EPK kaolin	18	
Zinc oxide	9	
Rutile	7	
Omit Flint	3	Stonier surface. Yellower color.
Omit Flint	6	Increase of stony surface. Increase of yellow color.

Cone 5–6 Oxidation

PORTCHESTER COPPER

Figure 1.6, p83.

Nepheline Syenite	55%	Satin-matt. Opaque.
Whiting	25	Green-black color.
Flint	10	
EPK kaolin	7	
Tin oxide	3	
Copper oxide	2	
Omit Flint	10	Dry, crawled surface. Black-green color.

Not only does the free silica increase the A/S ratio of this glaze so as to produce the satin-matt surface, but it also contributes to the fusion of the glaze.

For additional evidence of the eutectic power of silica, as well as silica's effect on the matt-shine ratio, see color tests on pp. 249–252. The surfaces of these tests show increased opacity and lowered gloss whenever free silica was omitted or reduced.

3. The addition of free silica in a glaze will combine with the alkaline earth melters to produce a satin-matt opacity.

Chapter 3, pp. 159–161, describes the formation of calcium-silicate crystals, which results in the jade-like, satin-matt surface. The section on zinc oxide shows how zinc oxide combines with silica to produce showy, flowerlike crystals on the surface of a clay form (pp. 217–218). The achievement of the macrocrystalline surface depends on the following factors:

a. A low-alumina, high-soda, high-gloss, fluid glaze.

b. The glaze must also contain a 2:1 ratio of zinc oxide to silica in order to form Willemite crystals. A combination of 2 and 2/3 parts zinc oxide to one part silica created a successful seeding agent for the growth of Willemite crystals (Sanders 1974, 21). The zinc-silica mixture was applied over or under the base glaze and, together with a colorant such as iron oxide or manganese dioxide, became the nucleus (seeding agent) for the growth of a few large Willemite crystals. (For a superb account of macrocrystalline glazes, complete with spectacular photographs of flowerlike crystals, detailed formulas, and precise firing instructions, see Sanders 1974, 13–39.)

c. It is necessary to have a long cooling period at the specific temperature best suited for the growth of macrocrystalline formations within a particular glaze base. The temperature will vary depending on the firing temperature and batch ingredients of the glaze. A common firing temperature for a macrocrystalline glaze is approximately 2102°F–2282°F; the temperature of the slow cooling period (48 hours) is 1832F°–932°F (Hamer and Hamer 1986, 344). We achieved macrocrystalline formations in a cone 4 firing (2167°F) by holding the cooling period for 5 hours at 2000°F. The seeding agent of Zinc oxide (60) Flint (25) was applied under a high-soda, low- alumina, transparent, glassy, fluid glaze (see p. 220).

d. The kiln atmosphere must be oxidation and not reduction, due to zinc oxide's early reduction to metal, and consequent volatilization, at temperatures above 1742°F in a reduction atmosphere (Hamer and Hamer 1986, 343).

4. The presence of unmelted particles of free silica in a glaze magma will create a frosty, white opacity. The opposite result can occur if all of the free silica enters into the melt. In such a case, additions of free silica form a eutectic bond with other glaze materials to increase the fusion of the glaze. The stronger melting power of both Wollastonite ($CaSiO_3$) and talc ($3MgO \cdot 4SiO_2 \cdot H_2O$) compared to Whiting ($CaCO_3$)

and magnesium carbonate ($MgCO_3$) is an example of the eutectic power of silica (see p. 241).

Cone 9/10 Reduction

The addition of 10% silica to Rhodes 32 glaze, which has no free silica in the formula (p.245), causes the stony, opaque surface to display glassy, shiny spots. The addition of 30% free silica transforms the opaque, stony-matt, white-orange-brown surface into a high gloss, transparent, blue-gray surface. The addition of 40% silica results in a frosty, white, glossy opacity; this surface is caused by the presence of unmelted silica particles, which lie suspended within the matrix of melted glass.

The addition of silica to the Temple White glaze (p. 245) demonstrates the consequences of increasing the amount of free silica in a semitransparent, semigloss glaze that already contains almost 19% free silica. Additions of 10%–30% silica increased the whiteness and opacity of the surface, but did not destroy its gloss. These tests, together with the other tests on p. 245, demonstrate the eutectic power of lower additions of free silica and the opposite effect of increased amounts of silica. Once again, proportion is the key; different amounts of silica in the same glaze base will create a broad spectrum of glaze surfaces.

Cone 5/6 Oxidation

At lower firing temperatures, with more complex glaze formulas, it becomes harder to generalize about the results of silica additions. At the cone 5–6 oxidation temperatures, even small additions of free silica in certain glazes will disturb the balance of the glassmaker, alumina, melter ratio and result in a dry, less-melted surface. Yet the same amount of free silica added to a different glaze may produce just the opposite effect and increase the fusion power of the melter. The particular balance of the numerous glaze ingredients in each glaze magma governs the final surface and produces variable results.

CORNWALL STONE

Crafts Students League White Stoneware.		White color.
Potash feldspar	44%	Semigloss.
Cornwall Stone	22	Semi-opaque.
Whiting	18	
Zinc oxide	8	
Titanium oxide	4	
EPK kaolin	5	
Add Flint	5	Higher gloss. Increased transparency. Increased flow.
Add Flint	15	Lower gloss. Increased opacity.

CARACAS WHITE I

(North River Pottery)		White matt.
G-200 Feldspar	49.4%	
Whiting	17.9	
Zinc oxide	11.5	
Flint	6.4	
EPK kaolin	14.7	
Omit Flint		Sugary surface. Uneven flow and coverage.

CARACAS WHITE II

North River Pottery   Off-white. Matt.

G-200 Feldspar	47.5%
Whiting	16.9
Zinc oxide	7.6
Flint	5.1
EPK kaolin	15.3
Rutile	7.6

Omit Flint   Shinier.
                                             Smoother surface.

Note: Free silica in Caracas Glaze I increases fusion and also assists the even disbursement of the rest of the glaze materials in the melt. A similar result of uneven disbursement occurred in a gloss, blue-green, semitransparent glaze when 20% free silica was omitted from the formula (see p. 244). Fluidity increased and coverage became uneven. Here we see evidence of the distinction between low-silica and high-silica magma. A key characteristic of melted silica is that it is slow moving, heavy, and viscous. Magma low in silica tends to flow rapidly. Heavy, viscous, molten silica surrounds the other materials in the melt and slows down their flow; the result is a more even coverage of the glaze surface.

As we have seen in some of the tests, additions of low amounts of free silica can increase the fusion power of the melters, and thus create a shinier, more transparent surface. On the other hand, because of the refractory nature of silica, increased additions of silica will eventually opacify the glaze surface. It is to be expected that this result would occur faster at the cone 5/6 temperatures than at the higher stoneware temperatures of cone 9/10. The Cornwall glaze demonstrates this result. The addition of 5% silica increased the shine and transparency of the surface. However, adding 10% more silica caused increased opacity and lowered gloss.

KATHERINE CHOY

Figure 4.6, p270.   Yellow color. Glossy,
Crafts Students League   white streaks. Opaque,
Stoneware   satin-matt. Clear gloss
                                             on Grolleg porcelain
                                             cone 4–6 claybody.

Nepheline Syenite	53.9%
Whiting	11.7
Lithium carbonate	4.7
Zinc oxide	10.3
Flint	1.5
EPK kaolin	17.9

Omit Flint	1.5	Glassier surface Whiter, less yellow color.
Add Flint	10	Whiter color. Increased opacity. Increased crystallization.

Whiter color suggests presence of unmelted particles in glaze magma, and despite the increased gloss, a lowering of transparency.

TRANSPARENT BLUE

Crafts Students League   Blue. Transparent.
Stoneware Glaze   Gloss.

Potash Feldspar	44 grams
Colemanite	20
Whiting	1
Zinc oxide	3
Flint	24
EPK kaolin	1
Cobalt oxide	0.5%
Copper oxide	4.0%

Omit Flint   Greener color.
                                             Shinier surface.
                                             Resembles celadon
                                             surface. Beautiful test!

The powerful melter and glassmaker colemanite (20%) has been held in check by a more than equivalent amount of viscous, sluggish silica magma. Once this check is removed, the surface shine increases and the color becomes greener.

JACKY'S CLEAR II

(Portchester claybody)
Figure Intro.6, p16.

Beige color. Transparent. Gloss.

Nepheline Syenite	50%
Colemanite	10
Wollastonite	10
Zinc oxide	5
Flint	20
Ball clay	5

Add Flint	10	Whiter color. Less transparent.
Add Flint	20	Whiteness increases. Semitransparent.

The addition of sizable amounts of silica opacifies the surface. Note that the same result does not occur at the higher stoneware temperatures, where the increased heat melts more of the silica into silica glass. Hence, many cone 9/10 high-gloss, transparent celadon glazes contain 30% silica (see Rhodes Porcelain, p. 245, Mirror Black II and Sanders Celadon, following).

5. *Silica has an extraordinary effect on the color of the glaze surface.* It is to be expected that a material that increases transparency and gloss would also increase the brilliance of the surface color. What is perhaps not so obvious is the dramatic effect that the addition of a relatively low amount of silica can have on a specific color. Here again sounds a refrain that echoes throughout the pages of this book: *The ratio of the base ingredients of a glaze (glassmaker, melter, and adhesive oxides) are as important for the creation of a specific color as are the colorants themselves.*

Cone 9–10 Reduction

SANDERS CELADON BASE

Transparent. Gloss.

Kona F-4	44%
Whiting	18
Flint	28
EPK kaolin	10

Add Red iron oxide

1/2%	2%	5%	9%
Blue-green. Gloss. Transparent.	Deep green. Gloss. Transparent.	Brown-yellow. Black; Mahogany where thin. Gloss. Transparent.	Mahogany flecked with black streaks where thick. Gloss. Transparent.

Add 20% Flint to above tests.

Milky-blue. Lower gloss. Opaque.	Blue-green. Brilliant. Gloss. Opaque.	Deep moss green. Black spots. Semi-opaque. Dense, even coverage.	Black. High gloss. Opaque. Dense, even coverage.

MIRROR BLACK I

Shiny. Opaque. Black.

Custer Feldspar	56.0%
Whiting	16.0
Flint	20.5
Ball clay	7.5
Red iron oxide	5.0

Omit Flint	Reduced gloss. Opaque. Yellow-brown.

MIRROR BLACK II

Shiny black

Custer Feldspar	25.8%
Cornwall Stone	12.9
Whiting	12.9
Flint	31.4
EPK kaolin	6.5
Red iron oxide	9.5
Bentonite	1.0%

Omit Flint	20	Metallic black.

PAM'S SATURATED IRON

		Black. Red-brown streaks. Gloss.
Custer Feldspar	52.4%	
Whiting	10.1	
Gerstley Borate	6.2	
Flint	17.0	
EPK kaolin	5.6	
Red iron oxide	8.6	
Omit Flint		Stony surface. Yellow color edged in brown.
Replace Flint	8.5	Stony surface. Red-brown-black metallic.
Add Flint	17.0	High gloss. Dense black.

Note: Substitution of lower silica glaze core produced similar results. *See Appendix B1.*

SANDERS SATURATED IRON

		High gloss. Red-brown background. Black streaks where thick.
Kona F-4	44%	
Whiting	18	
Flint	28	
EPK kaolin	10	
Red iron oxide	9%	
Add Flint	20	Higher gloss. Dense overall black.

These tests show that a brilliant, overall black color is dependent on a large amount of free silica. Whenever the silica was reduced or omitted, the strong black changed to a metallic, dull black or yellow-brown. In addition, these tests show once again that a reduction or total absence of free silica favors the yellow-brown surface color, hence the absence of free silica from the yellow-brown, stony, Rhodes 32 glaze. In general, note the reduction of shine and increased opacity of the surface whenever silica is substantially reduced. The brilliant black color that results from the increase of free silica seems to go hand-in-hand with an increased surface gloss.. Conversely, the change to a yellow-brown color is accompanied by loss of shine and transparency and the reversion to a stonier, more opaque surface, similar to the Rhodes 32 glaze. These color effects of black, brown, or yellow depend in part on the manner of light transmission, which depends on the amount of crystals and/or unmelted particles present in the glaze layers. Thus, it is the combined effect of the fusion, crystallizing, and glassmaking (transparency) properties of silica that are largely responsible for the color changes of the high iron-glazes in the above tests.

MAMO MATT

Figure 4.4, p268.		Orange-green-black. Mottled. Opaque. Stony.
G-200 Feldspar	49.0 grams	
Dolomite	19.0	
Whiting	4.0	
Grolleg	21.0	
Tin oxide	8.0	
Copper oxide	1.5	
Add Silica	20–30	Gloss. Blue-purple color.

SANDERS CELADON

Figure 1.5, p82.		Transparent. Gloss. Gray-blue-green. Noncrazed surface. Iron-spotting.
Kona F-4	44%	
Whiting	18	
Flint	28	
EPK kaolin	10	
Barnard clay	5%	
Omit Flint	28	Gray color. Opaque. Semi-matt.
Omit Flint	14	Pale, blue-gray. Gloss. Visible crazing.
Add Flint	10	Brighter blue; less gray. Fatter surface.
Add Flint	20	Paler, milky blue. Increased opacity. No iron-spotting.
Add Flint	30	Pale blue-gray. Opaque, stony, sugary surface. Low gloss. Drip marks evident.
Add Flint	40	White with bluish cast. Stony, unmelted surface. Sintering cracks visible.

PAVELLE GLAZE

Figure 3.7, p229.

Orange-yellow-brown.
Opaque, smooth, matt.

Bone ash	42.0 grams
Dolomite	30.0
Zinc oxide	4.0
Cornwall Stone	17.5
Nepheline Syenite	10.5
Rutile	3.0%

BYRON YELLOW

Custer Feldspar	52.7 grams
Whiting	21.3
Talc	4.0
Bone ash	2.0
EPK kaolin	25.0
Red iron oxide	4.0%

RHODES 32

Figure 4.2, p266.

Orange-brown-yellow (thin);
White (thick); opaque, smooth, matt.

Custer Feldspar	48.9%
Dolomite	22.4
Whiting	3.5
EPK kaolin	25.1

The yellow-brown-white surface of Rhodes 32 changed to blue-gray with the addition of free silica. *See tests, p. 245.*

Note: The absence of free silica in the above three glazes.

FELDSPAR GLAZE PLUS ADDITIONS OF SILICA

Nepheline Syenite	80%	Kona F-4	80%	Cornwall Stone	80%
Whiting	20	Whiting	20	Whiting	20
Red iron oxide	1/2%	Red iron oxide	1/2%	Red iron oxide	1/2%

ADD:
Silica

0%	Gray-yellow-green. Mottled surface; overall matt. Puddles of green glass. Craze.	Gray-blue-green. Gloss. Transparent. Craze.	Gray-blue. Gloss. Semitransparent. Craze.
10	Green. Less gray. Increased flow.	Slightly whiter.	More brilliant color.
20	Increased flow. More puddles of green glass. Less matt surface.	Paler color. Less crazing.	Increased whiteness. Lowered gloss.
30	Green-gray color. Overall gloss. Increased flow.	Bluish, milky cast. Crazing invisible. Stony surface.	Opaque white. Drip marks show. High craze.
40	Blue-green-gray. High gloss Transparent.	Similar to above. Increase of bluish milky color.	Dry-white, unmelted surface. Craze.
50	Similar to above. Crazing not visible.		
60	Similar to 50% test.		

Cone 8 Oxidation

ELENA'S SATIN-MATT

		Gray-white yellow.
		Mottled. Satin-matt.
Custer Feldspar	56 grams	
Whiting	20	
Flint	6	
EPK kaolin	18	
Zinc oxide	9	
Rutile	7	
Omit Flint		Overall yellow color.
Omit Flint	3	Paler yellow color.

The omission of 6% free silica caused the mottled gray-yellow-white color to change to an overall yellow. When 3% free silica was returned to the glaze, the yellow cast lessened. This test, however, was more yellow than the color of the original glaze and shows the sensitivity of surface color to the presence or absence of low amounts of free silica—in this case we are dealing with a mere 3% Flint!

Cone 5–6 Oxidation

RANDY RED

Figure 3.11, p233.		Red-brown-green
		mottled gloss.
Kona F-4	20%	
Gerstley Borate	32	
Flint	30	
Talc	14	
EPK kaolin	5	
Red iron oxide	15	
Omit Flint	30	Brown-yellow color.
		Increased fluidity.
		Glassy.

The powerful melter and glassmaker, Gerstley Borate (32%), has been held in check by an equivalent amount of viscous, sluggish silica magma. Once this check is removed, the glaze flows freely, and the color changes. See similar result in Transparent Blue glaze, p. 248.

PORTCHESTER COPPER

Figure 1.6, p83.		Blue-green satin-matt.
Nepheline Syenite	55%	
Whiting	25	
Flint	10	
EPK kaolin	7	
Tin oxide	3	
Copper oxide	1%	
Omit Flint.	10	Green color.
		Dry, crawled matt.

The variable results that occur in the above tests illustrate the complexity of the interaction of free silica with their various glaze components. Except for the obvious conclusion, that the amount and presence of free silica has a decided effect on the color, shine, and flow of the glaze surface, it is difficult to predict more specific rules of behavior that would hold true for all glazes. This is especially true of the lower-stoneware glazes, which contain many glaze ingredients. Reliable predictions can be made only within the context of each individual glaze formula.

6. *Silica magma is heavy, viscous, and slow moving. The addition of a large amount of free silica to a glaze can reduce fluidity without drastically changing its color or texture.* A large addition of silica (50%!) to a runny, highly crazed, shiny, green-yellow, cone 9/10 reduction glaze residue produced a glossy, smooth, uncrazed blue-green surface, which flowed into an even, fat roll at the bottom of the clay form. However, because free silica will affect the glaze fit, the addition of large amounts of free silica is advisable only in the case of runny, highly crazed glazes; the addition of a non-shrinking material such as silica in many glazes could disturb the glaze fit and cause shivering and cracking problems. Hence, this is a remedy that should be used with caution and depends entirely for its success on a disproportionate amount of volatile, high-shrinking melters.

SUMMARY OF TEST RESULTS

1. Both the omission and the addition of large amounts of silica will produce opacity, lowered gloss, and a dull, pale,

surface color, depending on the balance of the glaze ingredients in a particular glaze. Such is the extraordinary effect of too little or too much free silica. Exactly what constitutes too little or too much depends on the pre-existing ratio of the glassmaker, adhesive, and melter oxides in any one glaze. This result is shown clearly by comparative tests of Nepheline Syenite, Cornwall Stone, and Kona F-4, both alone and with increased additions of silica (see p. 251). The Cornwall Stone glaze core is relatively high in silica and low in soda and potash melters. The melter content was too low to bring all the silica into the melt. Thus, further additions of silica (20%) lightened the color and opacified the surface of the glaze. On the other hand, Nepheline Syenite is relatively low in silica and high in soda melters. This enables 50%–60% additions of free silica to be made without opacifying or dulling the glaze surface. The Kona F-4 glaze core is an intermediate glaze core, and additions of free silica begin to lighten the color and opacify the surface at about the 40% increment. Hence, differences in the silica and melter content of the various glaze cores must always be kept in mind whenever you make additions of glaze materials. It is possible to increase the silica content of a glaze by substitution of a high-silica glaze core such as Cornwall Stone, for a low or medium glaze core such as Nepheline Syenite, or Kona F-4. Conversely, a reduction in silica content may be accomplished by substituting a low-silica glaze core for a high-silica glaze core. Hence, the same kind of changes will occur in a glaze surface irrespective of whether the source of the additional silica is free silica (Flint) or the substitution of a high-silica glaze core such as Cornwall Stone or Custer feldspar.

2. Silica additions affect the fluidity and melt of the glaze surface. In many of the tests, lower amounts of silica (10%–15%) tended to increase fluidity, as the silica entered into a eutectic combination with the melters of the glaze. Thereafter, additional amounts of silica reduced the flow of the glaze. Once again, the exact amount needed to produce a given result will always depend on the ratio of the oxides in each particular glaze.

3. Additions of silica are crucial for the development of color. At the cone 9/10 reduction temperatures, the addition of 10%–15% silica caused a blue-green-gray color of a feldspathic, glossy, transparent glaze with a low iron content (1/2%–2%) to become bluer and more brilliant. Thereafter, further increments of silica bleached and whitened the surface color and eventually opacified the surface. Saturated, high-iron glazes changed from a mottled, red-brown-black color to an overall glossy black surface as a result of the additions of 20% more silica. The surface also became more opaque, dense, and viscous.

The addition or omission of silica to the cone 5/6 oxidation surface affected the brilliance and hue of the color and surface of the glaze. General rules at this firing temperature were harder to observe—it seemed always to depend on the existing ratio of the oxides in each particular glaze.

ALUMINA
Al_2O_3

Alumina is made up of the elements aluminum and oxygen. After oxygen and silicon, aluminum is the third most abundant element found on the earth. Aluminum is never found alone in nature; it exists mostly in combination with oxygen as aluminum oxide (alumina), and in this form it constitutes 15% of the crust of the earth. In ceramics, alumina is the stabilizing oxide of the glaze trinity. It forms a necessary bridge between the acidic glassmaker silica and the fast-flowing, volatile melter oxides.

Alumina (Al_2O_3) has an off-balance, electrical charge and regains stability by bonding with silica (SiO_2) and other melter oxides such as soda (Na_2O), potash (K_2O), calcia (CaO), and magnesia (MgO); this is the form in which alumina appears in our feldspars, clays, and other silicate minerals of the earth's crust.

The powerful weathering action of flowing waters loosens these sturdy, aluminosilicate-melter bonds. Hydrogen atoms escape from their watery bonds and slip into the spaces that were occupied by silica and melter atoms. One result is the aluminohydrate bauxite, which is the primary source for commercial aluminum.

Fortunately for potters, weathering forces were not able to completely destroy all of the bonds between alumina and silica. The weathering of granites and other igneous rocks dissolved the weaker-bonded melter oxides, but left in place the stronger silica and alumina bonds; in this way, the infinitely varied, aluminosilicate clay minerals came into being. The flat, platelet structure of the clay molecule, which creates the property of plasticity, is formed by aluminosilicate bonds. Of most importance is the fact that certain aluminosilicate clay minerals constitute the primary source of alumina for ceramic glazes.

SOURCES OF ALUMINA[2]
Figure 4.1, p265.

The mineral corundum and the gemstones of ruby and sapphire are crystalline forms of alumina. Bauxite, gibbsite, and diaspore are aluminohydrates and are the sources of commercial aluminum. These hydrated minerals, together with the mineral laterite, provide purified alumina for the ever-expanding field of high-alumina (corundum) ceramics.

"Of all the new ceramic materials, alumina probably has the widest diversity of uses and the greatest potentialities. For electronic and aerospace applications, its outstanding mechanical strength, excellent thermal shock resistance, excellent electrical properties (high dielectric strength, low power factor, etc.), and its chemical and abrasion resistance make it well suited in this field."

(*Ceramic Industry* 1998, 57)

The primary and least expensive source of alumina for potters is found in the kaolin clays in which the ratio of the alumina to the silica is about 38%–45%. Ball clays provide another prime source. Here the ratio of alumina to silica is approximately 30%–50%. Feldspar is also a source of alumina; however, the alumina content is much lower (15%–23.5%). Kaolins and ball clays are a preferred source of alumina, not only because of their higher alumina content, but also because their fine clay particles help to suspend the particles of the glaze materials evenly in water and thus facilitate even coverage and physical application of the glaze. Pure, powdered aluminohydrate is expensive and is not used in large amounts. However, in small doses, it is

used to prevent porcelain lids and feet from sticking to pots or kiln shelves during the firing process.

PROPERTIES OF ALUMINA

Alumina possesses many unique properties; a knowledge of these properties is essential for the successful achievement of ceramic surfaces.

1. *Alumina contributes hardness and durability to both the claybody and the glaze.* During the firing of the claybody, alumina and silica bonds are rearranged to form the aluminosilicate mullite crystal, which provides the strength of the fired ware. The presence of alumina in glazes strengthens and preserves the glaze coat and protects the ware from the destructive effects of acid and time.

2. *Alumina adheres the molten glaze to the sides of the clay form.* It contributes viscosity to the glaze magma; this property prevents the molten liquid from flowing off the form during the firing process. In most cases, subject to a few exceptions, glazes with low amounts of alumina are excessively fluid and will run off the walls of the form and onto the kiln shelf. Excess fluidity in a glaze is caused by weak and irregular bonds between the molecules of the glaze magma. Alumina strengthens and solidifies these bonds; it weaves the molecules of silica and melter oxides (soda, potash, calcia, etc.) into a strong, compact lattice structure that inhibits the movement of the molten glaze. This property is an essential requirement for most glaze surfaces. On the other hand, an excess amount of alumina in a glaze will produce a sluggish glaze flow and a crawled or pinholed glaze surface. The reason lies, once again, in the compact bonds formed by alumina with silica. A tightly constructed aluminosilicate structure prevents the liquid glaze from flowing into sintering or pre-firing cracks. It also will not heal the pinholes left by escaping gases. In addition, it may cause the glaze to crawl. High alumina magmas tend to have a high surface tension. Their surface molecules are strongly attracted to each other and encapsulate the liquid glaze into small spheres, leaving bare patches of exposed claybody. We shall see from the tests that follow that what constitutes too little or too much alumina depends primarily on the particular oxide ratio that exists in

each individual glaze. According to *The Potter's Dictionary,* most stoneware glazes have 4%–10% alumina; 10% kaolin renders about 4% alumina; 25% kaolin renders about 10% alumina (Hamer and Hamer 1986, 7). Bear in mind that the feldspathic glaze core can also contribute up to 10% alumina. For example, 57%–60% of Custer feldspar provides 10%–10.5% alumina (Zakin, 1978, 6).

3. *Alumina has an important effect on the opacity of the glaze surface.* Small amounts of alumina can dissolve in the glaze magma. Large amounts will speedily recrystallize into tiny, dense crystals. As stated previously, what constitutes a small or large amount will depend on the specific oxide ratio of each glaze. The speed of the crystallization ensures that the crystals will be tiny and dense. This microcrystalline surface absorbs the light and creates the visual effect of a stony-matt surface. In addition, alumina inhibits the tendency of silica to form large, visible crystals with other melter oxides such as calcia and zinc oxide. Hence, the ratio of the alumina to the silica in any particular glaze usually determines whether or not the surface will be transparent and glossy, or opaque and matt.

4. *Alumina raises the fusion point of a glaze.* The melting temperature of alumina is about 3722°F. Hence, at the stoneware and porcelain temperatures, alumina usually functions as a refractory material and effectively raises the firing temperature of a glaze. Whenever sizable amounts of alumina are added, the firing temperature of the glaze increases. However, at temperatures well above 2400°F, alumina functions as a melter. Even at lower temperatures, it can bring down the melting temperature of the acidic glass-maker silica and thus lower the fusion point of the glaze. The interaction of alkali, acid, and heat promotes fusion (Hamer and Hamer 1986, 6). Chameleon-like, alumina changes to an alkali in the presence of acidic silica and thus increases the fusion of the glaze. A few of the tests that follow show this result. Note that the amount of the added alumina is fairly low in all such instances.

5. *Alumina has a distinct effect on the color of the glaze surface.* At reduction stoneware temperatures, it tends to bleach color and reduces the brilliance and hue of the glaze surface. In the oxidation atmosphere, in combination with chrome or manganese, it encourages the production of reds and pinks (*Ceramic Industry* 1998, 56).

6. *Alumina has the third lowest expansion rate after boric oxide and fused silica* (Hamer and Hamer 1986, 356). Hence, it will lessen the crazing of the glaze surface.

7. *Like so many of our ceramic materials, breathing alumina dust can be harmful.* The toxic effect on the nerve cells from the inhalation of alumina is clearly set forth by Jane Brody, in her "Personal Health Column," *New York Times,* B14, 6 April 1989.

GLAZE FUNCTION

The glaze tests that follow demonstrate the various properties of alumina and graphically illustrate the changes in the glaze surface produced by separate additions or omissions of high-alumina clays. Note that in Chapter 1, similar changes in a glaze surface are described with the substitution of high-alumina glaze cores such as Nepheline Syenite or iron-bearing clays (see also Appendix B1).

1. *Alumina contributes the property of viscosity to the glaze magma and inhibits the excessive flow of the glaze magma.* This is the most common characteristic of alumina. Every beginning potter is advised to add high-alumina kaolin to an overly fluid glaze. The following tests illustrate this well-known principle and demonstrate that additions of both kaolin and ball clay will curb the fluidity of most stoneware glazes.

Cone 9–10 Reduction

LAU LUSTRE

Figure 1.11, p88.		Orange-gold-white. Opaque Gloss.
Nepheline Syenite	53%	
Whiting	2	
Lithium carbonate	5	
Soda ash	5	
Flint	10	
EPK kaolin	25	
Tin oxide	2	
Omit EPK		Less gold color. Runnier and glassier.
Repeat Test		Blue-green-gray. Transparent.

Cone 5–6 Oxidation

KATHERINE CHOY

Figure 4.6, p270. Crafts Students League Stoneware		Opaque satin-matt. Streaks of yellow-white gloss. Off-white background.
Nepheline Syenite	53.9%	
Whiting	11.7	
Lithium carbonate	4.7	
Zinc oxide	10.3	
Flint	1.5	
EPK kaolin	17.9	
Omit EPK		Whiter, glassy, runny.
Add EPK	10	Whiter and stonier.

CORNWALL STONE

Crafts Students League Stoneware		White. Semigloss. Opaque.
Potash feldspar	44%	
Cornwall Stone	22	
Whiting	18	
Zinc oxide	8	
Titanium dioxide	4	
EPK kaolin	5	
Omit EPK	5	High gloss. Transparent.
Add EPK	5	White. Pinholed. Opaque. Stony. Speckled.
Add EPK	10	White. Stony. Smoother. Overall speckled.

2. *A less well known characteristic of alumina is that in certain glaze magmas it actually promotes fluidity and glaze fusion and thus increases the gloss and transparency of the surface.* One reason for this may lie in the potter's source of alumina, which is the aluminosilicate clay minerals. Alumina forms strong, eutectic bonds with silica at high temperatures, and it stands to reason an aluminosilicate material would be less refractory than pure alumina (see p. 255). In addition, alumina inhibits silica's tendency to form visible crystal formations with other materials such as calcium oxide, and thus promotes noncrystalline transparency and shine. The following tests show that increased fluidity, gloss, and transparency resulted with the addition of low amounts of kaolin and ball clays. These additions were made with contrary expectations and demonstrate once again the bewildering tendency of ceramic materials to produce unexpected results. Only the experience of repeated testing will enable you to guide these materials toward the achievement of your desired surface.

Cone 9–10 Reduction

SHINO CARBON TRAP

Figure 1.10, p87.		Orange-red. White where thick. Semi-matt. Gloss where thick. Pitted surface.
Nepheline Syenite	45.0%	
Kona F-4	10.8	
Spodumene	15.2	
Soda ash	4.0	
Ball clay	15.0	
EPK kaolin	10.0	
Bentonite	2.0	
Omit Ball clay		
Test A .		White, shiny surface with carbon trap.
Test B (different kiln)		White, shiny, pinholed, and crawled surface.
Omit EPK		
Test A		Pale-orange color. Shiny fat surface.
Test B (different kiln)		Shiny, white, surface. Pinholed and crawled.
Omit Bentonite		Shiny white surface. Pinholed and crawled.
Omit EPK and Ball		Drier, redder surface. Thick layers are white and glossy.

Lowering the clay content consistently produced a shiny white, pinholed, and crawled surface, with a loss of the orange color. Note the dry surface, which resulted with the omission of both EPK and ball clays. The tests were repeated with similar results.

FRAZIER WHITE

		White.Semitransparent. Satin surface. Craze.
Custer feldspar	60%	
Dolomite	20	
EPK kaolin	20	
Omit EPK		Creamy surface. Gloss.
Omit 1/2 EPK		Greener cast to white color. Shiny, glossy surface.
Double EPK		Whiteness increases. Crawled surface.

Lowering the alumina content by reducing and then omitting EPK increased the gloss of the surface. Note its effect on the color of the glaze surface. Higher amounts of alumina increased surface tension and viscosity and resulted in a crawled surface.

TEMPLE WHITE

Figure 4.8, p272.		Gray-white. Semitransparent.
Custer feldspar	34.7%	
Dolomite	19.6	Semigloss; No crackle.
Whiting	3.1	
Flint	18.9	
EPK kaolin	23.6	
Omit EPK		Whiter. Less melted. Less transparent. Uneven coverage. Increased gloss.
Add EPK	10	Slight increase in opacity.
	20	Similar.
	30	Dry and unmelted surface.

The test with omission of EPK kaolin shows the contribution of alumina to the fusion of the glaze. The uneven coverage points to the absence of the suspending agent contained in the clay materials. Subsequent additions of alumina to a glaze already enriched by 23.6% kaolin predictably increase opacity and eventually produce a dried-up, unmelted surface.

Note: See Sanders Celadon and Rhodes porcelain tests (p. 260) in which the absence of high alumina kaolin changed the transparent, glossy, gray-green surface to a paler, satin-matt.

RHODES 32

Figures 4.2, p266; 4.8, p272.		Orange-yellow-brown. White where thick. Brown iron-spots. Opaque, stony.
Potash feldspar	48.9%	
Dolomite	22.4	
Whiting	3.5	
EPK kaolin	25.1	
Omit EPK		Gray-white flecked with tiny, yellow-white crystals. Opaque, stony, fluid, pitted.

The intense crystalline formation reflects the reduction of alumina by approximately 10% due to the absence of the kaolin. Dolomite and silica tend to form visible crystals. Alumina retards this kind of macrocrystalline growth and promotes the smooth, dense, stony surface of this glaze.

Omit EPK	12.5	Similar to above test. Crystal structure smaller. Denser, less-mottled surface. Less pinholing and fluidity.

MIRROR BLACK II

		High gloss. Opaque. Black.
Custer feldspar	25.8%	
Cornwall Stone	12.9	
Whiting	12.9	
Flint	31.4	
EPK kaolin	6.5	
Red iron oxide	9.5	
Omit EPK		Gold crystals on shiny black background where thick.

MAMO MATT

Figure 4.4, p268.		Mottled orange-green-brown-gray. Stony-matt.
G-200 Feldspar	49 grams	Crystalline formation where thick.
Dolomite	19	
Whiting	4	
Grolleg clay	21	
Tin oxide	8	
Copper oxide	1.5	
Add Grolleg clay	10	Denser surface with no breakup or crystalline formation.

3. *The ratio of alumina to silica in a glaze usually determines whether or not the surface will be gloss, transparent, opaque or matt* (see pp. 241, 244–246). At high stoneware temperatures, 5–6 molecules of silica to 1 molecule of alumina predicts a transparent gloss surface; 3–4 molecules of silica to 1 molecule of alumina suggests a satin-matt surface, and finally, less than 3 molecules of silica to 1 molecule of alumina results in the stony-matt opacity. Note that these predictions are subject to many exceptions and may be changed by the kind and amount of melters included in the glaze magma. For example, as the following chart shows, opaque, satin-matt Charlie D glaze has an alumina-silica ratio higher than high gloss, transparent Sanders Celadon!

PERCENTAGE ANALYSIS
CONE 9–10 REDUCTION

	Sanders* Celadon	Iron* Residue	Temple White	Jacky's Clear	Charlie D	Rhodes 32	Sam Haile	Y Body	G-200 Spar
SiO_2	68.9	67.8	63.0	59.8	67.0	54.3	55.3	66.8	66.81
Al_2O_3	13.8	13.0	18.0	13.7	12.1	22.2	18.5	27.5	18.4
Fe_2O_3								0.25	0.08
KNaO	5.5	8.5	6.0	8.4	6.0	8.5	8.0	2.0	13.67
CaO	11.8	8.8	10.0	6.0	8.4	11.6	12.8	0.11	0.81
MgO			3.0		6.4	3.3		0.33	TR
B_2O_3		1.8		1.5					
SnO							5.3		
ZnO			10.5						
TiO_2								1.4	
A/S	1:5	1:5	1:3.5	1:4	1:5.5	1:2.4	1:2.9	1:2.3	1:3.6
	Gloss	Gloss	Semi-Gloss	Gloss	Satin-Matt	Stony-Matt	Silky-Matt		

Base Glaze

CONE 5–6 OXIDATION

	White Matt*	Portchester Copper II	Jacky's Clear II
SiO_2	49.4	57.5	61.7
Al_2O_3	19.7	14.6	13.9
KNaO	9.4	10.0	8.8
CaO	10.9	14.3	7.2
B_2O_3			3.0
SnO_2		3.4	
ZnO	10.5		5.3
A/S Ratio	1:2.5	1:3.9	1:4.4
	Stony Matt	Satin Matt	Transparent Gloss

*Nepheline Syenite 54.10%

Whiting	17.37
Zinc oxide	9.40
Flint	5.20
EPK kaolin	13.90

4. *It is to be expected that a refractory material, such as alumina, would lighten color and reduce brilliance of a glaze surface, in direct contrast to silica.* Ten to twenty percent additions of Flint (silica) intensified the black of cone 9/10 reduction high-iron glazes and blued the celadon color of low-iron glazes (see Tests, pp. 249–250). In the following celadon glazes, additions of high-alumina kaolin have milked and lightened the gray-blue color. Note how the total omission of kaolin provides a similar result, for reasons discussed on pp. 256–257.

Cone 9–10 Reduction

Alumina produces even more dramatic color changes in certain glazes.

RHODES PORCELAIN

Figure 4.3, p267.		Gray-blue. Transparent Gloss. Craze.
(Porcelain claybody)		
Kona F-4	33%	
Whiting	20	
Flint	32	
EPK kaolin	15	
Omit EPK		Cloudy, milky white. Satin-matt.
Add EPK	6	Lighter gray-blue. Craze still visible.
	12	White-gray. Opaque. Semigloss. Crawled surface. No visible craze.
	18	Similar to above. Test too thinly applied.
	25	White. Dry. Unmelted. Sintering cracks.

SANDERS CELADON

Figure 1.5 p82.		Blue-green-gray. Transparent. Gloss.
Kona F-4	44%	
Whiting	18	
Flint	28	
EPK kaolin	10	
Barnard	5%	
Omit EPK		Paler color. Satin-matt.
Add EPK	20	Grayer, greener color.
	40	White-gray; yellow cast. Dry, sintered surface.
Add Flint	20	Milky, blue color.
	40	White-gray; blue cast. Sintered surface. Less dry than EPK test.

RON'S SATURATED IRON

		Dense black-brown gloss, edged in red-brown where thin.
Kingman feldspar	43%	
Whiting	13	
Flint	19	
EPK kaolin	15	
Red iron oxide	10	
Omit EPK		Mottled brown-black gloss with more brown than original glaze and more depth.

The following glaze contains an entirely different combination of glaze materials from the prior glazes. In this glaze, the opaque orange-gold-white surface lost its opacity and gold luster; it changed to a transparent, blue-gray surface (repeat test) with the omission of 25% EPK kaolin. *Hence, the color changes produced by the addition or omission of additional alumina may not be predictable. They depend on the particular oxide trinity produced by the combination of glaze materials in each specific glaze.*

LAU LUSTRE

Figure 1.11, p88.		
Nepheline Syenite	53%	Black-white-orange gold.
Flint	10	Lustrous sheen.
Whiting	2	Opaque.
Soda ash	5	
Lithium carbonate	5	
EPK kaolin	25	
Tin oxide	2	
Omit EPK		Not as gold. Runnier and glassier.
Repeat test.		Transparent. Blue-green-gray. Runnier and glassier.

Alumina has an equally strong effect on glaze surface color at lower stoneware, oxidation temperatures. A low alumina content in a cone 8, gray-white-yellow, satin-matt glaze resulted in a blue-yellow, gloss surface.

Cone 8 Oxidation

ELENA'S SATIN-MATT		
	Gray-white-yellow. Satin-matt.	
Custer feldspar	56 grams	
Whiting	20	
Zinc oxide	9	
Flint	6	
EPK kaolin	18	
Rutile	7	
Omit EPK	Blue-yellow, high gloss.	

And again, at the cone 5–6 oxidation temperatures, the omission of 5% kaolin resulted in a brighter, redder surface.

Cone 5–6 Oxidation

RANDY RED		
Figure 3.11, p233.	Red-brown-green. Opaque. Gloss.	
Kona F-4	20%	
Gerstley Borate	32	
Talc	14	
Flint	30	
EPK kaolin	5	
Red iron oxide	15%	
Omit EPK	Brighter red.	

5. *An excess amount of alumina in a glaze magma could inhibit the flow of the glaze and prevent the healing of fissures and ruptures caused by escaping gases.* The result would be a crawled and/or pinholed surface. There can be no general rule as to the exact amount of alumina that will cause a crawled surface. It all depends on the rest of the glaze materials in each specific glaze. The cone 9–10 reduction tests of the Frazier White glaze show how additions of 20% or more kaolin in this high-feldspar glaze create a pinholed and crawled surface. On the other hand, in the Temple White glaze, this kind of surface did not appear until the 30% additions (p. 257). At cone 5–6 oxidation temperatures, a crawled surface appeared in one test with the addition of only 10% alumina. Obviously, the amount of alumina that causes crawling depends on the rest of the glaze materials, as well as on the firing temperature of the glaze. As a general rule, fluid glazes that are high in sodium and potassium melters and low in clay and/or free silica will be able to absorb larger amounts of alumina without crawling. High surface tension, which is a characteristic property of alumina magma, is the primary cause of a crawled surface. Not only are sodium and potassium oxides powerful melters, but potassium materials also characteristically have low surface tension. Another significant factor is whether or not the glaze contains sizable amounts of silica. Silica promotes a nonfluid, viscous melt; hence, glazes high in silica will usually absorb less alumina than low silica glazes. For example, in the Rhodes Porcelain glaze described previously, a crawled surface results from 27% kaolin; yet other glazes, such as the smooth, stony-matt Rhodes 32, can absorb this amount of kaolin without producing a crawled surface. The difference lies in the silica content of these two glazes. Rhodes 32 contains no flint (free silica) in its formula. In Rhodes Porcelain, the high silica content (flint 32%) increases the viscosity of the flow, and this factor, combined with the high kaolin content, produces the crawled surface. *Note that different glaze cores contain varying amounts of silica and alumina in their oxide structure. Thus, the effect of the alumina additions on a glaze surface will depend partly on the already existing proportion of the alumina and silica in the glaze core.* In the following tests, Nepheline Syenite, which is high in alumina and low in silica, begins to pinhole at the 10% addition and crawls at the 20% addition. On the other hand, Cornwall Stone, which is lower in alumina and higher in silica, still presents an unflawed satin surface at the 20% addition; cracks and crawling do not begin until the 30% addition.

Cone 9/10 Reduction

NEPHELINE SYENITE		CORNWALL STONE	
Nepheline Syenite	80%	Cornwall Stone	80%
Whiting	20	Whiting	20
Blue-green glass. Yellow-brown matt. High craze. High flow.		Gray blue-green gloss. Transparent. Less (larger) craze. Viscous flow (fat roll at base).	
Add 10 EPK Kaolin			
White-yellow color. Stony-matt. Nonfluid. Pinholes. Crystalline ring in center.		Greener color. Transparent. Gloss. Less flow.	
Add 20 EPK Kaolin			
White (thick). Gray (medium). Brown (thin). Stony, matt. Nonfluid. Beginning crawl.		Paler, grayer green. Satin-matt. Decreased flow. Less visible craze. Semitransparent.	
Add 30 EPK Kaolin			
Yellow-gray-white-brown. Stony-matt. Increased crawl.		White-gray. Satin surface but stonier than 20% test. Opaque. Sintering cracks visible. Beginning crawl.	
Add 40 EPK Kaolin			
Yellow-white. Dry. Stony. Unmelted. Sintered. Cracked eggshell surface.		White. Dry and stony (less than Nepheline Syenite). Unmelted. Sintered. Cracked eggshell surface.	

6. *The source of alumina for glazes is primarily kaolin clays.* Ball clays also figure as a source, but as the following tests show, ball clays do not contain as much alumina as the kaolins. In addition, their higher iron and titanium content will cause color changes in a light-colored surface. Other clays such as fire clays also contain alumina, but here again, their unique characteristics will leave a mark on the surface. The following tests show some of these differences.

COMPARATIVE TESTS OF O.M-4 BALL CLAY AND EPK KAOLIN

Cone 9–10 Reduction

RHODES 32		
Figures 4.2, p266; 4.8, p272.		
		Orange-yellow-brown. White where thick. Iron spots. Stony-matt.
Potash Feldspar	48.9%	
Dolomite	22.4	
Whiting	3.5	
EPK kaolin	25.1	
Reduce EPK to	12.5	White-gray ringed with yellow-white crystals where thick; inside ringed with glassy-white-gray crystals. Stony-matt.
Substitute Ball clay	12.5	Yellow-white; interior ringed with glassy gray crystals. Satin-matt.
Add EPK	5	No visible change.
Add Ball clay	5	No visible change.
Add EPK	10	Slightly stiffer, and stonier. Whiter surface where thick. Still yellow-brown where thin.
Add Ball clay	10	Silkier, satin surface. Less white and less stony than Kaolin test and control glaze.
Add EPK	15	Similar to 10% test but slightly drier.
Add Ball clay	15	Browner cast and silkier surface than EPK test.
Add EPK	20	Similar to 10% test.
Add Ball clay	20	Similar to 15% test.

Note change in color with one-half the amount of EPK and change of both surface texture and color with Ball clay substitution for EPK kaolin. Ball clay has higher iron, titanium, and melter content than EPK.

CORNWALL STONE

I

Cornwall Stone	80%	Blue-gray.
Whiting	20	Gloss. Craze.
Add EPK	20	Gray-green. Satin-matt. Less visible craze. Less flow.
Add Ball clay	20	Increased green. Increased gloss. Increased craze. Increased flow.

II

Cornwall Stone	60%	Gray-green.
Whiting	20	Satin-matt.
EPK	20	
Substitute Ball clay for EPK.		Deeper gray-green; blue cast. Slight increase in gloss.

RHODES PORCELAIN

Figure 4.3, p267.

		Blue-gray. Transparent. Gloss. No craze.
Custer Feldspar	33%	
Whiting	20	
Flint	32	
EPK kaolin	15	
Substitute Ball for EPK.		Yellower cast to surface.

Once again, the deeper coloring and increased melting action of Ball clay is evident in these tests. For additional comparative tests see Chapter 2, Clays and Claybodies, pp. 114–116.

Notes

[1] Comparable results are obtained with the substitute of lower or higher silica glaze cores. *See Appendixes B1, B2.*

[2] *Ceramic Industry* 1998, 54, 70.

LAB VIII SILICA   ALUMINA

FLINT   BALL CLAY   EPK   KAOLIN

1. Mix your glaze without Flint.

2. Mix your glaze without the clay ingredient. In order to minimize the possibility of excess fluidity, glaze only the interior and top half of the exterior walls.

3. Add to your original glaze: 10%, 20%, 30%, 40% increments of:
   (a) Flint
   (b) The clay ingredient (EPK, or Ball clay).

4. Apply the test mixtures in 3(a) and 3(b) to test pots striped with cobalt, iron, and copper oxides on their outer and/or inner walls. Compare color of stripes in various tests.

5. One-half of each and every test should contain the original glaze as a control.

CONCLUSION

With the addition of alumina, the trip around the ceramic circle has been completed. It began with glaze core minerals (primarily feldspars and clays) that contain a compound oxide structure of glassmaker silica, adhesive alumina, and melters of potassium, sodium, calcium, and lithium oxides. These complex earth materials make up the heart and soul of our stoneware and porcelain claybodies and glazes. Additions of auxiliary melters—calcium-magnesium-zinc materials transformed the opaque, high gloss of these glaze cores into a transparent surface; larger additions of these melters opacified the glaze core's melted surface. Additions of the glassmaker silica (flint) and the adhesive alumina (kaolin and ball clays) influenced transparency, glaze fit, fluidity, and physical application of the glaze. Increased amounts of these alumina and silica minerals created their own unique form of opacity.

Note that the classification of a particular ceramic material as a glaze (or claybody) core, auxiliary melter, glassmaker, or adhesive, changes depending on the particular kind of surface desired and/or the firing temperature of the kiln. Thus, feldspars appeared as a glaze core in one kind of ceramic surface and firing temperature, and yet functioned in another as auxiliary melter, glassmaker, or adhesive. Multiple functions were also characteristic of other materials, such as the calcium-borate minerals. Hence, the point is

not to rely on rigid classifications, but rather to explore the general and most typical function of the materials at the various firing temperatures in both glazes and claybodies; take due note of all possible variations of function, and then utilize this knowledge to achieve the desired result.

Throughout this continuum, emphasis has been placed on the importance of the proportion of these materials in the creation of glaze surfaces. Thus, an entire spectrum of ceramic surfaces was created, ranging from transparent-gloss, to semitransparent gloss, to satin-matt semigloss, to stony-matt opaque, and finally, to dry-matt opaque, simply by adding or omitting varying amounts of these minerals. (A summary of the contrasting effects of silica, alumina, and melter materials on the glaze surface is diagrammed in Appendix B3). In all cases, the crucial relationship between this proportion and the achievement of surface colors has been stressed.

And finally, throughout this book, the ancient geological origins of ceramic materials and a sense of their ephemeral nature is an ever-recurrent theme. The aim has been to restore the lost connection between potters and the materials of their craft.

This completes our record of the long journey made by ceramic minerals as they leave the earth, enter the fire, and re-emerge in a new form. Understanding and respect for their kaleidoscopic journey will open the door to a new and joyous experience with the materials of your craft.

Figure 4.1. Rocks of silica and alumina (Collection of Department of Earth and Environmental Sciences, Columbia University, New York).

Bauxite (*front left and right*) and quartz (*front and back center*) surrounding obsidian (*center*) with two nodules of cristobalite.

Figure 4.2. Covered jar by author.

Stoneware, cone 9–10 reduction. Glaze: Rhodes 32 (Dolomite 22.4%; EPK kaolin 25.1%; no Flint).

Figure 4.3 Tests: Rhodes Porcelain.

Stoneware, cone 9–10 reduction.

Descending order:
1. Rhodes Porcelain glaze (33% Custer Feldspar).
2. Without Whiting (20%).
3. Without Flint (32%).
4. Without EPK kaolin (15%).

267

Figure 4.4 Stoneware shell by Sue Browdy.

Cone 9–10 reduction. Glaze: Mamo Matt (Dolomite 18.8%; Kaolin 20.8%; no Flint).

Figure 4.5 Tests.

Stoneware, cone 9–10 reduction.

Left: Temple White/Temple White without Flint (18.9%).

Center: Temple White/Temple White with substitution of Nepheline Syenite for G-200 feldspar (34.7%).

Right: Nepheline Syenite 80%, Whiting 20%, Flint 30/Nepheline Syenite 80, Whiting 20, no Flint.

Figure 4.6. *Left:* **Bowl** by Jacqueline Wilder.
Right: **Jar** by Sue Browdy.

Katherine Choy glaze. Cone 5–6 oxidation.

Bowl: Sheffield Grolleg Porcelain claybody.

Jar: Portchester red-brown claybody. Note extraordinary surface change from matt to gloss when claybody changes from the Portchester red-brown to the porcelain claybody.

Figure 4.7 Range of surface jars.

Cone 5–6 oxidation. Porcelain claybody.

Left: Satin-matt surface: Nepheline Syenite 80%, Wollastonite 20%.

Back: Gloss surface: Jacky's Clear.

Front: Matt surface: Ron's White Matt #5 (p.219).

Figure 4.8 Range of surface jars.

Cone 9–10 reduction.

Left: Gloss surface. Sanders Celadon (p. 24). Porcelain.

Front center: Semigloss surface. Temple White. Stoneware.

Right: Satin-matt surface. Charlie D Black (Black Mason Stain 3%, Manganese dioxide 3%, Cobalt oxide 5%). Stoneware.

Back center: Stony-matt surface. Rhodes 32. Stoneware (p. 158).

1. Using any of the glaze cores discussed in Chapters 1 and 2, together with any of the melter materials described in Chapter 3, and the free silica and alumina materials of Chapter 4, make up the following glazes:

(a) a gloss, noncrazed surface for cone 5–6 oxidation firing temperature.

(b) a gloss, noncrazed surface for the oxidation and/or reduction cone 9–10 firing temperature.

2. Change the above gloss surfaces into:

(a) a satin-matt surface (opaque-gloss).
(b) a stony-matt surface (opaque-nongloss).

You may change or add to any of the ceramic materials contained in your glaze except the glaze core, which should remain the same.

3. Adjust the cone 9–10 glaze that you have been working with for a cone 5–6 firing temperature and /or:

Adjust the cone 5–6 glaze that you have been working with for a cone 9–10 firing temperature.

You may add or substitute any of the glaze cores, and/or other ceramic materials covered in this book.

4. Make up a claybody for the cone 5–6 firing temperature. Try not to add any new materials, but rather, accomplish (a) through (c) below by altering the proportion of the materials in the cone 5–6 claybody.

(a) Adjust this claybody for the cone 9–10 firing temperature.

(b) Change the cone 5–6 claybody into a glaze for the cone 5–6 firing temperature.

(c) Change the cone 5–6 claybody into a glaze for the cone 9–10 firing temperature.

Use the charts in Appendixes B and C as a guide in solving problems 1–4.

5. Compute the molecular formula and the percentage oxide analysis of the glaze or glazes used as the control in your tests.

Follow the procedures outlined in Appendixes A.

6. Analyze the structure of your glaze in terms of the function of its component materials and their oxide structure.

Follow the analysis of the structure of Sanders Celadon in Appendix A2.

APPENDIX A1
GLAZE CALCULATION TECHNIQUES

Glaze calculation techniques are useful diagnostic tools for the solution of glaze and claybody problems. These techniques are particularly helpful when substitutions are sought for ceramic materials in glazes. The following methods provide molecular and ratio (percentage) analyses of glazes *without* the aid of a computer.

DEFINITIONS:

Batch Recipe: The ratio of the weight of the materials to the molecular weight.

Empirical Molecular Formula: The ratio of the molecules of each oxide contained in the materials of the recipe *after* kiln firing.

CONVERSION FROM BATCH RECIPE TO EMPIRICAL MOLECULAR FORMULA

Steps to convert a batch recipe to an empirical molecular formula are as follows:

1. List materials in the batch recipe and their percentage weight units.

2. Divide weight units of each material in batch recipe by the molecular weight[1] of that material, thus obtaining the number of molecules or percentage of one molecule of each material present in the glaze batch. This will establish ratio-multipliers of weights for all the materials in the glaze batch, in order to arrive at the ratio of the oxides.

Conversion from Batch Recipe to Empirical Molecular Formula
Sanders Celadon (Base)

Kona F-4	= 44 batch weight divided by 507 molecular weight = .087 ratio-multiplier
Whiting	= 18 batch weight divided by 100 molecular weight = .18 ratio-multiplier
EPK kaolin	= 10 batch weight divided by 258 molecular weight = .039 ratio-multiplier
Flint	= 28 batch weight divided by 60 molecular weight = .466 ratio-multiplier

Batch Recipe	Weight Units %	Molecular Weights	Ratio-Multiplier of Weights
Kona F-4	44	507	.087
Whiting	18	100	.18
EPK kaolin	10	258	.039
Flint	28	60	.466

[1] *Note: Current chemical data sheets, which contain the percentage oxide analyses, molecular formulas, and molecular weights of the ceramic materials in the batch recipe, can be obtained from your supplier. Similar but often less-current chemical analyses appear in the appendixes of general ceramic texts. Slight variations in molecular weights and oxide structures may occur in chemical data sheets of different years. See pp 279, 281. Note that the molecular formula of a ceramic material can be computed from its percentage oxide analysis (comparable to an oxide batch recipe) by following the procedure described on page 280.*

3. Make a chart, which we will call "Intermediate Molecular Formula", listing horizontally at the top, left to right, all the oxides introduced by the materials in the batch recipe. To the left on this chart list vertically all the materials in the batch recipe.

4. Multiply the ratio-multipliers by the values found in the oxide structure *(see Appendix A2, p281).*

Solution: Sanders Celadon

Consult the oxide analysis of materials in Sanders Celadon provided in APPENDIX A2, p.281, note 1. **Making use of the oxide structure information on Kona F-4, we multiply each oxide structure in Kona F-4 by the ratio-multiplier: .087.**

$$.087 \times 0.49 = .0428 \ Na_2O$$
$$.087 \times 0.31 = .0268 \ K_2O$$
$$.087 \times 0.20 = .0174 \ CaO$$
$$.087 \times 1.00 = .0878 \ Al_2O_3$$
$$.087 \times 5.5 \ = .4828 \ SiO_2$$

Next, using the ratio-multiplier 0.18, we multiply 0.18 x 1 for the oxide, CaO, in Whiting. We use the number 1 because Whiting is 97.05% $CaCO_3$.

To continue, using the ratio-multiplier 0.039, we multiply 0.039 x 1 for Al_2O_3 and multiply 0.039 x 2 for SiO_2, both components of EPK kaolin, because its structure for this glaze contains 2 silica for 1 alumina.

Finally, using the ratio-multiplier 0.466, we multiply 0.466 x 1, because Flint is 99.670% SiO_2.

Intermediate Molecular Formula[2]

	R_2O		RO	R_2O_3	RO_2
	Na_2O	K_2O	CaO	Al_2O_3	SiO_2
Kona F-4	.0428	.0268	.0174	.0878	.4828
Whiting			.18		
EPK kaolin				.039	.078
Flint					.466
	.0428			.0878	.4828
	.0268			.039	.078
	.0174				.466
	.18				
	.2670			.1268	1.0268

---

[2]*Note: Oxides contained in ceramic materials are divided into three main groups according to the amount of their oxygen atoms. Oxides, which contain one atom of oxygen, are classified as RO-R_2O. This group includes the primary melters, such as Na_2O, K_2O, and CaO. Oxides with a 1:2 oxygen ratio are classified as $RO_2(SiO_2)$; and finally, oxides with a 2:3 oxygen ratio (Al_2O_3) are known as the R_2O_3 group.*

5. To obtain the molecular formula:

a) Convert the RO-R$_2$O oxides to one or unity.

Total the RO-R$_2$O columns.

Divide each number in these columns by the RO-R$_2$O combined numerical total.

b) Express the remaining oxides as multiples of .2670.

      1. Express R$_2$O$_3$ in terms of RO-R$_2$O.

      2. Express RO$_2$ in terms of RO-R$_2$O.

This allows the total oxide structure to be expressed in terms of the sum RO-R$_2$O, namely .2670. Thus, each R$_2$O, RO, R$_2$O$_3$, and RO$_2$ is a multiple of the sum RO-R$_2$O (.2670).

Solution: We sum the values of Na$_2$O, K$_2$O, and CaO.

    Na$_2$O   .0428
    K$_2$O   .0268
    CaO   .0174
            .18
            .2670

Then, divide each *part* of the sum by the sum.

    .0428/.2670 = .16   (Na$_2$O)
    .0268/.2670 = .10   (K$_2$O)

Adding: .0174 + .18 = .1974
            .1974/.2670 = .74   (CaO)

Summing under Al$_2$O$_3$ and dividing again by .2670, we have:

    .0878 + .039 = .1268
    and: .1268/.2670 = .475   (Al$_2$O$_3$)

Finally, under SiO$_2$, we find:

    .4828 + .078 = 1.0268
    and: 1.0268/.2670 = 3.85   (SiO$_2$)

<u>The empirical molecular formula is, therefore:</u>

Na$_2$O	.16	
K$_2$O	.1	} 1 (unity)
CaO	.74	
Al$_2$O$_3$	.475	
SiO$_2$	3.85	

CONVERSION FROM EMPIRICAL MOLECULAR FORMULA TO BATCH RECIPE

Steps to convert empirical molecular formula to a batch recipe are as follows:

1. List each oxide of molecular formula horizontally, left to right, and arrange them in RO-R$_2$O, RO, R$_2$O$_3$, and RO$_2$ columns on left side of chart.

2. On right side of chart, place four columns headed as follows: molecular equivalent, molecular weight, weight units, and % batch recipe.

Molecular equivalent:

This is the percentage of the molecule of the raw material that you need to satisfy the oxide requirement in the molecular formula.

For compound materials set up an equation:

$$\frac{\text{Formula oxide requirement}}{\text{Oxide amount in raw material}} = \text{X (molecular equivalent)}$$

Example: Sanders Celadon molecular formula requires 0.16 Na$_2$O. Using Kona F-4 for its Na$_2$O content, you need only an amount of Kona F-4 that will give you 0.16 Na$_2$O. One molecule of Kona F-4 yields 0.492 Na$_2$O. To find what amount of Kona F-4 yields 0.16 Na$_2$O, set up the following equation:

$$\frac{.16*}{.492**} = \text{X}*** = .33 \times 100 = 33\%, \text{ the molecular equivalent}$$

Notes:
*Amount of Na$_2$O in molecular formula.
**Amount of Na$_2$O in Kona F-4 molecular formula (See p. 281, note 1).
*** % of Kona F-4 that yields 0.16 Na$_2$O.

Solution:

3. First, in the top row write the values of the oxides in Sanders Celadon (Base). Making use of Kona F-4,[3] (which has five oxides), multiply each of the oxide components of Kona F-4 by .33, the molecular equivalent.

$$.33 \times .48 = .16 \quad (\text{Na}_2\text{O})$$
$$.33 \times .32 = .1 \quad (\text{K}_2\text{O})$$
$$.33 \times .20 = .07 \quad (\text{CaO})$$
$$.33 \times 1 = .33 \quad (\text{Al}_2\text{O}_3)$$
$$.33 \times 5.6 = 1.83 \quad (\text{SiO}_2)$$

Thus, we get row 2 in Kona F-4[3]. Subtract row 2 from the top row. As a result Kona F-4 has a remainder of 0, 0, .67, .145, and 2.02 to be satisfied by the remaining three materials—Whiting, EPK kaolin, and Flint.

---

[3]**Note:** Kona F-4 Molecular formula used in the above solution.

Na$_2$O	.48	
K$_2$O	.32	} 1
CaO	.20	
Al$_2$O$_3$	1.0	
SiO$_2$	5.6	

4. Using the molecular equivalent 0.33, multiply the molecular weight 509 of Kona F-4, arriving at the product 167.9 weight units.

5. One molecule of Whiting ($CaCO_3$) for the purpose of this computation represents 100% CaO. Hence, 0.67 of Whiting satisfies the CaO requirement. Now 0.67 × the molecular weight 100 results in the product weight unit of 67.

6. For the EPK kaolin row, bring down from the Kona F-4 row the value 0.145 Al_2O. As previously stated, EPK kaolin with respect to SiO_2 and Al_2O_3 has a relation of 2 to 1 respectively. Hence, from the 0.145 EPK we get 0.29 SiO_2 and 0.145 Al_2O_3. The 0.145 is eliminated by subtraction.

7. Now multiply the molecular weight 258 by 0.145, arriving at the product 37.41.

8. Subtracting 0.29 within the EPK kaolin row from 2.02 in the Kona F-4 row, we are left with the molecular equivalent, 1.73 SiO_2 in the Flint row. Finally, 1.73 × 60 gives a product of 103.8. Sum all products in the weight units column. The sum is 376.11.

9. We now have:

167.9/376.11 = .446 or 44%   (We ignore .006)

67/376.11   = .18 or 18%

37.41/376.11 = .10 or 10%

103.8/376.11 = .28 or 28%

Conversion from Molecular Formula to Batch Recipe

Molecular Formula
Sanders Celadon (Base)

Na_2O	.16
K_2O	.1
CaO	.74
Al_2O_3	.475
SiO_2	3.85

	Na_2O	K_2O	CaO	Al_2O_3	SiO_2	Molecular Equivalent	Molecular Weight	Weight Units	% Batch Recipe
Kona F-4[3]	.16 (.16)	.1 (.1)	.74 (.07)	.475 (.33)	3.85 (1.83)	.33	509[4]	167.9	44
	0	0	.67	.145	2.02				
Whiting			.67			.67	100	67	18
			0						
EPK Kaolin				.145	.29	.145	258	37.41	10
				0	1.73				
Flint					1.73	1.73	60	103.8	28
					0				
Total								376.11	

---

[3]**Note:** Kona F-4 Molecular formula used in above computation.

Na_2O	.48
K_2O	.32
CaO	.20

} 1

Al_2O_3	1.0
SiO_2	5.6

[4]Prior computation using earlier data sheet gave 507 as the molecular weight. See p. 275, note 1.

PERCENTAGE ANALYSIS METHOD

This method is by far the easiest and most comprehensible of the three methods. It was introduced to Brookfield Craft Center students by Richard Zakin, as part of a proposed ceramics text (Zakin 1978). This method analyzes the materials in a glaze recipe in terms of ratio of the weight of their various oxide components, as opposed to the ratio-multiplier of each oxide. The chemical analysis sheets provided by your supplier for ceramic materials contain analyses of each material in terms of the percentage weight of its component oxides.

Steps to arrive at percentages weight of oxides:

1. In previous calculations we move vertically. Now we progress horizontally for variety. We sum the oxides horizontally. Here we consider a ratio of a sum of weight of individual oxides to the total sum of all the oxides in the material.

PERCENTAGE ANALYSIS Sanders Celadon (Base)						
Batch Recipe Glaze Material	**Kona F-4**	**Whiting**	**EPK Kaolin**	**Flint**	**Total**	**%Total**
Weight Units	44	18	10	28		
Oxides						
SiO_2	29.4		4.5	28	61.9	68.7
Al_2O_3	8.6		3.7		12.3	13.7
Na_2O	3.0				3.0	3.3
K_2O	2.2				2.2	2.4
CaO	.7	10.0			10.7	11.9
Total					*90.1*	*100*

A/S ratio is 1:5

$$\frac{Al_2O_3}{SiO_2} = \frac{13.7}{68.7} = \frac{1}{5}$$

The sum of the following oxides is:

SiO_2	=	29.4 + 4.5 + 28	=	61.9
Al_2O_3	=	8.6 + 3.7	=	12.3
Na_2O	=	3.0	=	3.0
K_2O	=	2.2	=	2.2
CaO	=	.7 + 10.0	=	10.7

The total number of oxides is 90.1.

Now we consider:

SiO_2	=	61.9/90.1	=	.687	= 68.7%
Al_2O_3	=	12.3/90.1	=	.137	= 13.7%
Na_2O	=	3.0/90.1	=	.033	= 3.3%
K_2O	=	2.2/90.1	=	.024	= 2.4%
CaO	=	10.7/90/1	=	.1187	= 11.9%

APPENDIX A2

ANALYSIS OF STRUCTURE OF SANDERS CELADON

	Glaze Core	Auxiliary Melters	Auxiliary Glassmaker	Auxiliary Adhesive	Colorant
Ceramic Material	Kona F-4 Feldspar	Whiting	Flint	EPK Kaolin	Barnard Clay
Batch Recipe	44%	18%	28%	10%	3%–5%
Glaze Function	Provides the core of the glaze surface.	Increases fusion. Encourages blue-green celadon color.	Promotes craze-free surface.	Improves physical application. Increases adhesion of glaze magma. Increases transparency.	Provides the iron colorant that creates blue-green celadon.
Oxide Structure[1]	SiO_2 5.6 Al_2O_3 1.0 Na_2O .48 K_2O .32 CaO .20	$CaCO_3$[2] 1	SiO_2 1	SiO_2 2 Al_2O_3 1	Fe_2O_3 SiO_2 Al_2O_3

[1] *The slight variations in Kona F-4 oxide structure (5.6 versus 5.549, 0.48 versus 0.492, and 0.32 versus 0.308) are due to Technical Data Sheets of different years. In order to get the intermediate values as shown in the Sanders Celadon solution, p. 276, for the ratio-multiplier, .087, we multiply the Kona F-4 oxide structure in the following manner:*

$$.087 \times .492 = .0428 \ Na_2O$$
$$.087 \times .308 = .0268 \ K_2O$$
$$.087 \times 5.549 = .4828 \ SiO_2$$

[2] *Whiting (calcium carbonate) is a pure form of calcia (calcium oxide).*

APPENDIX B1

ADDITIONAL GLAZE CORE SUBSTITUTION TESTS

Cone 9–10 Reduction

PAM'S SATURATED IRON

Black gloss.

Custer Feldspar	52.4%
Whiting	10.4
Gerstley Borate	6.2
Flint	17.0
EPK kaolin	5.6
Red iron oxide	8.6

Omit Custer Feldspar   Red-brown-metallic.
                             Opaque. Stony.

Substitute for Custer:
a. Kona F-4   Greener color.

Once again, a less black and greener color occurred with the substitution of Kona F-4. Its chemical analysis at the time of the test showed a lower glass content and a higher alumina and soda content than the original glaze core; this is a possible reason for the color change. The surface otherwise is not noticeably different.

b. Nepheline Syenite   Increased green-yellow.
                            Lower gloss.

The increased green color of the surface reflects the lower silica and higher alumina and soda content of this substituted glaze core.

c. G-200   Less black color.

The lower glass-silica content of this substituted glaze core could account for the change to a lessened black color.

d. Albany Slip clay   Yellowest color.

This glaze core has the lowest glass content of any of the above glaze cores.

This fact combined with its higher calcium-magnesium content promotes the yellow cast of the iron colorant.

LORD'S MATT

Gray. Transparent. Stony.

G-200 Feldspar	54.3%
Whiting	24.7
Ball clay	21.0

Omit G-200 glaze core   Dry, mottled.
                            Yellow-green.

Substitute for G-200:
a. Kona F-4 Feldspar   Similar surface.

Test confirms Kona F-4's similarity to potash feldspars and shows how substitution of one for the other can often be made without causing dramatic surface changes.

b. Nepheline Syenite   Yellow-brown.
                            Opaque. Stonier.

The changes in this surface are caused by the lower glass content and higher alumina content of Nepheline Syenite and is a predictable result whenever this glaze core substitutes for a higher-glass, lower alumina feldspar.

c. Cornwall Stone   Blue. Gloss. Transparent.

This substitution has just the opposite effect from the Nepheline Syenite substitution. A glaze core with increased silica has now been substituted for G-200. The increased gloss and bluer color of the surface are a characteristic result of increased glass content.

d. C-6 Feldspar (soda)   Slightly bluer color. Stony.

The slightly different surface of this test compared to the Kona F-4 test result (C-6 has but a slight increase in soda, potash, and glass content) could be due to its thicker application. *Differences in glaze application can often result in more variation than substitution of different materials.*

MAMO MATT

	Green, orange-black.
	Mottled. Opaque.
	Stony.
G-200 Feldspar (potash)	49.0 grams
Dolomite	19.0
Whiting	4.0
Grolleg	21.0
Tin oxide	8.0
Copper oxide	1.5

Omit G-200 Feldspar   Yellow-brown.
Rough. Dry.

Substitute for G-200:

a. Kona F-4   Similar.

Note similarity once again of Kona F-4 and G-200 feldspars when one is substituted for the other.

b. Nepheline Syenite   Yellow-brown-black.

Lower silica and higher alumina content of Nepheline Syenite produces a characteristic yellow color and drier surface.

c. Cornwall Stone   Lighter color.
White speckles.
Satin surface.

The higher silica content of Cornwall Stone has changed the normally stony surface of this glaze to a satin-matt

COMPARATIVE TESTS OF LITHIUM, SODA, AND POTASH GLAZE CORE SUBSTITUTIONS

Cone 9–10 Reduction

CASCADE

	Orange-red.
	White where thick.
	Stony-matt.
Custer Feldspar (potash)	30%
Spodumene (lithium)	20
Dolomite	22
Whiting	2
EPK kaolin	22
Tin oxide	6

For Custer Feldspar, substitute:

a. Kona F-4   Similar to control.
Slight increase in red.

Note that the least change once again was caused by the Kona F-4 substitution, and that its slightly increased soda content caused a somewhat redder shade.

b. Nepheline Syenite   Browner color.

The higher alumina and lower silica content of Nepheline Syenite, once again, moved the color toward the brown shade.

c. Cornwall Stone   Green-gray cast.

The higher silica content of Cornwall Stone produced increased transparency; this, in turn, produced a grayish, greener surface color. (The underlying claybody color is gray, and the change in surface color from red-orange to green-gray reflects the influence of the claybody and thus indicates transparency.)

d. Spodumene   Orange-brown.
e. Lepidolite   Green.
f. Petalite   Green.

Surprisingly, the same green color change of the Cornwall Stone substitution occurred with the Lepidolite substitution despite its low silica content. One possible explanation could be that increased transparency in this case was caused by increased fusion produced by the powerful lithium-potash-fluorine melters in Lepidolite. The same green color result appeared in the Petalite substitution. In this case, there is both increased glass content as well as the powerful melting action of the lithium melter. Note the contrary result in the Spodumene substitution. Although this glaze core contains a higher proportion of lithium than Lepidolite, it contains a lower total melter and silica content.

Whatever the reasons for these diverse results, these tests point to a single overwhelming conclusion—the oxide balance of this glaze is highly sensitive, and with the possible exception of Kona F-4, a change in glaze core produces visible changes in surface color.

Cone 8 Oxidation

SHINO CARBON TRAP

Reddish-orange edged in black. White, pinholed, crawled where thick. Opaque. Luster.

Nepheline Syenite	45.0%
Kona F-4	10.8
Spodumene	15.2
Soda ash	4.0
Ball clay	15.0
EPK kaolin	10.0

Substitute for Nepheline Syenite:

a. Kona F-4 — Shiny white. No red. No luster.

b. Custer — Pale orange-white. Stiffer surface. Bubbled texture. No luster. Repeat test similar.[1]

c. Cornwall Stone — Shiny white. No red. Stiffer. Crawled. No luster.

d. Lepidolite — White-brown-red milky surface. Pinholes. No luster.

e. Spodumene — Increased red-orange. Drier and more pitted. Crawled where thick. Repeat test similar to Lepidolite test but stiffer and more crawling.[1] Larger pinholes. Orange where thin; stiff, milky, white-brown-red where thick.

The oxide balance of this glaze is highly sensitive. Any increase in silica and decrease in soda/and or melter content causes a corresponding loss of the red-orange color and a stiffer, less-melted and more pinholed and crawled surface.

[1] *Tests were repeated in a different studio, using different supplies.*

ELENA'S SATIN-MATT

Gray-white-yellow. Satin-matt.

Custer Feldspar (potash)	56.0%
Whiting	20.0
Flint	6.0
EPK kaolin	18.0
Zinc oxide	9.0
Rutile	7.0

Omit Custer Feldspar	Yellow, dry surface.
Omit one-half Custer Feldspar	Paler yellow. Dry.
Add 24 Custer Feldspar	Speckled gray. High gloss.

Substitute for Custer Feldspar:

a. Kona F-4 — Drier surface.

b. Cornwall Stone — Gray-white. No yellow. White flecks.

c. Lepidolite — Light yellow. Increased dryness.

d. Spodumene — Brown-yellow. Stonier surface but silkier than Lepidolite test.

These tests clearly show the importance of the silica-alumina ratio with respect to color and surface.

Note the drier surface that appeared whenever the silica content was reduced by the substitution of a lower silica, higher alumina glaze core. Both the silica content and the type of melter in the glaze core clearly affect the final color of the surface. The Cornwall Stone substitution with its increased silica content eliminated the yellow color of the original surface and produced an overall gray-white, spotted color. Note the increased yellow-brown color with the substitution of the lower silica, lithium melter, Lepidolite, and Spodumene substitutions.

APPENDIX B2
CHART OF GLAZE CORE SUBSTITUTION TESTS

Cone 9–10 Reduction

		POTASH FELDSPAR	KONA F-4 FELDSPAR	NEPHELINE SYENITE
SANDERS CELADON BASE				
Kona F-4	44%	No visible difference.	*Original Glaze.*	Greener color.
Whiting	18	Slightly bluer color.	Gray-green-blue.	More iron spots.
Flint	28		Transparent.	Carbon trap (black spotting).
EPK kaolin	10		Gloss.	Crazed.
			No craze on stoneware.	
RON'S SATURATED IRON				
Kingman Spar (potash)	43%	*Original Glaze.* Black-brown.	Oranger color. Glossier.	Increased orange. Gloss.
Whiting	13	Gloss.		
Flint	19			
EPK kaolin	15			
Red iron oxide	10			
PAM'S SATURATED IRON				
Custer spar	52.4%	*Original Glaze.*	Greener cast.	Increased green and yellow.
Whiting	10.1	Black-brown.		Lower gloss.
Gerstley Borate	6.2	Gloss.		
Flint	17.0			
EPK kaolin	5.6	with G-200: more brown,		
Red iron oxide	8.6	less black.		
MIRROR BLACK I				
Custer spar (potash)	56.0%	*Original Glaze.* Dense, shiny black.	Greenish-black-brown. Pitting.	Greener color. Increased pitting.
Whiting	16.0		Increased flow.	Increased flow.
Flint	20.5			Matt-shine surface.
Ball clay	7.5			
Red iron oxide	5.0			
TEMPLE WHITE				
G-200 spar (potash)	34.7%	*Original Glaze.* White-gray.	Subtle difference. Slightly greener	Yellow-brown. Matter surface.
Dolomite	19.6	Semigloss.	cast. Lower gloss.	Lower gloss.
Whiting	3.1	Semitransparent.	More satin.	Increased iron-spotting.
Flint	18.9	Satin surface.		Similar to Rhodes 32.
EPK kaolin	23.6			
LORD'S MATT				
G-200 spar (potash)	54.3%	*Original Glaze.* Gray.	Similar.	Yellow-brown. Stonier surface.
Whiting	24.7	Stony.		
Ball clay	21.0			

CHART OF GLAZE CORE SUBSTITUTION TESTS, (CONTINUED)

CORNWALL STONE	LEPIDOLITE	SPODUMENE	PETALITE
Milkier, paler color. Less transparent. Lower gloss.	Yellower color. Milkier surface. Opaque. Increased flow.	Yellow-brown-green. More opaque. Increased flow. Shivering.	Greener and browner color. Shivering.
Blacker color. Satin surface. Reduced gloss.			
Green-black. Pitting. Nonfluid. High gloss. Closest to original glaze.			
Greener color. Gray-green. Increased gloss. Transparent.		Yellow-brown. Lower gloss.	
Bluer color. Increased transparency. Increased gloss.			

CHART OF GLAZE CORE SUBSTITUTION TESTS, (CONTINUED)

		POTASH FELDSPAR	KONA F-4 FELDSPAR	NEPHELINE SYENITE
RHODES 32				
Custer spar (potash)	48.9%	*Original Glaze.* White (thick). Orange-red spots (thin).	Oranger color. Similar surface.	Greener cast. Drier, stonier surface.
Dolomite	22.4			
Whiting	3.5			
EPK kaolin	25.1			
MAMO MATT				
G-200 spar (potash)	49.0 grams	*Original Glaze.* Blue-green-orange-black. Mottled. Opaque, stony surface.	Similar.	Yellow-brown-black color. Drier surface.
Dolomite	19.0			
Whiting	4.0			
Grolleg	21.0			
Tin oxide	8.0			
Copper oxide	1.5			
CASCADE				
Custer spar (potash)	30.0%	*Original Glaze.* Orange-red-White (thick). Stony matt.	Redder color. Similar surface.	Browner. Not as orange-white.
Spodumene	20.0			
Dolomite	22.0			
Whiting	2.0			
EPK kaolin	22.0			
Tin oxide	6.0			
SHINO CARBON TRAP				
Nepheline Syenite	45.0%	Pale orange-white. Stiffer surface. Bubbled. No luster.	White color only. No red. Gloss. No luster.	*Original Glaze.* Black on rim (carbon trap). Reddish-orange-white (thick). Pinholed and crawled. Opaque.
Kona F-4	10.8			
Spodumene	15.2			
Soda ash	4.0			
Ball clay	15.0			
EPK kaolin	10.0			

CHART OF GLAZE CORE SUBSTITUTION TESTS, (CONTINUED)

CORNWALL STONE	LEPIDOLITE	SPODUMENE	PETALITE
Bluer cast. Higher gloss. Reduced opacity.			
Lighter color. Speckle of white spots. Increased satin surface.			
Greener color.	Green.	Orange-brown.	Green.
White. Gloss. Stiffer. Crawled. No luster.	White-red-brown. Milky surface. Pinholed. No luster.	Increased red color.	Similar color to Lepidolite test. Stiffer surface.

CHART OF GLAZE CORE SUBSTITUTION TESTS, (CONTINUED)

Cone 5–6 Oxidation

		POTASH FELDSPAR	KONA F-4 FELDSPAR	NEPHELINE SYENITE
ALEXANDRA'S GLAZE				
Cornwall Stone	40%	Drier surface.		*Original Glaze.*
Nepheline Syenite	40	Loss of orange		Pearly-white.
Whiting	20	singe on edge of glaze.		Orange singe at edge.
				Satin-matt[1]
TRANSPARENT BLUE				
Potash spar	44.0 grams	*Original Glaze.*	Greener color.	Greener color.
Colemanite	20.0	Blue color.	No surface change.	Craze.
Whiting	1.0	Gloss.	Gloss remains.	No surface change.
Zinc oxide	3.0	Transparent.		Gloss remains.
Flint	24.0			
EPK kaolin	1.0			
Cobalt carbonate	1/2%			
Copper carbonate	4%			
KATHERINE CHOY SILKY MATT[2]				
Nepheline Syenite	53.9%	Shinier surface.	Drier surface.	*Original Glaze.*
Whiting	11.7			Off-white, yellow.
Lithium carbonate	4.7			Stony-satin matt.
Zinc oxide	10.3			
Flint	1.5			
EPK kaolin	17.9			
RANDY RED				
Kona F-4	20%	Overall green	*Original Glaze.*	Test I: Yellow-red spots.
Gerstley Borate	32	with red flecks.	Red-green. Brown where	Lowered gloss.
Talc	14	Lower red color.	thin. Redder where thick.	Test II: Brighter green.
Flint	30	Lower gloss.	High gloss.	Muddier red.
EPK kaolin	5			
Red iron oxide	15			
PORTCHESTER COPPER				
Nepheline Syenite	55%	Greener color.	Same as Custer feldspar	*Original Glaze.*
Whiting	25	Shinier surface.	substitution.	Black-green.
Flint	10	(Higher silica content		Satin-matt.
EPK kaolin	7	of Custer feldspar.)		
Tin oxide	3			
Copper oxide	2			
CORNWALL STONE GLAZE				
Custer Feldspar	43.6%	*Original Glaze.*	White. Stony.	White-yellow.
Cornwall Stone	21.8	White. Semigloss.	Pinholed.	Grainy texture.
Whiting	17.8	Semitransparent.		
Zinc oxide	7.9			
Titanium Dioxide	4.0			
EPK kaolin	4.9			

Cone 8 Oxidation
ELENA'S SATIN MATT

		POTASH FELDSPAR	KONA F-4 FELDSPAR	NEPHELINE SYENITE
Custer spar	56 grams	*Original Glaze.*	Drier surface.	
(potash)		Gray-white-yellow.		
Whiting	20	Satin-matt.		
Flint	6			
EPK kaolin	18			
Zinc oxide	9			
Rutile	7			

[1] *Later test of glaze on 5/6 porcelain produced a stony surface.*
[2] *Later test of glaze on 5/6 porcelain produced gloss surface.*

CHART OF GLAZE CORE SUBSTITUTION TESTS, (Continued)

CORNWALL STONE	LEPIDOLITE	SPODUMENE	PETALITE
Greener color. No other surface change. Gloss remains.	Deep green-black. Increased gloss.	Green-black. Satin-matt.	Blue-green-gray. Satin-matt.
Lower yellow. Increased flow (thick).	Yellow color. Increased flow.	Brown yellow. Cracked rim.	High gloss. Pinholes.
Overall mahogany red. Satin-matt. Lowered gloss.	Yellow cast. Fluid. *Repeat:* Blue red. *Repeat:* Greener.	Yellow cast. *Repeat:* Yellow-red-blue. *Repeat:* Purple-green-lime.	Brown gloss. *Repeat:* Mahogany. *Repeat:* Satin-matt.
Paler, bluer color. Less green. Lowered gloss. (Lower melter content of Cornwall Stone.)	Brilliant blue-green. Pitted.	Bluer color. Increased flow. Matt-shine.	
	Yellow. Dry. Stony	Light yellow. Drier surface	Pale yellow-white. Flecks of white. Stony but not as dry as other tests. Increased flow.
Gray-white. Flecks of white.	Light yellow. Very dry surface.	Brown-yellow. Silkier surface.	

APPENDIX B3

ANALYSIS OF STONEWARE GLAZE SURFACE (CONE 9–10)

GLAZE SURFACE DETERMINED BY THE RATIO OF:

- **GLAZE CORE•** (glass-glue-melter)
  - ■ ● ◆ Feldspar / Feldspathic rocks
- **AUXILIARY GLASS-MAKER •** (SiO$_2$)
  - ◆ Flint
- **AUXILIARY ADHESIVE GLUE •** (Al$_2$O$_3$)
  - ● Kaolin / Ball Clay
- **AUXILIARY MELTERS** CaO/MgO
  - ● Calcium-Magnesium Minerals*

	Smooth, Stony-Matt	Dry, Stony-Matt	Satin-Matt	Opaque-Gloss	Transparent-Gloss
Feldspar	50	35	40	34.5	35
CaO/MgO*	25	10	30	23	20
Flint	—	25	20	19	30
Kaolin	25	30	10	23.5	15
High	●	●	■	●	■
Medium	■	■	◆	■	●
Low	◆	◆	●	◆	◆

Smooth, Stony-Matt: Al$_2$O$_3$ dominates surface. Medium-high melter content melts some Al$_2$O$_3$ to form tiny dense crystals. Light trapped and absorbed by dense, micro-crystalline Al$_2$O$_3$ and remaining unmelted particles. The surface appears stony-matt opaque.

Dry, Stony-Matt: Unmelted particles dominate surface. Melter content insufficient to melt SiO$_2$ and Al$_2$O$_3$.

Satin-Matt: CaO/MgO-SiO$_2$ crystals dominate. Large crystals form due to low Al$_2$O$_3$ content. Light is reflected off crystal facets. Result is satin-matt opacity.

Opaque-Gloss: Medium-high melter content melts SiO$_2$ into glass and accounts for gloss. Dense, micro-crystalline surface of some melted Al$_2$O$_3$ plus remaining unmelted SiO$_2$ and Al$_2$O$_3$ particles create opacity and whiteness of surface. Higher fire or increased melter could render surface transparent.

Transparent-Gloss: Melted silica dominates. Melters high enough to melt large amount of Flint into SiO$_2$ glass. Sufficient Al$_2$O$_3$ to inhibit formation of large crystals which would inhibit transparency and gloss. All glaze materials are in correct balance for a transparent and gloss surface.

*Whiting and/or Wollastonite and/or Dolomite and Talc or Magnesium Carbonate.

APPENDIX C1

PRIMARY FUNCTION OF COMMON CERAMIC MATERIALS IN CLAYBODIES AND GLAZES

CERAMIC MATERIAL	GLAZE FUNCTION	CLAYBODY FUNCTION
Albany Slip clay	Glaze core	Colorant
A.P. Green clay	Texture (ST)	Open body Texture Core (ST)
Ball clay	Alumina Opacity	Plasticity
Barnard clay	Glaze core Color (ST,P)	Color
Bone ash	Opacifier	Melter (4–6)
Borax	Melter (5–6, E) Glassmaker Carbon Trap (ST, P)	Melter (4–6, E)
Boric Acid	Melter (5–6, E) Glassmaker	Melter (4–6, E)
Boron Frits	Glaze core (5–6, E) Melter Colemanite (s) Gerstley Borate (s)	Melter (4–6, E)
Colemanite	Glaze core (5–6, E) Melter Gerstley Borate (s) Boron Frits (s)	Melter (4–6, E)
Cornwall Stone	Glaze core (ST,P) (Low melter, high SiO_2)	Melter (P)
Dolomite	Melter(ST) Opacifier Whiting (s)	Melter (ST)
EPK kaolin	Alumina Opacity (ST, P)	Core (P, W)
Flint (silica)	Glassmaker	Glassmaker Glaze-fit

PRIMARY FUNCTION of COMMON CERAMIC MATERIALS
in CLAYBODIES and GLAZES (Continued)

CERAMIC MATERIAL	GLAZE FUNCTION	CLAYBODY FUNCTION
Fluorspar	Melter	
Gerstley Borate	Glaze core (4–6, E) Melter Colemanite (s) Boron Frits (s)	Melter (4–6, E)
Gold Art clay		Core (ST)
Kentucky Stone		Core (ST)
Lepidolite	Lithium glaze core	Melter (FL)
Magnesium Carbonate	Melter (ST, P, W) Opacifier	Melter
Nepheline Syenite	Glaze core (low SiO_2) (high Na_2O) (high Al_2O_3)	Melter (ST, P)
Ocmulgee Red clay	Color(ST)	Color Melter Core (E 4–6)
Petalite	Lithium glaze core (ST 9–10)	Melter (FL)
POTASH SPARS	*GLAZE CORE (ST, P)*	*MELTER (ST, P)*
Custer G-200 K200	G-200, K200 (s) Custer, K200 (s) Custer, G-200 (s)	
Red Art	Color	Melter Color Core (E)
Rotten Stone	Glaze core (ST, P)	Melter Color (ST)
SODA SPARS	*GLAZE CORE (ST, P)*	*MELTER (ST, P)*
Kona F-4	C–6 (s)	
C–6	Kona F-4 (s)	

PRIMARY FUNCTION of COMMON CERAMIC MATERIALS
in CLAYBODIES and GLAZES (CONTINUED)

CERAMIC MATERIAL	GLAZE FUNCTION	CLAYBODY FUNCTION
Spodumene	Lithium glaze core (ST, P)	Melter (FL)
Talc	Melter Opacifier	Melter (E, 4–6, W)
Volcanic ash	Glaze core (ST, P) Cornwall Stone (s)	
Whiting	Melter (ST, P) Opacifier Wollastonite (s) Dolomite (s)	Melter (ST)
Wollastonite	Melter (ST, P) Opacifier Whiting (s) Dolomite (s)	Melter (ST, P)
Wood Ash	Glaze core (ST, P) Melter (ST, P) Colorant	
Zinc	Melter (ST, P) Opacifier (ST, P)	

KEY: (s)=substitute option
(E)=earthenware claybody
(ST)=stoneware claybody
(P)=porcelain claybody
(FL)=flameware claybody, c/9-11
(W)=White-burning claybodies, c/4-10

295

APPENDIX C2

MODEL OF MOLECULAR STRUCTURE OF IDEAL FELDSPAR, CLAY, CLAYBODY, AND GLAZES

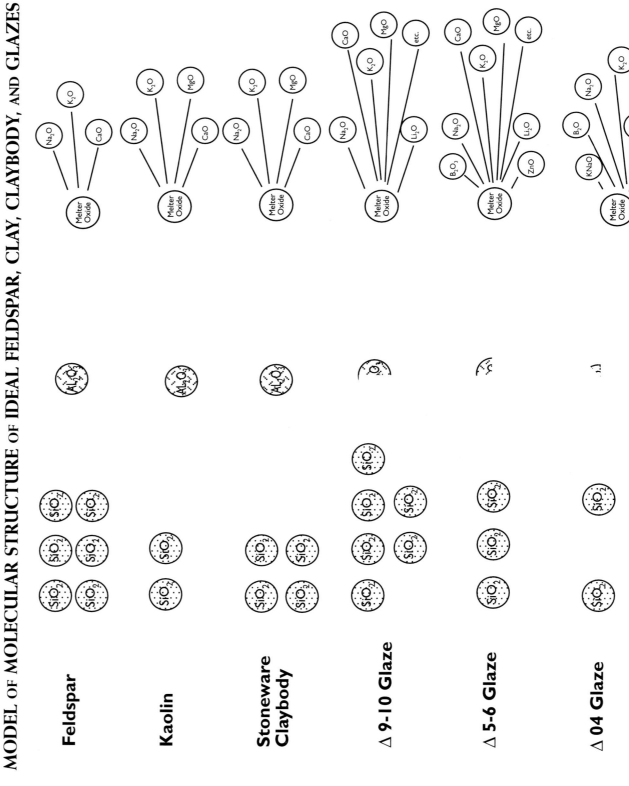

WORKS CITED

BOOKS

Bates, Richard L. 1969. *Geology of the industrial rocks and minerals.* New York: Dover Publications, Inc.

Bartlett, John. 1955. *Familiar quotations.* 13th ed. Boston and Toronto: Little, Brown and Company.

___.1980. *Familiar quotations.* 15th ed. Boston: Little, Brown and Company.

Birkeland, Peter. 1984. *Soils and geomorphology.* Oxford University Press.

Blatt, H. 1992. Weathering and soils. In H. Blatt, *Sedimentary petrology* (2nd ed., Chapter 2, pp. 30-45). New York: W. H. Freeman and Company.

Brodie, Regis C. 1982. *The energy-efficient potter.* New York: Watson-Guptill Publications.

Cairns-Smith, A. G., and H. Hartman 1986. *Clay minerals and the origin of life.* London and New York: Cambridge University Press.

Cardew, Michael. l973. *Pioneer pottery.* London: Longman Group Limited.

Chapell, James. 1991. *The potter's complete book of clay and glazes.* New York: Watson-Guptill Publications.

Conrad, John W. l973. *Ceramic formulas: The complete compendium.* New York: Macmillan Publishing Co., Inc. London: Collier Macmillan Publishers.

Cooper, Emmanuel. 1980. *The potter's book of glaze recipes.* New York: Charles Scribner & Sons.

Craig, James R., David J. Vaughan, and Brian J. Skinner. 1988. *Resources of the earth.* Upper Saddle River, N.J.: Prentice Hall.

Cypress Specialty Metals Company. Lithium in glasses, Sec.4:49-73. Lithium in glazes and enamels, Sec.5:76-89. Lithium in glass ceramics, Sec.6:90. *Applications of Lithium in Ceramics.*

Duda R., and L. Rejl. 1989. *Minerals of the world.* New York: Arch Cape Press.

Encyclopedia Britannica 1968 Vol. 21. S.v. "Syenite." C.E.T. 555-556.

Frye, Keith (Ed.). l98l. *Encyclopedia of mineralogy.* Vol. IVB of *Encyclopedia of Earth Sciences Series* (Ed. Rhodes W. Fairbridge). Stroudsburg, Pa.: Hutchinson Ross Publishing Company.

Gary, Margaret, Robert McAfee, Jr., and Carol L. Wolf (Eds.). 1974. *Glossary of Geology.* Washington, D.C.: American Geological Institute.

Grimshaw, Rex W. l980. *The chemistry and physics of clays.* 4th ed. rev. New York: John Wiley & Sons.

Grebanier, Joseph. 1975. *Chinese stoneware glazes.* New York: Watson-Guptill Publications. London: Pittman Publishing.

Gross, M. Grant. 1987. *Oceanography, a view of the earth.* 4th ed. Upper Saddle River, N.J.: Prentice Hall.

___. 1990. *Oceanography, a view of the earth.* 5th ed. Upper Saddle River, N.J.: Prentice Hall.

Hamer, Frank, and Janet Hamer. 1977. *Clays.* Ceramic Skillbooks (Ed. Murray Fieldhouse). London: Pittman Publishing LTD. New York: Watson-Guptill Publications.

___. 1986. *The potter's dictionary of materials and techniques.* New York: Watson-Guptill Publications.

Hetherington, A. L. 1948. *Chinese ceramic glazes.* 2nd rev. ed. South Pasadena, Calif.: P. D. and Ione Perkins.

Hopper, Robin. 1984. *The ceramic spectrum.* Radnor, Pa.: Chilton Book Company.

Kingery, W. David, and Pamela B. Vandiver. 1986. *Ceramic Masterpieces.* New York: The Free Press.

Kistler, Robert B., and Ward C. Smith. 1983. Boron and borates. *Industrial Minerals and Rocks.* 5th ed. (Ed. S. J. Lefond). New York: AIME.

Koyama, Fujio, and John Figgess. *Two thousand years of oriental ceramics.* New York: Harry N. Abrams, Inc.

Kuenen, P. H. 1950. *Marine geology.* New York: John Wiley & Sons.

Lawrence, W. G. 1972. *Ceramic science for the potter.* Philadelphia: Chilton Book Company.

Leach, Bernard. 1962. *A potter's book.* Albuquerque, N.M.: Transatlantic Arts Inc.

___. 1975. *Hamada potter.* Tokyo: Kodansha International LTD, distributed in U.S. through Harper & Row.

Nelson, Glenn C. 1984. *Ceramics.* 5th ed. Fort Worth, Tex.: Harcourt Brace Jovanovich College Publishers.

Parmelee, C. W. 1937. *Clays.* Ann Arbor, Mich.: Edwards Bros. Inc.

___. 1951. *Ceramic glazes.* 2nd ed. (Rev. E. D. Lynch and A. L. Friedberg). Chicago: Industrial Publications, Inc.

___. 1973. *Ceramic glazes.* 3rd ed. (Rev. C. G. Harmon). Boston: Cahners Books.

Peterson, Susan. 1977. *The living tradition of Maria Martinez.* Tokyo: Kodansha International LTD, distributed in New York through Harper & Row.

___. 1992. *The craft and art of clay.* Upper Saddle River, N.J.: Prentice Hall.

Rhodes, Daniel. 1958. *Clay and glazes for the potter.* New York: Greenberg.

Sanders, Herbert H. 1974. *Glazes for special effects.* New York: Watson-Guptill Publications.

___. 1982. *The world of Japanese ceramics.* Tokyo: Kodansha International LTD.

Tarbuck, Edward J., and Frederick K. Lutgens. 1984. *The earth. An introduction to physical geology.* 2nd ed. Columbus, Ohio: Merrill Publishing Company.

Tichane, Robert. 1990. *Clay bodies.* Painted Post, N.Y.: The New York Glaze Institute.

Webster's New Universal Unabridged Dictionary. 1979. 2nd. ed. S.v. "clay," "montmorillonite," "smectite."

Wood, Nigel. 1978. *Oriental glazes.* Ceramic Skillbooks. (Ed. Murray Fieldhouse). London: Pittman Publishing LTD. New York: Watson-Guptill Publications.

PERIODICALS AND OTHER SOURCE MATERIALS

Behrens, Richard. 1974. Colemanite and Gerstley Borate. *Ceramics Monthly* (February): 40-41.

Beumee, David. 1994. Porcelain bodies for potters. *Ceramics Monthly* (January): 38-42.

Bikita Minerals (Pvt) LTD. 1998. Quality Manager Report.

Ceramic Industry *Materials Handbook.* (January) 1966, 1992 138-1, 1993 140-1, 1994 142-1. Business News Publishing Co. Michigan.

Counts, Charles. 1981. True to the seam. *Studio Potter* 10-1 (December): 16-18.

Cushing, Val. l986. Notes on glaze calculation. *Glaze class. Alfred University. Claybody book.*

Davidson, D., A. Elliott, and D. J. Kingsnorth. 1989. Spodumene: A mineral source of lithium. *The AusIMM Annual Conference* (May): 103-109.

ECC International. 1987. *China and ball clays for the ceramic industry.*

____. 1989. China clay production. *Test methods for ceramic materials.*

Englund, Ivan. 1984. Glazes from granite. *Ceramic Review* (May-June): 87:13.

Eppler and Robinson. 1998. New Opportunities for Wollastonite in Traditional Whitewares. *Ceramic Industry* (April).

Fronk, Dale A., and Milan Vukovich, Jr. 1974. Deformation behavior of pyrometric cones and the testing of self-supporting cones. *Ceramic Bulletin.* Edward Orton Jr. Ceramic Foundation. 53-2. 156-157.

Foote Mineral Company. Technical Data.

Petalite, *Bulletin 301.* Lithium Carbonate, *Bulletin 312.* Spodumene, *Bulletin 313A.* Lithia in glasses, *Bulletin 315.* Lithia in whiteware, *Bulletin 317.* Lithia in glazes, *Bulletin 319.* Lithia in enamels, *Bulletin 320.*

Gertsley Borate, Notes. http://www.ceramicssoftware.com/ education/material/gerstley:htm,July1998.

[1]Haigh, M. 1990. *Providing a low cost source of Lithia to the glass and ceramic industries* (17 January).

[1]Haigh, M., and D. J. Kingsnorth. 1990. The lithium minerals industry. *Glass Magazine* (January).

Hennessy, David. 1982. Wollastonite and old celadons. *Ceramics Monthly* (March): 95-96.

Hosterman, John W. 1984. Ball clay and Bentonite deposits of the Central and Western Gulf of Mexico Coastal Plain,

United States. *Geological Survey Bulletin.* 1558-C 2-22.

Hunt, William. 1978. Frit formulas. *Ceramics Monthly* (May) 48-54.

____. 1998. Bone China, *Studio Potter* (June): 33–35.

Indusmin Inc. *Indusmin Syenite in ceramic whitewares.*

____. 1967. Mining and milling nepheline syenite. *Canadian Clay and Ceramics.*

Joanides, Thomas. 1976. Lithium clay bodies. *Ceramics Monthly* (February): 38-40.

Kaplan, Jonathan. 1981. Variables in raw materials. *Studio Potter* 10-1 (December): 19-22.

Keeling, P. S. 1962. 64. The Wollastonite deposit at Lappeenranta (Willmanstrand), S.E. Finland. Research Paper 442, *British Ceramic Research Association* (15 November): 877-894.

Kingsnorth, D. J. Marketing Manager Lithium Australia LTD. 1987. *Lithia in Glass.* Presented at The American Ceramic Society, San Diego, Calif. 3 November.

____. 1988. Adding lithium minerals can reduce melting costs. *Glass Industry* (March): 18-25

Kusnik, J., and K. W. Terry. *Final report on physical properties of spodumene-kaolin mixtures in the firing range 1270-1300 C.* Katee Enterprises.

Malmgren, R. 1992. Glaze Calculation software. *Ceramics Monthly* (January): 29-33.

____. 1998. A Look at Glaze Calculation Software. *Ceramics Monthly* (June July August): 38-45.

McDowell, Michael. 1981. Glazing with Mount Saint Helens ash. *Ceramics Monthly* (January): 34-37.

Murtagh, Martin J. *A comparison review of the geological origins of the Sheffield glacial clay deposit (Massachusetts) with the Albany Slip glacial clay deposit (New York).* Sheffield, Mass.: Sheffield Pottery.

Orton, Edward Jr. Ceramic Foundation. Standard pyrometric for the potter. 1-12. Fronk, Dale A., and Milan Vukovich, Jr. 1973. Deformation behavior of pyrometic cones and the testing of self-supporting cones. *Ceramic Bulletin,* (March) 156-158.

Power, Tim (Ed.). 1986. Wollastonite performance filler potential. *Industrial Minerals* (January): 19-34.

Powers, Ann. 1981. Mining and processing Pine Lake clay. *Studio Potter* 10-1 (December): 23-25.

Presidential Address of C.V. Smale, Technical Manager, The Goonvean & Rostowrack China Clay Co. LTD. 1977. To Cornish

Institute of Engineers. (May).

Rieger, K. C. R. T. Vanderbilt, Norwalk, Ct. Talc, pyrophyllite, and Wollastonite. 37-44. Adapted from *Industrial Minerals:* 1979 No. 144; 1981 pp.125-129; 1981 No. 167.

Robinson, Jim. 1981. Fear of silica: An approach to stable stoneware. *Studio Potter* 9-2 (June): 76-80.

____. 1988. Body building for potters: A clay-blending formulary. *Studio Potter* 16-2 (June): 73-82.

Rowan, Gerald, Michael Smyser, and Tony Hansen. 1988. Living without Albany. *Ceramics Monthly* (October): 48-50.

Spinks, H. C. Clay Company Inc. 1989. *Spinks clays for the ceramic arts.*

Studio Potter 1981. 10-1 (December) 4-33. Ralph S. Mason, Prospecting for clay. Charles Counts, True to the seam. Jonathan Kaplan, Variables in raw materials. Ann Powers, Mining and processing Pine Lake clay. Paul Buckles, Procedures and testing for potters ceramic clays. D. E. Duewel and W. D. Fitzpatrick, Quality control program for refractory raw materials. John T. Callahan, Testing Kaolin: Standard methods for testing raw materials. Richard P. Isaacs, Buyer beware.

Tantalum Mining Corporation of Canada LTD. TANCO. *Tantalum Mining Corporation—an overview.*

Taylor, Gerald H. 1989. Nepheline Syenite. *Ceramic Bulletin.* Vol. 68-5.

Terry, K. W. *The use of spodumene in ceramic bodies and glazes.* Katee Enterprises. 2-23.

United States Borax & Chemical Corporation. 1985. *The story of borax.* 6th ed. Glendale, Calif.: Macson Printing & Lithography.

U.S.G.S. Minerals Information. 1998. Mineral Industry Surveys, Boron, 1997

Weltner, George H. l986. MFA Alfred University. Research report, *Val Cushing 1986 Claybody book.*

Zakin, Richard. 1978. Glaze analysis. *An atlas of ceramic technology.* Oswego, N.Y.: 1-17.

Zamek, Jeff. 1989. Economics and raw materials. *Ceramics Monthly* (January): 22.

____. 1998. Gerstley Borate and Colemanite *Ceramics Monthly* (July-August).

[1]Courtesy of F & S Alloys and Minerals Corporation, Lithium Australia LTD

SUBJECT, AUTHOR, and TITLE INDEX

An italicized *f* indicates a photograph.

Crawling, 23
Craze-crackle 80*f*, 224*f*, 28, 29–30
    cornwall, 37
    on melted feldspar, 23
    nepheline syenite, 33
    test of feldspathic glaze cores, 42–43,
      80*f*, 85*f*
Cristobalite, 103, 104, 118, 242
Crystalline formation, 155
Crystallization (fractional), 2
Curie-Joliot, I., 5
Custer feldspar, 27, 294
C-6, 294. *See also* soda feldspar

D

D.F. Stone, 36
Diaspore, 254
Dolomite
    chart, 214
    claybody function, 199, 293
    color effects, 204–206
    compared to calcium carbonate, 198
    Dolocron, 164
    firing cycle, 198
    glaze function, 199, 202–213, 293
    lab, 221
    melting power, 202–203
    opacity and texture, 203–204
    origin, 198
Drybranch Kaolin Company, xi, 108

E

Earth
    crust, elements of, 5
    magma, 2
    melted stone, 13*f*
    minerals of, 2
Edward Orton Jr. Ceramic Foundation
    Equivalents for Orton Standard Pyrometric
    Cones, 9
Empirical Molecular Formula to Batch Recipe,
    Conversion, 278–279
Encyclopedia of Mineralogy, The, 99
EPK Kaolin, 107, 293. *See under* Kaolins
Epsom Salts, 23, 32, 193
Etibank, 185
Eutectic
    bonds, 64, 157
    power, 6
    ratio, 6

F

Feldspars, 4, 224*f*
    alumina, 254
    categories of, 25

characteristics of, 22–23
    crystalline, 22
    Custer, 23, 27
    function in claybody, 38, 104
    G-200, 23
    model of molecular structure, 296
    origins of, 22
    oxide structures of, 24
    problem and solution, 23
    role of, 21
    stoneware glaze, 80*f*, 102
    tests
      borax, 191
      considerations when evaluating,
        40–41
      glaze cores, comparative, 42–43,
        80*f*, 224*f*
      glaze dissection, 24–25
      with nepheline syenite and
        cornwall stone, 39
      Rhodes Porcelain Glaze, 24, 267*f*
      soda ash, 191
    wood ash with, 170
Feldspathic glaze, correcting problem with, 23
Feldspathic rocks, 21, 80*f*
    Cornwall Stone, 34–38, 85*f*
    function in claybody, 38–40
    nepheline syenite, 31–33, 81*f*
    origins of, 22
    tests, evaluating, 40–41
Fern ash, 170
Ferro Corporation, Frit and Color Division,
    222
Fireclays, 144*f*, 148*f*, 151*f*
    brown-black speckles, 116
    characteristics, 116
    chart, 117
    glaze function, 119
    mineralogical changes, 118
    particle size, 118
    sandstone and iron, 118
    test (chart), 119
Firing
    conditions, 19, 230*f*
    speed of, 9
    variations, 3, 236*f*
Flint, 293. *See also* Silica
    mineral, 242
    potters, 112, 242, 264
Fluorspar (Fluorite)
    chart, 179
    claybody function, 181, 293
    disadvantages of, 179–189, 181
    glaze function, 180–181, 293
    lab V, 182
    melter and opacifier, 179
    origin, 179

Foundry Hill Creme clay, 122, 123
Franklin Industrial Minerals, 27
Frits, 26
F&S International, Inc., xi, 54, 59, 60, 184

G

Gell-Mann, M., 5
Geological Survey Bulletin, The (1558-C), 110
Gerstley Borate, 171, 183, 211*f*, 294
    chart, 188
    hydrate, 192–193
    lab, 196
    origin, 189
    test
      colemanite, 189–190
      frits and colemanite, 191–195
G-200 feldspar, 23, 27, 294
G-200 Glaze Tests I–III, 42–44
Gibbsite, 254
Glass
    definition, 5–6
    phosphorus, 174
    silica, 6
Glauconites, 101
Glaze
    alumina in, 6, 163
    application, 161, 163
    calculation techniques, 1, 275–280
    classic, 5
    definition, 5, 6, 7
    firing conditions, 163
    internal structures, 10, 14*f*, 281
    model of molecular structure of, 296
    stoneware, 4, 80*f*
Glaze calculation techniques, 1, 275–280
Glaze cores
    characteristics of, 21
    clays and claybodies as, 105
    cornwall stone, 36–37, 85*f*, 253
    definition, 41, 155
    feldspars and feldspathic rocks.
      See Feldspars
    glaze application, 94
    kona F-4, 43, 47, 253
    lab II, 94
    lithium, 53–64
    Nepheline Syenite, 32, 253
    oxide analysis (chart), 43, 47
    rotten stone, 68–75
    substitutions
      batch recipes of (charts), 45
      oxide analysis of (charts), 45
      oxide percentage analyses of
        (charts), 43, 46, 47

GLAZE INDEX